MEDICAL AND EXPERIMENTAL MAMMALIAN GENETICS: A PERSPECTIVE

RECENT TITLES PUBLISHED BY ALAN R. LISS, INC. FOR MARCH OF DIMES BIRTH DEFECTS FOUNDATION

Birth Defects Compendium, Second Edition, Daniel Bergsma, *Editor*

BIRTH DEFECTS: ORIGINAL ARTICLE SERIES

1983 — Volume 19

No. 1 **Dentition: Genetic Effects,** Ronald J. Jorgenson, *Editor*

No. 3 **Primary Immunodeficiency Diseases,** Ralph J. Wedgwood, Fred S. Rosen, and Natalie W. Paul, *Editors*

No. 4 **Nervous System Regeneration,** Bernard Haber, J. Regino Perez-Polo, George A. Hashim, and Anna Maria Giuffrida Stella, *Editors*

No. 5 **Annual Review of Birth Defects, 1982, Birth Defects: Clinical and Ethical Considerations,** Sara C. Finley, Wayne H. Finley, and Charles E. Flowers, Jr., *Editors*

1984 — Volume 20

No. 1 **CMV: Pathogenesis and Prevention of Human Infection,** Stanley A. Plotkin, Susan Michelson, Joseph S. Pagano, and Fred Rapp, *Editors*

No. 3 **Craniofacial Mesenchyme in Morphogenesis and Malformation,** Kenneth S. Brown and Carlos F. Salinas, *Editors*

1985 — Volume 21

No. 2 **Diagnostic Accuracy: Effect on Treatment Planning,** Robert J. Shprintzen and Natalie W. Paul, *Editors*

No. 5 **The Patient Within the Patient: Problems in Perinatal Medicine,** Bruce K. Young, *Editor*

1986 — Volume 22

No. 1 **Bone Marrow Transplantation for Treatment of Lysosomal Storage Diseases,** William Krivit and Natalie W. Paul, *Editors*

No. 3 **Prospective Studies on Children With Sex Chromosome Aneuploidy,** Shirley G. Ratcliffe and Natalie W. Paul, *Editors*

1987 — Volume 23

No. 1 **Genetic Aspects of Developmental Pathology,** Enid F. Gilbert and John M. Opitz, *Editors*

No. 2 **Genetic Risk, Risk Perception, and Decision Making,** Gerry Evers-Kiebooms, Jean-Jacques Cassiman, Herman Van den Berghe, and Géry d'Ydewalle, *Editors*

No. 3 **Medical and Experimental Mammalian Genetics: A Perspective,** Victor A. McKusick, Thomas H. Roderick, Joe Mori, and Natalie W. Paul, *Editors*

Please contact the publisher for information about previous titles in this series published by Alan R. Liss, Inc.

March of Dimes Birth Defects Foundation
Birth Defects Original Article Series, Volume 23, Number 3, 1987

MEDICAL AND EXPERIMENTAL MAMMALIAN GENETICS: A PERSPECTIVE

Editors

Victor A. McKusick, MD
The Johns Hopkins Medical Institutions
Baltimore, Maryland

Thomas H. Roderick, PhD
The Jackson Laboratory
Bar Harbor, Maine

Joe Mori
March of Dimes
Birth Defects Foundation
White Plains, New York

Natalie W. Paul
March of Dimes
Birth Defects Foundation
White Plains, New York

ALAN R. LISS, INC., NEW YORK

To enhance medical communication in the birth defects field, the March of Dimes Birth Defects Foundation publishes the *Birth Defects Compendium (2nd Edition),* an *Original Article Series,* a *Reprint Series, Nursing Staff Development Modules, Genetics in Practice* (a quarterly newsletter), and provides a series of films and related brochures.

Further information can be obtained from:

Professional Education Department
March of Dimes Birth Defects Foundation
1275 Mamaroneck Avenue
White Plains, New York 10605

Published by:
Alan R. Liss, Inc.
41 East 11th Street
New York, New York 10003

Views expressed in articles published are the authors', and are not to be attributed to the March of Dimes Birth Defects Foundation or its editors unless expressly so stated.

Library of Congress Cataloging-in-Publication Data

Medical and experimental mammalian genetics.

(Birth defects, original article series; v. 23, no. 3)
"14 papers . . . collected to celebrate the first 27 years of "The Bar Harbor Course"—Pref.
Includes index.
1. Medical genetics—Congresses. 2. Genetics—Congresses. 3. Mammals—Genetics—Congresses. I. McKusick, Victor A. (Victor Almon), 1921- II. Bar Harbor Course. III. Series. [DNLM: 1. Genetics—congresses. 2. Genetics, Medical—congresses. W1 BI966 v.23 no.3 / QZ 50 M4883 1960–86]
RG626.B63 vol. 23, no. 3 616′.043 s 87-3754
[RB155] [616′.042]
ISBN 0-8451-1065-9

Contents

Contributors

Stylianos E. Antonarakis, MD, The Johns Hopkins University School of Medicine, Baltimore, MD 21205 **[65]**

Jane E. Barker, PhD, The Jackson Laboratory, Bar Harbor, ME 04609 **[221]**

Wesley G. Beamer, PhD, The Jackson Laboratory, Bar Harbor, ME 04609 **[221]**

Edward H. Birkenmeier, MD, The Jackson Laboratory, Bar Harbor, ME 04609 **[209]**

Samuel H. Boyer, MD, Department of Medicine, The Johns Hopkins University School of Medicine, and the Howard Hughes Medical Institute Laboratory for Biochemical Genetics, Baltimore, MD 21205 **[191]**

C. Thomas Caskey, MD, Howard Hughes Medical Institute/Institute for Molecular Genetics, Baylor College of Medicine, Houston, TX 77030 **[297]**

Webster K. Cavenee, PhD, Department of Microbiology and Molecular Genetics, University of Cincinnati College of Medicine, Cincinnati, OH 45267; present address: Ludwig Institute for Cancer Research, Montreal, Quebec H3A 1A1, Canada **[93]**

Stephen M.W. Chang, PhD, Institute for Molecular Genetics, Baylor College of Medicine, Houston, TX 77030 **[297]**

Neal G. Copeland, PhD, Mammalian Genetics Laboratory, BRI-Basic Research Program, NCI-Frederick Cancer Research Facility, Frederick, MD 21701 **[109,123]**

Nancy A. Jenkins, PhD, Mammalian Genetics Laboratory, BRI-Basic Research Program, NCI-Frederick Cancer Research Facility, Frederick, MD 21701 **[109,123]**

Haig H. Kazazian, Jr., MD, Department of Pediatrics, Genetics Unit, The Johns Hopkins University School of Medicine, Baltimore, MD 21205 **[77]**

Priscilla W. Lane, MS, The Jackson Laboratory, Bar Harbor, ME 04609 **[221]**

Edward H. Leiter, PhD, The Jackson Laboratory, Bar Harbor, ME 04609 **[221]**

Victor A. McKusick, MD, The Johns Hopkins Medical Institutions, Baltimore, MD 21205 **[xi,1]**

The numbers in brackets are the opening page numbers of the contributors' articles.

Orlando J. Miller, MD, Department of Molecular Biology and Genetics, and Department of Obstetrics and Gynecology, Wayne State University School of Medicine, Detroit, MI 48201 **[19]**

Joe Mori, The March of Dimes Birth Defects Foundation, White Plains, NY 10605 **[323]**

John A. Phillips III, MD, Division of Genetics, Vanderbilt University School of Medicine, Nashville, TN 37232 **[259]**

Thomas H. Roderick, PhD, The Jackson Laboratory, Bar Harbor, ME 04609 **[xi,1]**

Leonard D. Shultz, PhD, The Jackson Laboratory, Bar Harbor, ME 04609 **[221]**

Douglas C. Wallace, PhD, Departments of Biochemistry, Pediatrics, and Anthropology, Emory University, Atlanta, GA 30322 **[137]**

Preface

The 14 papers presented here were collected to celebrate the first 27 years of "The Bar Harbor Course." Initiated in 1960 as the Short Course in Medical Genetics, it has been an intensive, 2-week, summer learning-teaching experience held annually since that time on the Maine coast. For six alternate odd-numbered years from 1967 to 1977, inclusive, it was transformed into the Short Course in Experimental Mammalian Genetics. For the last eight years, beginning in 1979, it has been the hybrid Short Course in Medical and Experimental Mammalian Genetics—symbolic of the recent remarkable commonality, not only of research objectives (which has always been the case), but of methodology as well.

This collection is a sampling of the far-ranging topics of the Course. It is clearly not a textbook, and was never intended as such. Many opportunities have been offered by publishers to develop a multiauthored textbook on the basis of the Course. Textbook publication was always resisted, however, out of the feeling that the fast-moving nature of the field covered in the Course would make annual update necessary. Although from the beginning the Course has been first and foremost a pedagogic undertaking, an engaging and important feature of the Course has always been the introduction of exciting work in progress in the laboratories and clinics of the lecturers who are at the forefront of the particular research area. We resisted a "cast-in-concrete" impression of the Course (and the field) that might have been created by a published textbook.

To repeat: this is not a textbook; it is a collection of papers aimed, as its title indicates, at giving a perspective on medical and experimental mammalian genetics and particularly the interrelation of the two.

We—the 296 persons who have constituted the faculty of the Course through the 27 years, the 2299 persons who have been "students" in the Course, and the thousands more who as students of the "students" have profited indirectly from the Course—owe a tremendous debt to the March of Dimes Birth Defects Foundation. In late 1959, the then National Foundation (which had just chopped Infantile Paralysis off its name and embarked on work in birth defects and arthritis) committed funds for the Course. Dr. Tom Rivers and Mr. Basil O'Connor were personally responsible for the support at the beginning. The March of Dimes support has continued uninterrupted since then. In the last two years supplementary support for the Course has come from the National Institutes of Health and from the Lucille P. Markey Charitable Trust. To them also we express our gratitude. Finally, to the March of Dimes we are indebted for putting together this perspective on medical and experimental mammalian genetics.

Victor A. McKusick, MD
Baltimore, Maryland

Thomas H. Roderick, PhD
Bar Harbor, Maine

Twenty-Five Years in Medical Genetics and Experimental Mammalian Genetics With Particular Reference to the Gene Map of Mouse and Man

Victor A. McKusick, MD,[1] and Thomas H. Roderick, PhD[2]

The Johns Hopkins Medical Institutions, Baltimore, MD 21205[1]; The Jackson Laboratory, Bar Harbor, ME 04609[2]

The first 25 years of the Bar Harbor Course saw three methodologic revolutions in human genetics, and in mammalian genetics in general. Each methodologic revolution brought scientific advances with important relevance to medicine. The Course had its conception (in 1959) and its inception (in 1960) near the beginning of the era of clinical chromosomology. In the middle period (about 1970) came banding methods, as well as gene mapping by interspecies somatic cell hybridization, which the banding techniques made possible (or at least more accurate and precise). Toward the end of the first quarter century, methods of molecular genetics were brought to bear on scientific questions in mammalian genetics.

The first 25 years of the Course also saw the medicalization (Motulsky's term) of human genetics and indeed of mouse genetics as well. It saw the development, almost from scratch, of clinical genetics as a discrete discipline[1]

[1]*Clinical genetics* is concerned with diagnosis and management of genetic disorders in individual patients and their families. *Medical genetics* is concerned with the entirety of genetic factors in human health and disease. It is a broader discipline than human genetics, which does not necessarily concern itself with all of hereditary disease, especially the clinical aspects (although these days leading contributions to human genetics come from the study of disease, and the human geneticist who knows nothing of human genetic disease is scarcely worthy of the designation). *Medical genetics* is concerned with all of clinical genetics and all of human genetics, because understanding of the normal is fundamental to understanding disease.

Birth Defects: Original Article Series, Volume 23, Number 3, pages 1–17

[McKusick, 1975]. To no small extent, the developments in clinical genetics can be credited to the neo-Vesalian basis (Charles Scriver's term and concept) it acquired through chromosome methods and gene mapping, which reveal the "anatomy of the human genome." The clinical geneticist acquired "his organ," just as the cardiologist had the heart and the neurologist the nervous system. Elsewhere [McKusick, 1982], attention has been called to the parallelism between the influence of Vesalius's *Fabrica* (1543) on the evolution of medicine and the influence of studies in genomic "anatomy" on the development of clinical genetics.

Progress in mammalian genetics, particularly in the formal genetics of the mouse, was reviewed in considerable detail in "Mammalian Genetics and Cancer" [Russell, 1981], a 1979 symposium commemorating the 50th anniversary of The Jackson Laboratory. The important role that the inbred mouse has played was well documented [see also Russell, 1985]. Progress in medical genetics in the first 20 years of the Bar Harbor Course was also detailed in five areas [McKusick, 1981]: 1) chromosome abnormalities in disease, 2) molecular (enzymatic and other) defects in hereditary disease, 3) immunogenetics, especially HLA and disease, 4) the formal genetics of hereditary disease in man, and 5) clinical genetics. Most of these advances in medical genetics and experimental mammalian genetics will not be reviewed again here.

In the years since the 1979 symposium, molecular genetics using methods of recombinant DNA has come into full use in studying the normal and abnormal genome of both man and mouse. Applications include demonstration of restriction enzyme polymorphisms ("restriction fragment length polymorphisms," RFLPs) and their exploitation for gene mapping by the linkage principle; mapping of Huntington disease to chromosome 4 was an early triumph followed by Duchenne muscular dystrophy, adult polycystic kidney disease, and cystic fibrosis. They include DNA diagnosis by direct methods such as demonstration of nucleotide change through alteration in cleavage by a site-specific endonuclease (eg, *Mst*II and sickle cell anemia) and hybridization with specific oligonucleotide probes (eg, in sickle cell anemia and certain forms of thalassemia). It includes demonstration of deletion through absence of a given band(s) from the Southern blot, eg, the various deletion types of thalassemia and the Illig form of growth hormone deficiency [Phillips et al, 1981]. The molecular genetic methods have defined the variety of mutation as illustrated by the thalassemias [Antonarakis et al, 1985] and permitted identification of the basic defect in some disorders through the "reverse genetics" and "candidate gene" approaches.

During the history of the Bar Harbor Course, human genetics has become an experimental science, for example, by the study of cultured cells as a surrogate for the whole organism and by the molecular approaches that permit the medical scientist to go directly to the DNA. Human genetics has become less exclusively dependent on observation and statistical methods. The medical geneticist has become an experimentalist rather than purely a naturalist. There is not space for a listing of all the topics that were fresh and exciting—in the genetics of man and mouse—each year of the Course. Probably at least 95% of the material presented in the latest sessions of the course was, for good and obvious reason, not included in the 1960 course.

Linkage and chromosome mapping, which is its objective, have been central themes of the Bar Harbor Course from the beginning. They have been the main motif of the Course just as they are of genetics in general. It was a proposed project for studying linkage in man that brought Earl L. Green, then Director of the Jackson Laboratory, to Johns Hopkins as a project site visitor in early 1959. Out of that grew an invitation to one of us to visit the Laboratory in the summer of 1959, at which time the Course was conceived.[2]

Although the primary objective of the Course, especially in the early years, was educational, research inevitably was fostered (and, of course, clinical application).[3] As stated above, linkage was an area of particular research emphasis; it profited by comparison of approaches and results in mouse and man. Comparative mapping of the X chromosome was of particular interest in the 1960s. Opportunity for comparisons increased greatly after 1970 (with use of somatic cell hybridization for mapping genes to specific autosomes in both man and mouse) and after 1977 (with mapping by

[2]"Exactly 20 years ago this month July 1975, I first visited the Jackson Laboratory; and it was at lunch at a well-known Bar Harbor restaurant that the idea of a course first came up in conversation with John L. Fuller, then Assistant Director for Training. Those of us in medical genetics and the scientists at the Jackson Laboratory have all along had similar goals; both are interested in determining the genetics and basic gene-determined defect of the mutations they encounter as well as in determining the role of genetic factors in cancer, longevity, and behavior. The methodology then used by each was quite different, but each had much to learn from the other" [McKusick, 1981, p 127].

[3]"From the beginning, the objective of the course was to 'teach the teachers.' A great vacuum then (1960) existed in genetic instruction in medical schools; and the largest part of the students in the short course consisted of members of medical school faculties, although the faculties of dental, veterinary, and other professional schools and persons teaching human genetics at the university level were also represented. We insisted that this was a course and not a seminar or symposium. An attempt was made to cover the field in a systematic manner. In addition, new material was introduced each year, much of which had not yet been published, thereby conveying a sense of immediacy and excitement" [McKusick, 1981, pp 127–128].

molecular genetic methods) simply because the chromosomal locations of more loci were known in both species. The syntenic homology for the X chromosomes became apparent in the 1960s and was the basis of Ohno's law of the evolutionary conservatism of the mammalian X chromosome. The extent of *autosomal* homology of synteny between mouse and man came as something of a surprise. It is our purpose here to describe the progress in delineating the genetic anatomy of mouse and man in the first quarter century of the Course.

Although the educational objective of the Course—"to teach the teachers"—was still primary, in the latter part of its first quarter century the character of the student body shifted to some extent; in recent years there has been a larger representation of graduate students and postdoctoral and active younger researchers. Recently, fostering of research has been a particularly evident outcome of the Course. We can point to many collaborations that developed out of the contacts among students, between student and faculty, and among faculty during the 2 weeks in Bar Harbor.

In the first 7 years, the Course was called "Medical Genetics." Beginning in 1967, the Course, in alternate years through 1977, was entitled "Experimental Mammalian Genetics" with "Medical Genetics" continuing in even-numbered years. Since 1977, the course has been one—"Short Course in Medical and Experimental Mammalian Genetics."[4] It was no longer neces-

[4]The 2-week course, which had its 25th annual session in 1984, was described in *Science* [McKusick, 1972] and in the 50th Anniversary Symposium of the Jackson Laboratory held in 1979 [McKusick, 1981]. The faculty of the Course has consisted, in about equal numbers, of members of the Johns Hopkins University faculty, staff of the Jackson Laboratory and faculty members and researchers from other universities and research institutes. Through the first 25 years of the Course, 56 persons from the Hopkins faculty, 53 persons from the Jackson staff, and 154 persons from other institutions have lectured or otherwise taught in the Course, some for many years. The only person with a perfect record, performing in the Course every one of the first 25 years, is Leroy C. Stevens of the Jackson Laboratory. Over 20 years of faculty performance were put in: from The Jackson Laboratory by Elizabeth S. Russell, director of the Experimental Mammalian Genetics Course (22 years); from Hopkins, by Victor A. McKusick, cofounder and codirector of the Medical Genetics Course (23 years). Fifteen or more years service on the faculty was provided by the following: from the Jackson Laboratory, Seldon E. Bernstein, a codirector of the Course for a few years (17 years); Earl L. Green, former director of Jax (18 years); Margaret C. Green (18 years); Thomas H. Roderick, codirector of the Course beginning in 1968 (15 years); George D. Snell, Nobel laureate (19 years); and, from Hopkins: Helen Abbey (18 years); Samuel H. Boyer (19 years); Howard M. Dintzis (15 years); and Edmond A. Murphy (17 years). Park S. Gerald with 21 years at the Bar Harbor Course out of the first 25 holds the record for guest faculty. Through the first 25 years, the student body numbered, in all, 2,227—a count that includes a few "repeaters." Some "students" have returned subsequently as faculty and vice versa.

sary or desirable to have separate alternating medical genetics and experimental mammalian genetics courses. The mouse was—and still is—a powerfully useful surrogate for man in the inquiry as to whether principles and phenomena found in "simpler" systems apply to the human. The new parasexual methods (somatic cell studies, molecular genetic studies) are being used in both man and mouse, and similarities and differences can be recognized. As much as or even more than before, new mouse mutations of physiologic interest provide useful ways to study mechanisms of gene expression, and mouse mutants of pathologic interest are more reliable models of human disease because of assurance, in many cases, that they are mutations in homologous genes.

The new "parasexual" methods, targeted for application primarily in human genetics, have also made mouse genetics less dependent exclusively on breeding methods. Useful as the latter methods are—and development in 1971 of the recombinant inbred strains method for mapping indicates that further genetic exploitation of the inbred mouse is still possible—somatic cell genetics and molecular genetics are important additions to the armamentarium of the mouse geneticist, just as they are for the medical geneticist.

During the history of the Bar Harbor Course, also, mouse models of human genetic disorders have become increasingly important to biochemical and clinical researchers—not only as accurate homologs of human disease but also as tools for elucidating biochemical, physiologic, and developmental mechanisms of genetic diseases (Leiter et al, this volume). It is no longer facetious to say that biomedical researchers are now equally interested in human models of mouse diseases that have been studied in great detail. Many presentations at the Bar Harbor Course in recent years have provided data from both mouse and man, with emphasis on comparative genetics, physiology, and development.

Indeed, a convergence of the fields of medical and mouse genetics has occurred in the last 25 years—in terms of the methods used, the objectives of research, the topics of key interest, and even the researchers engaged in the investigations. Evolution of the Bar Harbor Course reflected the convergence. In a sense, interspecies somatic cell hybridization between mouse and man is symbolic of the fusion of the two fields.

Since gene mapping is central to genetics and has been the most important single motif of the Bar Harbor Course throughout its history, progress in human and mouse genetics since 1960 can be surveyed no better than by focusing on this area of research. Table 1 presents a summary of progress before the initiation of the Course; Table 2 summarizes and compares progress in mouse and man during the years of the Course.

TABLE 1. Progress in Gene Mapping up to 1959 (when the Bar Harbor Course was conceived)

Year	Man	Mouse
1911	First chromosomal assignment in a mammal: color blindness to human X [Wilson, 1911]	1915—First demonstration of linkage *pink-eyed dilution* and *albino* in mouse [Haldane et al, 1915]
1937, 1947	First estimation of map interval in man: color blindness vs hemophilia [Bell and Haldane, 1937; Haldane and Smith, 1947]	1946—Centromere mapping of two linkage groups by adjacent-2 segregation in translocation heterozygotes [Snell, 1946]
	1944—DNA and the genetic material [Avery, McLeod, and McCarty, 1944]	
1951	First autosomal linkage in man: Lutheran vs secretor [Mohr, 1951]	
	About 35 human traits (or disorders) known from pedigree pattern to be X-linked	
1953	Elliptocytosis/Rh linkage [Chalmers and Lawler, 1953; Lawler and Sandler, 1954]	
	1953—Double helix model of DNA [Watson and Crick, 1953]	
1955	ABO/nail-patella linkage [Renwick and Lawler, 1955]	
1956	Linkage heterogeneity (genetic heterogeneity) of elliptocytosis [Morton, 1955]	
	Correct chromosome number of man [Tjio and Levan, 1956; Ford and Hamerton, 1956]	
	Suggestion of prenatal diagnosis by linkage principle, using amniocentesis [Edwards, 1956]	
1959	Male determinants (testis-determining factor; TDF) on the Y chromosome demonstrated by findings in XO [Ford et al, 1959], XXY [Jacobs et al, 1959], and other sex chromosome anomalies	Same in mouse with description of XO [Russell et al, 1959] and XXY [Cattanach, 1961]; 15 autosomal linkage groups known [Eicher, 1981, table I]

TABLE 2. Progress in Gene Mapping During the First 25 Years of the Bar Harbor Course†

Year	Man	Mouse
1960	Three pairs of linked autosomal loci known	Linkage known for about 100 loci
	Phytohemagglutinin for easy chromosome preparations from peripheral blood [Moorhead et al, 1960]	
	Denver conference for standardization of cytogenetic nomenclature; autosomes into 7 groups, A–G	
1961	Computer program for linkage analysis [Renwick, 1961]	
1962	Lyon hypothesis related to ocular albinism and other human traits [Lyon, 1962]	Lyon hypothesis for interpretation of *mottled* heterozygous female [Lyon, 1961]
	Catalog of X-linked traits: "On the X Chromosome of Man" [McKusick, 1962]: total number = 59 definite, 21 possible	
	X-linked blood group, Xg^a, discovered [Mann et al, 1962]	
	Map interval of color blindness vs G6PD estimated [Adam, 1961; Porter et al, 1962]	
1963	Chromosome 13 deletion first noted in retinoblastoma [Lele et al, 1963]	
	Clonal proof of lyonization and of X-linkage of G6PD [Davidson et al, 1963]	
	Linkage disequilibrium (Lewontin's term) used as evidence of linkage of the δ-globin and β-globin loci [Boyer et al, 1963]	
1964	HAT medium introduced for selection in human somatic cell hybridization studies [Littlefield, 1964]	
	Lesch-Nyhan syndrome described [Lesch et al, 1964]	
1965—The genetic code completely elucidated (Cold Spring Harbor Symposium, 1965)		
1965	Recombination often greater in females than males [Cook, 1965]	

†For references not given here, see "Mendelian Inheritance in Man" [McKusick, 1986a].

(continued)

TABLE 2. Progress in Gene Mapping During the First 25 Years of the Bar Harbor Course (continued)

Year	Man	Mouse
1966	Chicago conference on standardization of nomenclature in human cytogenetics; adoption of p and q for chromosome arms First edition of "Mendelian Inheritance in Man" [McKusick, 1966; later editions, 1968, 1971, 1975, 1978, 1983, 1986]	"The Biology of the Laboratory Mouse," 2nd Ed. [Green, 1966]; linkage known for 162 loci [Eicher, 1981, Fig. 2]
1967	HGPRT deficiency in Lesch-Nyhan syndrome [Seegmiller et al, 1967] Ohno's law of the evolutionary conservation of the mammalian X [Ohno, 1967] Description of gene assignment by study of somatic cell hybrids; preferential loss of human chromosome from man-mouse hybrids [Weiss and Green, 1967]	
1968	First autosomal assignment in man, Duffy blood group locus to chromosome 1 near centromere by linkage to heteromorphism [Donahue et al, 1968] Introduction of term *synteny* [Renwick, 1968, 1971]	Extensive use of electrophoretic polymorphisms (isozymes) for mapping [reviewed by Hutton, 1978]
	1969—First restriction enzymes	
1969	Differential staining ("banding") of chromosomes by quinacrine fluorescence [Caspersson et al, 1971] Second autosomal gene assignment, haptoglobin to 16, by linkage with 2 different translocations involving 16 [Robson et al, 1969] Nonsynteny of LDH-A and LDH-B shown by somatic cell hybridization [Nabholz et al, 1969] First description of marker X, later called *fragile X* [Lubs, 1969]	Location of A-T-rich satellite DNA near centromere in all chromosomes but Y [Pardue and Gall, 1970; Jones, 1970] by in situ hybridization

(continued)

TABLE 2. Progress in Gene Mapping During the First 25 Years of the Bar Harbor Course (continued)

Year	Man	Mouse
1970	Linkage of haptoglobin with heteromorphism of 16 called for the first time fragile site [Magenis et al, 1970] Giemsa banding of chromosomes Use of synteny test in somatic cell hybrid studies [Ruddle et al, 1970]	Testicular feminization (*tfm*) in mouse X-linked [Lyon and Hawkes, 1970] Proof of germ cell origin of ovarian teratomas and use in centromere mapping Eighteen known autosomal linkage groups [Green, 1971] Centromere location known for nine linkage groups by segregation in mice heterozygous for Robertsonian translocation or reciprocal translocation [Green, 1971]
	1971—Restriction enzyme mapping in SV40 [Nathans et al, 1971]	
1971	C-banding of centromeric heterochromatin Reverse, or R, banding Paris conference on standardization of nomenclature in human cytogenetics, with introduction of banding numerology Definitive assignment of thymidine kinase gene to 17 by somatic cell hybridization [Migeon and Miller, 1968; Miller et al, 1971] Prenatal diagnosis of hemophilia A by linkage to G6PD [McCurdy, 1971]	First assignment of an autosomal linkage group to a specific chromosome (XII to 19) by linkage to secondary constriction [Eicher, 1971] First assignment of autosomal linkage groups to specific chromosomes by translocation-karyotype analysis [Miller et al, 1971] Recombinant inbred strains for linkage analysis [Bailey, 1971; Taylor, 1978]
1972	Ancient tetraploidization suggested as mechanism of homeology of nonhomologous chromosomes [Comings, 1972] Mapping to morphologically identified chromosome, without selection, in somatic cell hybrids [Ruddle et al 1972]	Standardization of nomenclature for mouse karyotype [Committee, 1972; Nesbitt and Francke, 1973]; "Gene Mapping in Laboratory Mammals" [Robinson, 1972]
1973	First Human Gene Mapping Workshop, New Haven [Ruddle et al, 1973] Deletion mapping of acid phosphatase to 2p [Ferguson-Smith et al, 1973] Regional mapping of G6PD and PGK to Xq by use of KOP 14;X translocation [Ricciuti and Ruddle, 1973b]	Use by Chapman of subspecies of *Musculus* (eg, castaneus and molossinus) to provide variation at isozyme loci missing in inbred strain [reviewed by Chapman, 1973]

(continued)

TABLE 2. Progress in Gene Mapping During the First 25 Years of the Bar Harbor Course (continued)

Year	Man	Mouse
	Chromosome-mediated gene transfer in cultured cells Use of ovarian teratomas for centromeric mapping	
1974	Deletion in 11p in Wilms/aniridia syndrome [Ladda et al, 1974] Multipoint linkage analysis Second HGM Workshop, Rotterdam [Bootsma et al, 1974]	
1975	Proof of X-linkage of testicular feminization by cloning cultured cells from heterozygotes [Meyer et al, 1975] Third HGM Workshop, Baltimore [McKusick et al, 1975] Mapping by radiation-induced gene segregation [Goss and Harris, 1975, 1977]	Linkage groups IV and X same [Cacheiro and Russell, 1975] First comparative mapping committee at HGM Workshop [Warburton and Pearson, 1975]
1975—Southern blotting for DNA analysis of restriction fragments [Southern, 1975]		
1976	At least one gene assigned to each human chromosome; last assignment = β-galactosidase-2 (GLB2) to chromosome 22	At least one gene assigned to each chromosome; last assignment = *mahoganoid* to 16 by Roderick et al [1976]; intervening nontranslated sequences (introns) in mouse globin genes
1977—Methods for sequencing DNA [Maxam and Gilbert, 1977; Sanger et al, 1977; see also Sanger, 1981; Gilbert, 1982]		
1977	Fourth HGM Workshop, Winnipeg [Hamerton et al, 1977] Cloning of human genes and creation of complete library of human genomic fragments [Maniatis, 1977] High-resolution banding of prometaphase chromosomes [Yunis, 1977] Mapping of β-globin and α-globin gene clusters [Deisseroth et al, 1977, 1978] Linkage of 21-hydroxylase deficiency form of adrenal hyperplasia with HLA and use in prenatal diagnosis	Introns in immunoglobulin genes Nonsyntenic α-globin pseudogene(s) Origin of ovarian teratoma from germ cell after first meiosis [Eppig et al, 1977]

(continued)

TABLE 2. Progress in Gene Mapping During the First 25 Years of the Bar Harbor Course (continued)

Year	Man	Mouse
1978	First DNA polymorphism, detected with restriction enzyme *Hpa*I, closely linked to β-globin gene [Kan and Dozy, 1978] Centromere mapping of HLA and other loci on 6p [Bakker et al, 1979]	"Origin of Inbred Mice" [Morse, 1978] Regional mapping by study of duplication/deficiencies among segregants from translocation carriers [Eicher and Washburn, 1978]
1979	Fifth HGM Workshop, Edinburgh [Evans et al, 1979] Mapping of Duchenne muscular dystrophy by study of female cases with X-autosome translocation [Verellen, 1979]	450 loci mapped [M. Green in Russell, 1981] Assignment of IgK to mouse 6 Mapping of mouse IgH and linked prealbumin gene to 12 50th Anniversary Symposium Jackson Laboratory [Russell, 1981]
1980	Mapping of malformation syndromes by specifically localized chromosome anomalies: Prader-Willi syndrome; Langer-Giedion syndrome Mapping of genes in the hemoglobin gene clusters [Proudfoot et al, 1980] Prospectus for genetic mapping using restriction fragment length polymorphisms [Botstein et al, 1980]	
1981	Multiple dispersed pseudogenes of argininosuccinate synthetase [Su et al, 1981] In situ hybridization for assignment of immunoglobulin genes to 2p, 14q, and 22q; α-globin to 16p, insulin to 11p, etc [eg, Harper et al, 1981] Mitochondrial chromosome completely sequenced [Anderson et al, 1981] Sixth HGM Workshop, Oslo [Berg et al, 1981] Assignment of first human autosomal linkage group, Lutheran/secretor, to 19 Use of multiple linked DNA RFLPs to "haplotype" thalassemias [Kazazian, Orkin, et al, 1981]	
1982	Assignment of first polymorphic anonymous DNA segment to 14 [Wyman et al, 1982]	RFLPs in mice [Arnheim et al, 1982; Bennett et al, 1982]

(continued)

TABLE 2. Progress in Gene Mapping During the First 25 Years of the Bar Harbor Course (continued)

Year	Man	Mouse
	Assignment of many oncogenes: SRC to 20, MYC and MOS to 8q, FES to 15q, ABL to 9q	
1983	Seventh HGM Workshop, Los Angeles [Sparkes et al, 1983] Assignment of Huntington disease gene to 4p by linkage to RFLP [Gusella et al, 1983] Correlation of location of oncogenes and chromosomal aberrations in Burkitt tumor and chronic myeloid leukemia	
1984	Chromosome sorting and spot-blot analysis for assignment for muscle glycogen phosphorylase gene to 11p [Lebo et al, 1984] Use of close linkage of esterase D and RFLPs to demonstrate homozygosity or hemizygosity of the retinoblastoma locus on 13q	Mutant *gray tremor* (*gt*), determined by recessive gene on mouse 17, filtrable agent transmissible like kuru and Creutzfeld-Jacob disease [Sidman et al, 1985] Conservation of synteny in mouse and man [Nadeau and Taylor, 1984] The Oxford Grid for displaying mouse/man homology of synteny [Buckle et al, 1984]
1985	Adult polycystic kidney disease mapped to 16 [Reeders et al, 1985] Wilson disease, a recessive, mapped to 13 [Frydman et al, 1985] Eighth HGM Workshop, Helsinki [de la Chapelle et al, 1985] Deletion mapping of Duchenne muscular dystrophy [Francke et al, 1985] Cystic fibrosis, another recessive, mapped to 7 Use of Jeffreys' hypervariable DNA segments as markers [Jeffreys et al, 1985a]	
1986	"Reverse genetics" identification of basic defect from map location: CGD and Duchenne muscular dystrophy [Monaco et al, 1986] Candidate gene elucidation of elliptocytoses [McKusick, 1986b] Molecular mapping of cone pigment ("color blindness") genes [Nathans et al, 1986a,b] About 950 genes assigned to specific chromosomes in man [McKusick, 1986b]	Heritable fragile sites in mice [Sanz et al, 1986] 879 genes mapped to specific chromosomes in mouse [Roderick and Davisson, 1986]

REFERENCES

Adam A: Linkage between deficiency of glucose-6-phosphate dehydrogenase and colour blindness. Nature 189:636, 1961.

Antonarakis SE, Kazazian HH Jr, Orkin SH: DNA polymorphism and molecular pathology of the human globin gene clusters. Hum Genet 69:1–14, 1985.

Arnheim N, Treco D, Taylor B, Eicher EM: Distribution of ribosomal gene length variants among mouse chromosomes. Proc Natl Acad Sci USA 79:4677–4680, 1982.

Bailey DW: Recombinant-inbred strains. Transplantation 11:325–327, 1971.

Bakker E, Pearson PL, Meera Khan P, Schreuder GMT, Madan K: Orientation of major histocompatibility (MHC) genes relative to the centromere of human chromosome 6. Clin Genet 15: 198–202, 1979.

Bell J, Haldane JBS: The linkage between the genes for colour-blindness and haemophilia in man. Proc R Soc Med 123:199–250, 1937.

Bennett KL, Lalley PA, Barth RK, Hastie ND: Mapping the structural genes coding for the major urinary proteins in the mouse: Combined use of recombinant inbred strains and somatic cell hybrids. Proc Natl Acad Sci USA 79:1220–1224, 1982.

Boyer SH, Rucknagel DL, Weatherall DJ, Watson-Williams EJ: Further evidence for linkage between the beta and alpha loci governing human hemoglobin and the population dynamics of linked genes. Am J Hum Genet 15:438–448, 1963.

Buckle VJ, Edwards JH, Evans EP, Jonasson JA, Lyon MF, Peters J, Searle AG, Wedd NS: Chromosome maps of man and mouse II. Clin Genet 26:1–11, 1984.

Cacheiro NLA, Russell LB: Evidence that linkage Group IV as well as linkage group X of the mouse are in chromosome 10. Genet Res 25:193–195, 1975.

Caspersson T, Lomakka G, Zech L: The 24 fluorescence patterns of the human metaphase chromosomes—Distinguishing characters and variability. Hereditas 67:89–102, 1971.

Cattanach BM: XXY mice. Genet Res 2:156–158, 1961.

Chalmers JNM, Lawler SD: Data on linkage in man: Elliptocytosis and blood groups. I. Families 1 and 2. Ann Eugen 17:267–271, 1953.

Committee on Standardized Genetic Nomenclature for Mice: Standard karyotype of the mouse, *Mus musculus*. J Hered 63:69–72, 1972.

Davisson M, Roderick TH: Linkage map of the mouse. Mouse Newslett 61:19, 1979.

Davisson MT, Roderick TH: Status of the linkage map of the mouse. Cytogenet Cell Genet 22:552–557, 1978.

Deisseroth A, Nienhuis AW, Lawrence J, Giles RE, Turner P, Ruddle FH: Chromosomal localization of the human beta-globin gene to human chromosome 11 in somatic cell hybrids. Proc Natl Acad Sci USA 75:1456–1470, 1978.

Deisseroth A, Nienhuis A, Turner P, Velez R, Anderson WF, Ruddle RH, Lawrence J, Creagan RP, Kucherlapati RS: Localization of the human alphaglobin structural gene to chromosome 16 in somatic cell hybrids by molecular hybridization assay. Cell 12:205–218, 1977.

Dickie MM: The expanding knowledge of the genome of the mouse. J. Natl Cancer Inst 15:607–684, 1954.

Donahue RP, Bias WB, Renwick JH, McKusick VA: Probable assignment of the Duffy blood group locus to chromosome 1 in man. Proc Natl Acad Sci USA 61:949–955, 1968.

Edwards JH: Antenatal detection of hereditary disorders (letter). Lancet 1:579, 1956.

Eicher EM: Foundations for the future: Formal genetics of the mouse. In Russell ES (ed): "Mammalian Genetics and Cancer: The Jackson Laboratory Fiftieth Anniversary Symposium." New York: Alan R. Liss, Inc., 1981, pp 7–50.

Eicher EM: The identification of the chromosome bearing linkage group XII in the mouse. Genetics 69:267–271, 1971.

Eicher EM: Murine ovarian teratomas and parthenotes as cytogenetic tools. Cytogenet Cell Genet 20:232–239, 1978.

Eicher EM, Taylor BA, Riblet R: Assignment of *Igh* and *Pre-1* to chromosome 12. Mouse Newslett 61:42, 1979.

Eicher EM, Washburn LL: Assignment of genes to regions of mouse chromosomes. Proc Natl Acad Sci USA 75:946–950, 1978.

Eppig JT, Eicher EM: Application of the ovarian teratoma mapping method in the mouse. Genetics 103:797–812, 1983.

Eppig JJ, Kozak LP, Eicher EM, Stevens LC: Ovarian teratomas in mice are derived from oocytes that have completed the first meiotic division. Nature 269:517–518, 1977.

Ferguson-Smith MA: X-Y chromosomal interchange in the aetiology of true hermaphroditism and of XX Klinefelter's syndrome. Lancet 2:467–476, 1966.

Ferguson-Smith MA, Johnston AW, Handmaker SD: Primary amentia and microorchidism associated with an XXXY sex-chromosome constitution. Lancet 2:184–187, 1960.

Ferguson-Smith MA, Newman BF, Ellis PM, Thomsom DMG, Riley ID: Assignment by deletion of human red cell acid phosphatase gene locus to the short arm of chromosome 2. Nature 243:271–273, 1973.

Ford CE, Hamerton JL: The chromosomes of man. Nature 178:1020–1023, 1956.

Ford CE, Jones KW, Polani PE, de Almeida JC, Briggs JH: A sex-chromosome anomaly in a case of gonadal dysgenesis (Turner's syndrome). Lancet 1:711–713, 1959.

Francke U, Ochs HD, de Martinville B, Giacalone J, Lindgren V, Disteche C, Pagon RA, Hofker MH, van Ommen G-JB, Pearson PL, Wedgwood RJ: Minor Xp21 chromosome deletion in a male associated with expression of Duchenne muscular dystrophy, chronic granulomatous disease, retinitis pigmentosa, and McLeod syndrome. Am J Hum Genet 37:250–267, 1985.

Frydman M, Bonne-Tamir B, Farrer LA, Conneally PM, Magazanik A, Ashbel S, Goldwitch Z: Assignment of the gene for Wilson disease to chromosome 13; linkage to the esterase D locus. Proc Natl Acad Sci USA 82:1819–1821, 1985.

Green EL (ed): "The Biology of the Laboratory Mouse," 2nd Ed. New York: McGraw-Hill, 1966.

Green MC: Linkage map of the mouse. In Whitten WK (ed): "42nd Annual Report." Bar Harbor, ME: The Jackson Laboratory, 1971, p 56.

Haldane JBS, Smith CAB: A new estimate of the linkage between the genes for colour-blindness and haemophilia in man. Ann Eugen 14:10–31, 1947.

Haldane JBS, Sprunt AD, Haldane NM: Reduplication in mice. J Genet 5:132–135, 1915.

Hutton JJ: Biochemical polymorphisms—Detection, distribution, chromosomal location, and applications. In Morse HC III (ed): "Origin of Inbred Mice." New York: Academic Press, 1978, p 235.

Jacobs PA, Strong JA: A case of human intersexuality having a possible XXY sex-determining mechanism. Nature 183:302–303, 1959.

Jeffreys AJ, Wilson V, Thein SL: Hypervariable "minisatellite" regions in human DNA. Nature 314:67–74, 1985a.

Jeffreys AJ, Wilson V, Thein SL: Individual-specific 'fingerprints' of human DNA. Nature 316:76–79, 1985b.

Jones KW: Chromosome and nuclear location of mouse satellite DNA in individual cells. Nature 225:912–915, 1970.

Kunkel LM, Monaco AP, Middlesworth W, Ochs HD, Latt SA: Specific cloning of DNA fragments absent from the DNA of a male patient with an X chromosome deletion. Proc Natl Acad Sci USA 82:4778–4782, 1985.

Ladda R, Atkins L, Littlefield J: Computer-assisted analysis of chromosomal abnormalities: Detection of a deletion in aniridia Wilms tumor syndrome. Science 185:784–787, 1974.

Lalley PA, McKusick VA: Report of the committee on comparative mapping, HGM 8. Cytogenet Cell Genet 40:536–566, 1986.

Lathrop GM, Lalouel JM, White RL: Construction of human linkage maps: Likelihood calculations for multilocus linkage analysis. Genet Epidemiol 3:39–52, 1986.

Lawler SD, Sandler M: Data on linkage in man: Elliptocytosis and blood groups. IV. Families 5, 6 and 7. Ann Eugen 18:328–334, 1954.

Lebo RV, Gorin F, Fletterick RJ, Kao F-T, Cheung M-C, Bruce BD, Kan YW: High-resolution chromosome sorting and DNA spot-blot analysis assign McArdle's syndrome to chromosome 11. Science 225:57–59, 1984.

Lele KP, Penrose LS, Stallard HB: Chromosome deletion in a case of retinoblastoma. Ann Hum Genet 27:171–174, 1963.

Lesch M, Nyhan WL: A familial disorder of uric acid metabolism and central nervous system function. Am J Med 36:561–570, 1964.

Littlefield JW: Selection of hybrids from matings of fibroblasts in vitro and their presumed recombinants. Science 145:709–710, 1964.

Lubs H: A marker X chromosome. Am J Hum Genet 21:231–244, 1969.

Lyon MF: Gene action in the X-chromosome of the mouse (*Mus musculus* L.). Nature 190:372–373, 1961.

Lyon MF: Sex chromatin and gene action in the mammalian X-chromosome. Am J Hum Genet 14:135–148, 1962.

Lyon MF, Hawkes SG: X-linked gene for testicular feminization in the mouse. Nature 227:1217–1219, 1970.

Magenis RE, Hecht F, Lovrien EW: Heritable fragile site on chromosome 16: Probable localization of haptoglobin locus in man. Science 170:85–87, 1970.

Mann JD, Cahan A, Gelb AG, Fisher N, Hamper J, Tippett P, Sanger R, Race RR: A sex-linked blood group. Lancet 1:8–10, 1962.

McCurdy PR: Use of genetic linkage for the detection of female carriers of hemophilia. N Engl J Med 285:218–219, 1971.

McKusick VA: "On the X Chromosome of Man." Washington, DC: Am Inst Biol Sci, 1962.

McKusick VA: The Bar Harbor course in medical genetics. Science 176:820–821, 1972.

McKusick VA: Presidential address: The growth and development of human genetics as a clinical discipline. Am J Hum Genet 27:261–273, 1975.

McKusick VA: The last twenty years: An overview of advances in medical genetics. In Russell ES (ed): "Mammalian Genetics and Cancer: The Jackson Laboratory Fiftieth Anniversary Symposium." New York: Alan R. Liss, Inc., 1981, pp 127–144.

McKusick VA: The human genome through the eyes of a clinical geneticist. Cytogenet Cell Genet 32:7–23, 1982.

McKusick VA: "Mendelian Inheritance in Man: Catalogs of Autosomal Dominant, Autosomal Recessive and X-linked Phenotypes." Baltimore: Johns Hopkins University Press, 1986a.

McKusick VA: Human genomics 1986: Toward a complete gene map and nucleotide sequence of the human genome. Proc Seventh International Cong Hum Genet 1986b (in press).

Meyer WJ III, Migeon BR, Migeon CJ: Locus on human X chromosome for dihydrotestosterone receptor and androgen insensitivity. Proc Natl Acad Sci USA 72:1469–1472, 1975.

Migeon BR, Miller CS: Human-mouse somatic cell hybrids with single human chromosome (Group E): Link with thymidine kinase activity. Science 162:1005–1006, 1968.

Miller OJ, Miller DA, Kouri RE, Allderdice PW, Dev VG, Grewal MS, Hutton JJ: Identification of the mouse karyotype by quinacrine fluorescence and tentative assignment of 7 linkage groups. Proc Natl Acad Sci USA 68:1530–1533, 1971.

Miller OJ, Allderdice PW, Miller DA: Human thymidine kinase gene locus: Assignment to chromosome 17 in a hybrid of man and mouse cells. Science 173:244–245, 1971.

Mohr J: A search for linkage between the Lutheran blood group and other hereditary characters. Acta Pathol Microbiol Scand 28:207–210, 1951.

Monaco AP, Neve RL, Colletti-Feener C, Bertelson CJ, Kurnit DM, Kunkel LM: Isolation of candidate cDNAs for portions of the Duchenne muscular dystrophy gene. Nature 323:646–650, 1986.

Moorhead PS, Nowell PC, Mellman WS, Battips DM, Hungerford DA: Chromosome preparations of leucocytes cultured from human peripheral blood. Exp Cell Res 20:613–616, 1960.

Morton NE: Sequential tests for the detection of linkage. Hum Genet 7:277–318, 1955.

Nadeau JH, Taylor BA: Lengths of chromosomal segments conserved since divergence of man and mouse. Proc Natl Acad Sci USA 81:814–818, 1984.

Nathans J, Piantanida TP, Eddy RL, Shows TB, Hogness DS: Molecular genetics of inherited variation in human color vision. Science 232:203–210, 1986a.

Nathans J, Thomas D, Hogness DS: Molecular genetics of human color vision: The genes encoding blue, green, and red pigments. Science 232:193–202, 1986b.

Nesbitt MN, Francke U: A system of nomenclature for band patterns of mouse chromosomes. Chromosoma 41:145–158, 1973.

Ohno S: "Sex Chromosomes and Sex-linked Genes." Berlin, New York: Springer Verlag, 1967.

Ott J: Estimation of the recombination fraction in human pedigrees: Efficient computation of the likelihood for human linkage. Hum Genet 26:588–597, 1974.

Pardue ML, Gall JG: Chromosome localization of mouse satellite DNA. Science 168:1356–1358, 1970.

Paris Conference (1971): Birth Defects VIII(7), 1972.

Phillips JA III, Hjelle BL, Seeburg PH, Zachmann M: Molecular basis for familial isolated growth hormone deficiency. Proc Natl Acad Sci USA 78:6372–6375, 1981.

Porter IH, Schulze J, McKusick VA: Genetical linkage between the loci for glucose-6-phosphate dehydrogenase deficiency and colour-blindness in American Negroes. Ann Hum Genet 26:107–122, 1962.

Reeders ST, Breuning MH, Davies KE, Nicholls RD, Jarman AP, Higgs DR, Pearson PL, Weatherall DJ: A highly polymorphic DNA marker linked to adult polycystic kidney disease on chromosome 16. Nature 317:542–544, 1985.

Renwick JH, Bolling D: A program-complex for encoding, analyzing and storing human linkage data. Am J Hum Genet 19:360–367, 1967.

Renwick JH, Lawler SD: Linkage between the ABO and nail-patella loci. Ann Hum Genet 19:312–331, 1955.

Riccardi VM, Sujansky E, Smith AC, Francke U: Chromosomal imbalance in the aniridia-Wilms tumor association: 11p interstitial deletion. Pediatrics 61:604–610, 1978.

Ricciuti F, Ruddle FH: Assignment of nucleoside phosphorylase to D-14 and localisation of X-linked loci in man by somatic cell genetics. Nature [New Biol] 241:180–182, 1973a.

Ricciuti F, Ruddle FH: Assignment of three gene loci (PGK, HGPRT, G6PD) to the long arm of the human X-chromosome by somatic cell genetics. Genetics 74:661–678, 1973b.

Robinson R: "Gene Mapping in Laboratory Mammals, Part B." New York: Plenum Press, 1972.

Robson EB, Polani PE, Dart SJ, Jacobs PA, Renwick JH: Probable assignment of the alpha locus of haptoglobin to chromosome 16 in man. Nature 223:1163–1165, 1969.

Roderick TH, Davisson MT: The linkage map of the mouse. Jackson Lab Scient Rep, pp 39–45, 1985–86.

Roderick TH, Davisson MT, Lane PW: Mouse Newslett 55:18, 1976.

Ruddle FH, Chapman VM, Chen TR, Klebe RJ: Linkage between human lactate dehydrogenase A and B and peptidase B. Nature 227:251–257, 1970.

Ruddle FH, Ricciuti F, McMorris FA, Tischfield J, Creagan R, Darlington G, Chen T: Somatic cell genetic assignment of peptidase C and the Rh linkage group to chromosome A-1 in man. Science 176:1429–1431, 1972.

Russell ES: History of mouse genetics. Annu Rev Genet 19:1–28, 1985.

Russell ES (ed): "Mammalian Genetics and Cancer: The Jackson Laboratory Fiftieth Anniversary Symposium." New York: Alan R. Liss, Inc., 1981.

Russell WL, Russell LB, Gower JS: Exceptional inheritance of a sex-linked gene in the mouse explained on the basis that the X/O sex chromosome constitution is female. Proc Natl Acad Sci USA 45:554–560, 1959.

Sanz MM, Jenkins EC, Brown WT, Davisson MT, Kevin MJ, Roderick TH, Silverman WP, Wisniewski HM: Mouse chromosome fragility. Am J Med Genet 23:491–509, 1986.

Seegmiller JE, Rosenbloom FM, Kelley WN: Enzyme defect associated with a sex-linked human neurological disorder and excessive purine synthesis. Science 155:1682–1684, 1967.

Sidman RL, Hannah KC, Hope SO: Transmissible spongiform encephalopathy in the gray tremor mutant mouse. Proc Natl Acad Sci USA 82:253–257, 1985.

Siniscalco M, Motulsky AG, Latte B, Bernini L: Indagini genetiche pulla predisposizione al favismo. II. Dati familari associazione geneica con il daltonismo. Atti Accad Maz Lincei Rc 28:1–7, 1960.

Snell GD: An analysis of translocations in the mouse. Genetics 31:157–180, 1946.

Staff of the Roscoe B Jackson Memorial Laboratory: Contributions in mouse genetics. J Hered 36:271–274, 1945.

Taylor BA: Recombinant inbred strains: Use in gene mapping. In Morse HC III (ed): "Origins of Inbred Mice." New York: Academic Press, 1978, p 43.

Tjio JH, Levan A: The chromosome numbers of man. Hereditas 42:1–6, 1956.

Weiss MC, Green H: Human-mouse hybrid cell lines containing partial complements of human chromosomes and functioning human genes. Proc Natl Acad Sci USA 58:1104–1111, 1967.

Wilson EB: The sex chromosomes. Arch Mikrosk Anat Entwicklunsmech 77:249–271, 1911.

Structure and Organization of Mammalian Chromosomes: Normal and Abnormal

Orlando J. Miller, MD

Department of Molecular Biology and Genetics, and Department of Obstetrics and Gynecology, Wayne State University School of Medicine, Detroit, MI 48201

NORMAL CHROMOSOMES

In mammals, as in all eukaryotes, almost all the DNA in each cell is within the nucleus, packaged in the form of chromosomes. Although chromosomes are usually visible as distinct entities only during cell division, they remain more or less intact throughout the cell cycle. Methods have been devised to visualize chromosomes at any stage of the cycle by fusing resting or asynchronously growing cells with mitotic cells. As a result, the nuclear membranes of the interphase nuclei break down, and the chromosomes undergo some degree of condensation and become visible (Fig. 1). Those of G1 cells are single, long, thin chromosomes; those of S are fragmented, partially single- and partially double-stranded, and those of G2 are long, paired sister chromatids. The factors that mediate both the breakdown of the nuclear membrane and chromosome condensation during mitosis, meiosis, or premature chromosome condensation appear to be highly conserved in evolution, because those from human cells act quite effectively even in amphibia [Sunkara et al, 1979]. The phenomenon of precocious chromosome condensation (PCC) emphasizes the continuity of the chromosome throughout the cell cycle and the ease with which changes in degree of condensation or compaction occur. These changes clearly have a functional significance, with the chromosomes reaching a maximal state of compaction at the meta-

Birth Defects: Original Article Series, Volume 23, Number 3, pages 19–63

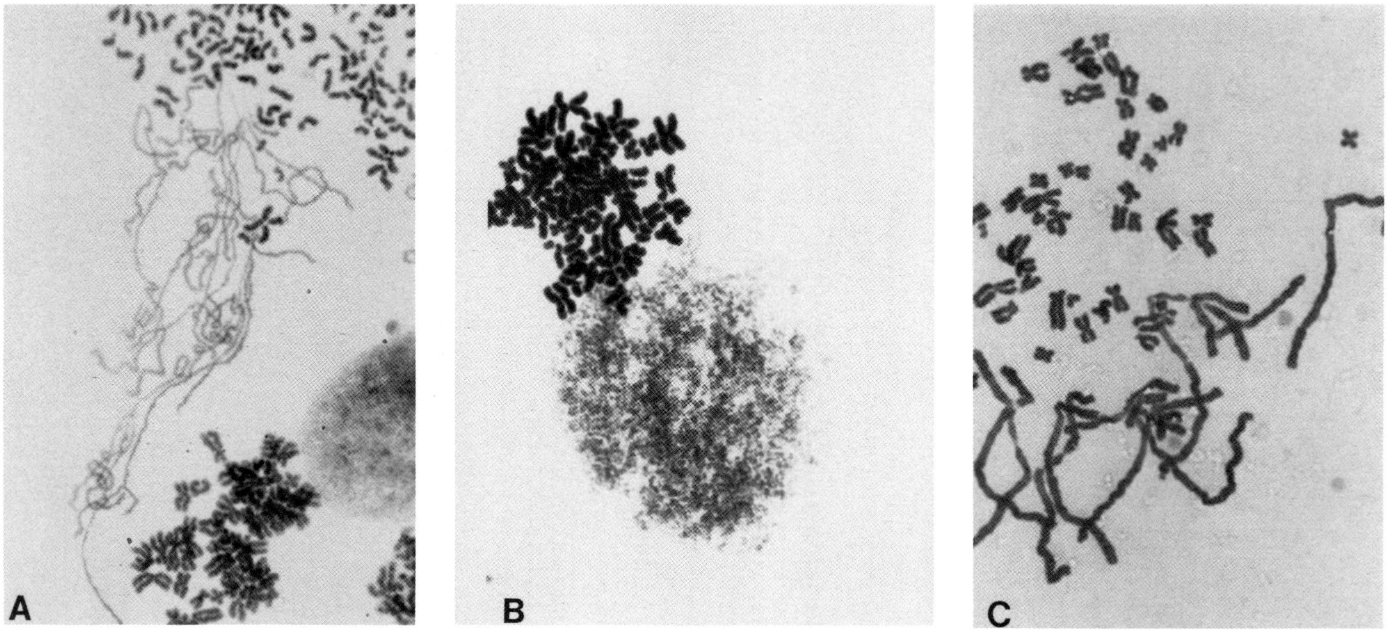
A
B
C

phase-anaphase boundary, when ease of movement toward the poles is essential, and rapidly decondensing after cell division is completed, when specific gene expression is required. (Even more extreme compaction occurs in the sperm head, with rapid decondensation after fertilization.) PCC chromosomes of resting (G0) human lymphocytes are the same length as mitotic chromosomes; those of G1 cells are generally two to three times as long or even longer, reflecting the decondensation associated with transcriptional activity [Schmiady and Sperling, 1981].

Hierarchic Organization of the Chromosome

The mechanisms by which changes in chromosome organization occur during the cell cycle and during specialized cell differentiation are easy to understand in terms of the underlying structure of chromosomes. Chromosomes are composed of DNA, histone proteins, nonhistone proteins, RNA, and various cations. These are organized in a hierarchic fashion (Table 1). Approximately 140 base pairs (bp) of DNA are wound in two turns around a block of eight core histones to produce a nucleosome. An 11–80 bp segment of DNA serves as spacer between nucleosomes and, in the presence of histone H1 and cations, participates in the production of a supercoil or solenoid containing six tightly packed nucleosomes per complete turn [McGhee et al, 1983]. This is the basic chromosome fiber that can be seen by electron microscopy of whole-mounted chromosomes [Woodcock et al, 1984]. This solenoid, in conjunction with various nonhistone proteins, is organized into a large number of domains of superhelicity that are independent of one another. Each contains 30–150 or more kb pairs of DNA in 0.3–1.4 μm of chromosome fiber that is attached at both ends to matrix proteins or nuclear cage. Transcribing genes are in domains that contain nonhistone proteins such as high-mobility-group (HMG) proteins 14 and 17, whereas nontranscribing genes contain other nonhistone proteins. Higher levels of organization are involved in producing the chromosome bands of prometaphase and metaphase mitotic chromosomes, the chromosomes of meiotic prophase, and the fully condensed chromosomes of early anaphase. There is

---

Fig. 1. Prematurely condensed chromosomes (PCC) of rat kangaroo cells (*Potorous tridactylis*, cell line Pt K_2) obtained after fusion with mitotic HeLa cells. **A:** G_1-PCC consisting of a single chromatid. **B:** S-PCC with "pulverized" appearance. **C:** G_2-PCC consisting of two chromatids, which are relatively longer than prophase chromosomes. The darkly stained chromosomes are of the mitotic HeLa cells. (Reproduced from Sperling and Rao [1974] with permission of the publishers. Photographs furnished by Potu N. Rao, The University of Texas M.D. Anderson Hospital and Tumor Institute, Houston.)

TABLE 1. Levels of Chromosome Organization

DNA—2 nm × 2–12 cm per chromosome
Nucleosome—10 nm diameter
DNA: 140 bp (core) + 11–80 bp (spacer)
Histones: two each of H2A, H2B, H3, and H4
Solenoid (chromosome fiber)—25–30 nm diameter
Nucleosomes: six per each 11 nm length
Histone H1, NMP A24, Mg^{2+}
Domain—30–150 kb supercoiled DNA per domain
Solenoid
Nonhistone proteins (HMG14 and -17 for active genes)
Band or chromomere—0.2–2 μm, 1–10,000 kb DNA per band
Phosphorylated, deacetylated histones and nonhistones
R bands: early-replicating, G + C-enriched DNA
Q or G bands: late-replicating, A + T-enriched DNA
C bands: late-replicating, highly repetitive DNA
Chromosome—0.8–1.2 × 2–12 μm
Arms
Centromere
Telomere
Nucleolus organizer
Suprachromosomal—nuclear cage
Transcription and replication
Telomere—nuclear membrane attachments
Clumped heterochromatin
Nucleolar fusion

also some degree of suprachromosomal organization of the nucleus, with a highly organized, proteinaceous nuclear matrix or cage, whose proteins are synthesized on mRNAs transcribed almost exclusively during S phase [Bludau et al, 1986]. This matrix plays an important role in transcription and replication [Vogelstein et al, 1980; Jackson et al, 1982]. It is also important in the attachment of chromosome ends (telomeres) to the nuclear membrane, at least in meiotic prophase, and with the clumping of transcriptionally inactive heterochromatin at the nuclear membrane and around the nucleolus. It is presumably involved in nucleolar fusion, which brings nucleolus organizer regions of different chromosomes together and enhances their activity. Chromosomes appear to occupy clearly defined domains within the interphase nucleus [Manuelidis, 1985]. Chromosomes may have a nonrandom arrangement with respect to each other within the cell, although the evidence for this in mammals is weak.

The hierarchic organization of chromosomal components is related to the functions of transcription, DNA replication, chromosome duplication, DNA damage and repair, and the excision or integration of DNA fragments or proviruses. It is also true of functions that require special structures, such as telomeres, centromeres, possibly centromeric heterochromatin, nucleolus organizer regions, and synaptonemal complexes.

The appearance of chromosomes at various stages of the cell cycle makes it apparent that condensation and decondensation are key mechanisms involved in chromosome function. A temperature-sensitive mutant mouse cell line that is unable to phosphorylate histone H1 at the nonpermissive temperature is unable to undergo chromosome condensation or to enter mitosis [Matsumoto et al, 1980]. Each chromosome appears to condense independently of the other chromosomes, in the sense that treatment with a strongly mutagenic agent (200–800 r irradiation) can block condensation of zero, one, or two chromosomes in a cell while others condense normally [Bianchi, 1982]. This suggests that chromosome-specific factors regulate chromosome condensation.

Condensation and decondensation of chromosomes are mediated by at least two rapidly reversible chemical modifications of chromosomal proteins: phosphorylation and acetylation. Phosphorylation of histones H1 and H3 is maximal during mitosis and occurs during the precocious condensation of interphase chromosome upon fusion of an interphase with a metaphase cell [Hanks et al, 1983]. Phosphorylation of at least eight nonhistone chromosomal proteins is found at the time of transition from G2 to mitosis [Sahasrabuddhe et al, 1984].

Decondensation of chromosomes in cycling cells is mediated by dephosphorylation of histone and nonhistone proteins and also by histone acetylation. Histone acetylation is one of the earliest events associated with the passage of cells from G0 to G1 [Mukherjee and Fabricant, 1978]. Chromosome organization influences these protein modifications: The nonhistone HMG14 and HMG17 proteins, which are associated with potentially or actively transcribing genes [Weisbrod and Weintraub, 1979], block histone deacetylation [Reeves and Candido, 1980]; so does n-butyrate, and this may be the mechanism by which it enhances gene expression in many systems.

Chromosome Duplication

The process of DNA replication is fairly well understood [Kornberg, 1980]; however, the closely coordinated assembly of new nucleosomes and other chromosomal components characteristic of the higher order of chromosome structure is only beginning to be clarified. Replication usually proceeds bidirectionally along the DNA fiber, starting from more than 10,000 origins of replication (ori) in mammalian cells. Replication of each strand of the double helix occurs only in the 5′ to 3′ direction, with one strand continuously replicated and the other put together in reverse as short Okazaki fragments, each replicated 5′ to 3′ and then spliced together. The old nucleosomes may be conserved during this process and travel with the continuously replicating strand, which is also the coding strand [Seidman et al, 1979]. New nucleosomes are assembled from the histones, which are newly synthesized during the S period. Presumably, specific nonhistone proteins and other components of higher-order domains are assembled not long after, although precise information on this is lacking.

Species-Specific Differences in G11-Staining Components

Most chromosome banding or staining procedures yield comparable results in humans and animals, but G11 staining is a striking exception. Mouse or hamster chromosomes stained with Giemsa at pH 11 show a rather intense reddish stain (resulting from protein or possibly RNA) along the entire chromosome, except for the centromeric heterochromatin, which has a faint bluish stain. Human chromosomes stained with this alkaline Giemsa show almost the reverse pattern, with faint bluish staining of the entire chromosome except for the centromeric heterochromatin of chromosome 9 and that of the acrocentric chromosomes of groups 13–15 and 21–22. This differential staining is maintained in rodent-human hybrids, where it provides a useful check for the presence of human chromosomes. Even a single human chromosome, or a tiny fragment of one, maintains its characteristic differential staining. We have found the same to be true of chimpanzee chromosomes in chimp-mouse hybrids. Clearly, no chromosomal proteins encoded by primate genes are required for the primate G11-staining pattern that must, instead, be a property of the primate DNA. Since the relevant sequences are found in every chromosome, they are presumably made up of one or more classes of the interspersed repeats that make up a sizable percentage of total DNA. These sequences can evolve quite rapidly; the Alu family of short interspersed repeats (SINES) and the Kpn family of long interspersed repeats (LINES) that are characteristic of humans and higher primates are sufficiently different from the corresponding SINES or LINES in rodents that differential protein binding to them might account for the differences seen in staining by the alkaline Giemsa technique.

Telomere

Each end of a chromosome is capped by a special structure, the telomere, whose existence was postulated by Muller more than 40 years ago on the basis of his radiation studies in *Drosophila*. Consideration of the detailed mechanism of DNA replication and its need for an RNA primer has also led to the recognition of the need for a special structure at the ends of each chromosome. Such structures have been identified in simple eukaryotes such as *Tetrahymena*, trypanosomes, protozoa, and yeast [Blackburn, 1984]. In *Tetrahymena*, for example, the telomere consists of $\begin{bmatrix} \text{CCCCAA} \\ \text{GGGGTT} \end{bmatrix}_{n}$, when n = 20–70. There is a short gap at the 5′ end of the sequence, and multiple chromatid gaps, providing a structure that permits replication right to the end of the chromosome. Loss of the telomere, as Muller and McClintock showed many years ago, is followed by formation of a dicentric chromosome, a breakage-fusion-bridge cycle, and loss of the chromosome as well as the telomeric fragment. Rarely, there appears to be an exception. In one mentally retarded individual, the distal region of a chromosome 1 had been separated from the remainder of the chromosome, yet both the major component and the fragment were stable in mitosis, being present in every cell examined [Kucerova et al, 1983]. Presumably, the break exposed a sequence that could function as a telomere, or new telomeres were constructed on both broken ends, as happens routinely in macronuclear differentiation in protozoa. In addition, the fragment must have gained centromeric function unless it was using an unknown and remarkably stable noncentromeric mechanism of transmission to daughter cells.

The tandemly repetitive structure of each telomere renders it susceptible to unequal crossing-over, leading to an increase or decrease in the number of copies. The presence of the same repetitive sequence at the ends of every chromosome can lead to recombination between nonhomologous chromosomes. The effect of this would be to homogenize the telomeric sequences, tending to keep the tandemly repetitive telomeric sequences on one chromosome identical to one another and to those on all the other chromosomes. Pairing between telomeres of nonhomologous chromosomes could lead to end-to-end translocations, which are fairly common.

In some cases, genes or other DNA sequences located near a telomere show a high level of genetic recombination. This is true in yeast and trypanosomes as well as in humans. Several genes or other DNA segments that lie within 5,000 kb of the telomere in the pairing segments of the X and Y chromosomes show very high levels of recombination in human males, roughly tenfold higher than they show in females [Rouyer et al, 1986].

Centromere

The centromere, or kinetochore, is the portion of the chromosome involved in spindle fiber attachment and chromosome movement during cell division. Virtually nothing is known about the centromeric DNA in mammals or any other eukaryote except the yeast *Saccharomyces cerevisiae*, which must therefore serve as a model. Centromeres from ten different yeast chromosomes have been isolated and cloned in recombinant plasmids [Hieter et al, 1985]. The shortest segment of the centromere of chromosome 3, CEN3, that will convert an autonomously replicating plasmid into a growth-regulated minichromosome is 627 bp long, whereas the shortest segment from CEN11 is 900 bp long. All ten centromeric sequences have a conserved 8 bp CDE I short region, a 78–86 bp A + T-rich, but not sequence-conserved, CDE II region, and a 25 bp CDE III region with dyad symmetry. These sequences are long enough to establish nucleosome phasing around a central DNA segment, which is free of nucleosomes but is protected from nuclease digestion by conformational constraints imposed by nonhistone protein binding [Bloom and Carbon, 1982]. This protected stretch of DNA or chromatin is presumably responsible for the direct attachment of a single microtubule to each yeast chromosome [Peterson and Ris, 1976].

Mammalian chromosomes are far larger than yeast chromosomes and their movement in anaphase requires attachment of many more microtubules; the number generally is proportional to the length of each chromosome. These microtubules are attached in a cluster to each centromere, usually at a special DNA plus protein (chromatin)-containing structure that is visible by electron microscopy [Ris and Witt, 1981]. Thus in mammals, as in yeast, microtubules attach directly to chromatin in the centromere, and it seems likely that they attach directly to the DNA. With the yeast model of one centromeric DNA sequence per chromosome serving as one microtubule attachment site, it seems likely that mammalian centromeres, with their multiple microtubule attachment sites, will be found to consist of a DNA sequence that is tandemly repetitive. Tandem duplication of a sequence is a common event in evolutionary terms. Tandemly repetitive sequences can undergo unequal crossing-over to produce either a larger or smaller number of copies of the sequence, providing a mechanism for selection of the number of centromeric sequences sufficient to mediate the movement of a chromosome whose size has been altered by deletion, duplication, translocation, or other means. This mechanism might also lead to the production of chromosomes with a diminished number of centromeric sequences, ie, too few for normal chromosome movement. Such a chromosome might tend to be lost or undergo nondisjunction, especially at the first meiotic division, in which each centromere has to move twice the chromosome mass it does in mitotic divisions. When human and murine centromeric sequences have been molecularly cloned, this idea will be open to experimental test.

A candidate human centromeric sequence has recently been identified. Mitchell et al [1985] found a 2.4 kb cloned fragment of human DNA that hybridizes even at high stringency to the centromere of every human chromosome, including the Y. One of the 14 somewhat divergent tandem repeats of 170 bp alphoid sequences making up this fragment contains a 119 bp sequence with multiple similarities to yeast centromeric sequences: a virtually identical 8 bp CDE I element, an 86 bp A + T-rich CDE II element, and a 25 bp CDE III element with dyad symmetry. Further tests are needed to determine whether this sequence or similar sequences can indeed function as part of a centromere.

Tandemly repetitive sequences, such as ribosomal RNA genes and satellite DNA, are subject to cycles of amplification (eg, by unequal crossing-over), diversification (by mutation, deletion, or insertion), and deletion. This can generate families of tandemly repetitive sequences that are related to the parental sequence but may be quite different in nucleotide sequence. If indeed centromeric DNA is tandemly repetitive, it should behave the same way, generating altered sequences, some of them no longer capable of centromeric function. The centromeres of most mammals are flanked by greater or lesser amounts of centromeric heterochromatin, which is composed of families of tandemly repetitive sequences of unknown origin and no apparent function. These may have been generated from the centromeric sequences themselves in the manner proposed. If so, they might retain vestiges of this relationship. Perhaps such an explanation accounts for the fact that microtubule-associated proteins bind more strongly to highly repetitive sequences near the centromeres of mouse chromosomes than to bulk DNA [Villasante et al, 1981].

Abnormalities of centromeric function are important causes of numerical chromosome abnormalities in that they can lead to nondisjunction and chromosome loss. Trisomy 21 and the other autosomal trisomies so common among spontaneous abortuses arise by meiotic or, rarely, premeiotic nondisjunction in the germ line. Many of these trisomies, including trisomy 21, show a marked maternal age effect, with an exponential rise in frequency after age 30 years or so. Mitotic chromosome loss and nondisjunction show a similar increase in women over 50 or 60 years of age, but this is pretty much restricted to the X chromosomes, with considerably more X0 cells than XXX cells observed. Suppression of the X-chromosome centromeric function, with premature separation of the X chromatids, has been observed in older women [Nakagome et al, 1984] and also in five of seven sibs aged 43–53 years [Fitzgerald et al, 1975], suggesting a role for chromosome-specific factors in the regulation of centromeric function. Evidence for repeated primary nondisjunction of chromosome 21 as a result of premature centromere division was reported in a 28-year-old woman who had had three 21-trisomic pregnancies [Fitzgerald et al, 1986].

Dicentric and Pseudodicentric Chromosomes

The behavior of centromeres in translocation chromosomes containing two centromeres is of particular interest and concern. When the translocation occurs in somatic cells of a mature individual, both centromeres remain functional, with the result that the two centromeres may be pulled toward opposite poles of the mitotic spindle and produce an anaphase bridge. Breakage-fusion-bridge cycles may follow, leading to an unbalanced complement and eventual chromosome loss. However, if a dicentric chromosome is produced as a result of events in the germ line or early embryo and passes through the differentiation events of embryogenesis, one of the centromeres is usually suppressed. The resultant pseudodicentric chromosome has a single active centromere and a latent centromere. Centromeric dot or CD staining of specific, unidentified proteins produces red dots at the sites of active centromeres but fails to stain latent centromeres. Sera from patients with the CREST variant of scleroderma contain antibodies that bind to one or more specific proteins at the centromere. One of these sera, which detects three specific centromeric proteins bound to the active centromere but not to the inactive centromere of an isodicentric X chromosome [Earnshaw and Migeon, 1985], is thus a good marker of functional centromeres. Some CREST sera show detectable, though reduced, binding to the inactive centromere of some dicentric chromosomes [Peretti et al, 1986]. Until the proteins detected by these sera are identified, we will not know whether this reflects some residual activity at these centromeres or whether some CREST sera detect proteins not involved in centromeric function. In one case, an end-to-end t(6;19) translocation chromosome with the centromere from 19 active and that from 6 latent tended to separate at the breakpoint, producing a normal 19 and a functionally acentric 6 [Drets and Therman, 1983]. This suggests that suppression of a centromere during embryogenesis is the result of a differentiation process that is not readily reversible. It would be interesting to determine whether demethylation by 5-azacytidine treatment would restore centromeric activity, analogous to the reactivation of genes on the inactive X chromosome.

Centromeric Heterochromatin

In the human, as in most animals and plants, centromeres tend to be surrounded by a variable amount of heterochromatin, material that remains condensed throughout interphase and is made up of highly repetitive DNA sequences that are not transcribed and are replicated late in the S period. Centromeric heterochromatin varies markedly in amount and distribution. It is usually present in greatest amounts on human chromosomes 1, 9, 15, and

16, but on each of these chromosomes, too, the amount varies greatly and in a directly heritable manner, with the result that one homolog can be considerably longer than the other. Arm ratios can change because the large block of centromeric heterochromatin on the long arm of, say, chromosome 9 is sometimes present on the short arm side of the centromere, presumably as a result of pericentric inversion.

Centromeric heterochromatin is made up of tandemly repetitive sequences, which, lacking any apparent function, can undergo rapid evolutionary change by sequential rounds of amplification, divergence, and deletion. Two general types of repeating sequences are found: 1) short, simple sequences, eg, TTCCA, which are tandemly repeated with many variations induced by mutation, insertion, or deletion and 2) more complex sequences of a specific length in a particular species, genus, or family [Singer, 1982]. The repeating unit is usually in the range of 100–300 bp long and is sometimes internally repetitive. These repetitive sequences are sometimes relatively AT-rich, as in the human and the mouse, or relatively GC-rich, as in the cow and rat kangaroo. In such cases, they have a buoyant density different from the main band of DNA such that they form one or more satellite DNA bands after isopyknic centrifugation. There are at least four distinguishable satellite DNA bands in human DNA, made up of very short repeating sequences, and they are abundant in the centromeric heterochromatin of about half the chromosomes. In addition, more complex sequences with a periodicity of about 170 bp are present in the centromeric heterochromatin of every chromosome. Restriction enzyme digestion of chromosomes has revealed heterogeneity for these sequences among chromosomes [Miller et al, 1983]. Cloned representatives of two 170 bp alphoid sequences, so called because of some conserved sequence homology with the alpha-satellite sequences of the African green monkey, have been studied by Jabs et al [1984] and Willard [1984]. The cloned probes hybridize at high stringency mainly to a single chromosome: one to the X and the other to chromosome 6; at reduced stringency, they hybridize far more widely although with marked quantitative differences in the extent. Interestingly, both these cloned probes detect restriction fragment length polymorphisms of sequences near the centromeres of chromosomes 1 and 6 [Jabs et al, 1986] and of chromosome 7 and the X [Willard et al, 1986]. Dorothy Miller has studied the chromosomal localization of another tandemly repetitive sequence with a 170 bp unit length from a library prepared by Arthur Mitchell from human DNA partially digested with *Eco*RI and *Hind*III [Mitchell et al, 1985]. This probe hybridizes to the centromeric region of every human chromosome and of every gorilla chromosome, too. In both cases, there is some interchromosomal variation in the extent of hybridization, but not a great deal, and the amount of signal is not correlated with the total amount of centromeric heterochromatin. These characteristics are what one would predict for a cloned probe that contains a centromeric sequence rather than just another alphoid sequence.

Centromeric heterochromatin in the mouse is more homogeneous than that in the human, consisting of more than 1 million slightly variant copies of a 234 bp repeat and about one-tenth as many copies of a different 117 bp repeat. Together they make up nearly 10% of the total DNA and are localized in the centromeric heterochromatin (C bands). The major satellite can be visualized by digesting chromosomal DNA with a restriction enzyme such as *Hae*III, which cuts DNA at a site that is common in main-band DNA but absent from the satellite DNA, and then staining with Giemsa. The C bands show unreduced staining, whereas the chromosome arms are much paler, sometimes with a faint G-banding pattern. Alternatively, the chromosomes can be digested with a restriction enzyme such as *Ava*II, which cuts most of the satellite DNA into 234 bp monomers and 468 bp dimers but cuts main-band DNA into much larger fragments. This greatly reduces subsequent Giemsa staining of the satellite DNA and tends to leave a small amount of centromeric heterochromatin stained. If this is the minor satellite, it is closer to the centromere than are the major satellite sequences [Kaelbling et al, 1984]. Minor satellite sequences are present on at least half of the mouse chromosomes and may, like the major satellite, be present on virtually all the chromosomes. The multichromosomal location of so many satellite DNAs indicates that genetic recombination can occur between sequences on non-homologous chromosomes, and extensive efforts are being made to understand the underlying mechanism and its evolutionary impact.

Despite the common presence of centromeric heterochromatin in close approximation to centromeres, there is no good evidence that it plays any role in centromeric function. It may, however, be involved in the pericentromeric puffing seen in some patients with the lethal Roberts-SC phocomelia syndrome [Tomkins et al, 1979]. A centromere can be separated from the great bulk of the heterochromatin, as in the T199H mouse translocation, and still function perfectly normally. In some strains of mice, one or more normal chromosomes have extremely little centromeric heterochromatin and yet show no deleterious effects.

Nucleolus Organizing Regions

Another specialized structure on the chromosome is the nucleolus organizer region (NOR), which is the site of transcriptionally active, tandemly repetitive, ribosomal RNA genes. NORs are visible by light microscopy as a secondary constriction, but by electron microscopy they are seen to be a nonconstricted region of different chromatin structure. NORs can also be visualized by ammoniacal silver staining. Such AgNORs are present on the short arms of some or all of the five pairs of human acrocentric chromosomes. Rarely, they are carried on the distal end of the Y as a familial trait, possibly predisposing to translocation [Genest et al, 1983]. These tandemly repetitive rRNA gene clusters are the site of one of the breakpoints in some inversions [Kaiser et al, 1984]. They are rather frequently involved in chromosome translocations [see, eg, Verellen-Dumoulin et al, 1984], and this can be associated with activation of an oncogene [Takahashi et al, 1986].

The sizes of AgNORS are highly variable and are generally correlated with the amount of rDNA at the site. Since rRNA genes are organized in long, tandemly repetitive arrays, it is not surprising that they show the same variability in copy number as satellite DNA and other sequences so organized. In some cases, copy number is strikingly amplified, a trait that can be transmitted in mendelian fashion. In such cases, most of the additional rRNA copies are transcriptionally silent, are hypermethylated, and do not stain with silver [Miller et al, 1981]. Suppression of rRNA gene expression is also seen in mouse-human hybrid cells, in which rRNA transcription and silver staining of the rRNA genes of only one species are present [Miller et al, 1976], that is, AgNORs are restricted to active or potentially active rRNA gene clusters, and AgNOR staining is a test for gene activity, not just gene presence.

Synaptonemal Complex

The exact pairing of homologous chromosomes in meiotic prophase is associated with the formation of a special structure, the synaptonemal complex [Solari, 1980]. This protein-rich axis, whose lateral elements are only 200–300 nm apart, extends the full length of each pachytene bivalent, with more than 98% of the chromosomal DNA surrounding it radiating out in lampbrush-like loops of chromatin. Tiny amounts of DNA and RNA are present in and between the lateral elements of the synaptonemal complex, and apparently only this tiny fraction of the total DNA is closely enough paired between homologs to initiate chiasma formation and genetic recombination. There is some evidence that specific DNA sequences are involved in the pachytene nicking of DNA, a prelude to genetic recombination. Thus, in mice as in the more thoroughly studied lily plant, a tiny fraction of genomic DNA undergoes repair replication during the pachytene stage of exact pairing and is therefore called P-DNA. Chromatin containing this P-DNA is rendered especially susceptible to cleavage by complementary binding of small nuclear RNA molecules (P-snRNA), and the approximately 10^5 P-DNA molecules in the genome may be the preferred sites of chiasma formation [Hotta and Stern, 1981].

If this model is correct, it presents a striking contrast to the standard model of recombination that was developed largely on the basis of Benzer's study of recombination at the *rII* locus of a bacterial virus (a nonmeiotic system) and in yeast. According to that model, recombination can take place at any point along the DNA strand. In organisms with a synaptonemal complex, on the other hand, recombination may be restricted to a relatively small number of recombinational hot spots. In this regard, it is interesting that all nine independent recombination events observed between the Aβ and Eα genes of the mouse major histocompatibility complex occurred within a 9.8 kb "hot spot," possibly at the same site, in the 85 kb stretch of DNA separating these loci [Steinmetz et al, 1982].

Analysis of synaptonemal complexes has yielded further information regarding the special events of meiosis, eg, the points of initiation of crossing-over. Tiny, 90 nm bars or granules are found between the lateral elements of the synaptonemal complex. These have been called recombination nodules. They occur in about the same number as the chiasmata observable by light microscopy and have much the same distribution along the chromosomes [Solari, 1980]. They may therefore mark the sites of crossing-over between chromatids of homologous chromosomes. The number of chiasmata per cell (roughly 32 in the mouse and 56 in the human) provides an estimate of the genetic length, or percent recombination, in each organism: 16 morgans, or 1,600 centimorgans, in the mouse, and 2,800 centimorgans, or 2,800 map units, in the human. Chiasmata, like recombination nodules, are not distributed at random but occur at higher frequencies near the ends of the chromosomes; eg, chiasmata are quite uncommon near the centromere of the metacentric human 16 but are quite common near the centromere of the acrocentric chromosome 15 [Saadallah and Hulten, 1983]. Comparison of the physical location of specific genes on the human X chromosome with their location on the genetic map confirms that there is a greater frequency of genetic recombination between genes near either end of the chromosome than between genes near the center of the chromosome [Hartley et al, 1984]. In the mouse, the distribution of known genes fails to show the Poisson distribution expected were the genes evenly distributed along the chromosome and crossing-over occurred at random [Lyon, 1976].

One final point should be made about the synaptonemal complex. The centromere, or kinetochore, is visible by electron microscopy as a localized differentiation of the lateral element. This permits measurement of the overall length and arm ratio of the synaptonemal complex of each chromosome.

Surprisingly, these are the same as those calculated from measurements on mitotic chromosomes [Moses, 1977]. This finding supports the view that a largely proteinaceous core or axis is a significant component of the structure of both meiotic and mitotic chromosomes.

CHROMOSOMAL ABERRATIONS IN HUMAN AND MOUSE

Euploid Aberrations

Chromosome aberrations can be numerical or structural (Table 2). Let us first consider numerical aberrations involving sets. Haploidy is rare. It exists for a limited time in parthenogenetic embryos in the mouse, but these usually diploidize rapidly. Nearly haploid leukemic cells have been reported, but they are the end result of a process of continuing chromosome loss from originally diploid cells.

Aberrant diploid states can sometimes be diagnosed in the human by chromosome studies. Chimerism is the presence of a mixture of cells as the result of two fertilization events. It can arise in various ways, eg, by placental anastomoses in a twin pregnancy, and can be recognized by the presence of independent sets of genetic and chromosome markers in normal individuals.

TABLE 2. Types of Chromosome Aberration

Numerical
Whole sets—euploid aberrations
n: Haploid
2n: Chimerism
2n: Gynogenetic diploid—ovarian teratoma
2n: Androgenetic diploid—hydatidiform mole
3n: Triploid
4n: Tetraploid
Single chromosomes—aneuploidy
2n − 1: Monosomy
2n + 1: Trisomy
2n + 2: Double trisomy or tetrasomy
Sex chromosome aneuploidy: one to five Xs, zero to four Ys
Mosaicism: 2n/2n − 1; 2n + 1/2n − 1; 2n/2n + 2
Structural
Breaks: fragile sites, breakage syndromes
Deletions: terminal, interstitial, ring
Duplication: tandem, inverted
Isochromosome
Translocation: Robertsonian, reciprocal, insertion
Duplication/deficiency (partial trisomy/deletion)
Amplification

Sometimes, however, one cell line is XX and the other XY, and in such cases intersexual development may occur.

Benign ovarian teratomas (dermoids) arise from germ cells, with an incidence of about one in 3,000 females. They usually contain two sets of maternal chromosomes but no paternal chromosomes and are called gynogenetic diploids. Nearly 60% are completely homozygous, arising from duplication of a mature ovum; about 20% are completely heterozygous, arising from failure of meiosis I; and the remaining 20% have a low level of heterozygosity, arising from a failure of meiosis II with genetic recombination of some distal markers [Parrington et al, 1984].

Hydatidiform moles, on the other hand, which occur about once in 500 pregnancies in Southeast Asia and considerably less often elsewhere, usually contain two sets of paternal chromosomes but no maternal chromosomes and are thus androgenetic diploids [Wake et al, 1984]. Studies in the mouse aid in understanding why this chromosome combination produces such devastating effects on embryogenesis. Using microinjection and enucleation procedures in fertilized eggs, McGrath and Solter [1984] have shown that reconstructed zygotes with a male and a female pronucleus survive and develop when implanted into the uterus of a pseudopregnant mouse, whereas those with two male or two female pronuclei die. This implies that the genomes of sperm and egg differ in a fundamental way that is critical for correct expression of genes essential in early embryogenesis. This could be due to differential modification of the DNA. It is known that a number of genes have sites that are methylated in sperm and must be demethylated in order to function in differentiated tissues. If these genes are not equally methylated in the ovum, then reconstructed zygotes with two paternal chromosome sets might be unable to produce certain critical gene products, whereas those with two maternal sets might produce too much.

Triploidy, the presence of three chromosome sets, occurs at high frequency and is usually lethal during embryogenesis, accounting for about 8% of human spontaneous abortions and thus over 1% of all recognized pregnancies. A few triploids survive until shortly after birth, but they have severe, multiple malformations. Genetic and chromosome marker studies indicate that triploidy can arise by double fertilization, fertilization by a diploid sperm, or suppression of the second meiotic division in the oocyte. This division usually occurs soon after sperm penetration of the ovum. Tetraploidy appears to arise primarily as a result of suppression of cytokinesis, with the cell failing to cleave despite duplication of the chromosomes. This is almost always an embryonic-lethal condition, although tetraploid cells themselves may be perfectly normal in certain tissues; they are common in the extraembryonic membranes and in certain differentiated cells such as liver and cerebellar neurons.

Aneuploid Aberrations

The major class of numerical chromosome aberrations involves only one or a few chromosomes, with loss leading to monosomy and gain to trisomy, double trisomy, tetrasomy, and so on. Sex chromosomes can be involved, and chromosome complements with one to five Xs and zero to four Ys have been observed in individuals with surprisingly mild developmental errors. Chromosomal mosaicism, the presence of two or more cell lines with different chromosome complements, is not uncommon, especially for the sex chromosomes. When two cell lines are present, they usually differ by either one or two chromosomes depending on their mechanisms of origin: anaphase lag or nondisjunction, respectively. Sometimes a much larger difference is seen, as in the recent report of 2n/2n + 2 mosaicism, the one line having two isochromosomes 21 for the equivalent of hexasomy 21; the individual had multiple malformations, mental retardation, and a non-Down syndrome phenotype [Ketupanya et al, 1984], which is surprising only in that both trisomy 21 and tetrasomy 21 are associated with Down syndrome.

Autosomal chromosome loss is so poorly tolerated that it is exceedingly rare, even in spontaneously aborted fetuses. Sex chromosome loss, too, has a severe effect, with Y0 embryos inviable by the two-cell stage, according to studies in the mouse, and X0 embryos severely disadvantaged, at least in the human. The incidence of X0 conceptions exceeds 1% in both human and mouse. Most X0s in the mouse develop into healthy adults, whereas about 97% of human X0s abort in the first trimester, so that the incidence at birth is less than one in 3,000. Adults with a 45,X karyotype have a characteristic phenotype, Turner syndrome, with sexual infantilism resulting from the absence or near-absence of germ cells from their streak gonads. Short stature, cubitus valgus, webbed neck, and other features make this syndrome easy to recognize. What is less easy to understand is the high embryonic mortality. Also uncertain is the mechanism of loss of one sex chromosome. It is unlikely to be meiotic nondisjunction, because older women do not have a higher incidence of X0s. In fact, women under age 20 appear to have the highest risk. Studies in the mouse may be relevant here. Russell [1976] found that the incidence of X0 mice could be raised almost sixfold (from 1% to 6%) by X-irradiation of mated female mice after fertilization but before the first cleavage division; the paternal X or Y was lost, presumably as a result of suppression of paternal X or Y centromere activity and anaphase lag. A similar process might account for human X0s and for the relatively high frequency of X0/XX and X0/XY mosaics.

Two other findings in the mouse should also be mentioned here. Females heterozygous for In(X)1H, an inversion with breakpoints near the two ends of X, produce a high frequency of X0 progeny [Evans and Phillips, 1975]. Since almost the entire chromosome will be included in a meiotic inversion loop, almost every oocyte with an odd number of chiasmata will form an anaphase bridge, leading to the loss of the maternal X. But that may not be the whole story. Recently, Searle et al [1983] found that females heterozygous for an X-autosome translocation, whose breakpoint, like that of the In(X)1H inversion, was close to the centromere of the X, produced many X0 daughters. (Remember that the mouse X chromosome, like the other chromosomes of that organism, is acrocentric.) Is it possible that a region near the centromere of the X, or of an autosome, is important in regulating centromeric function of the chromosome?

A consideration of other numerical aberrations of the sex chromosomes can aid in understanding mechanisms leading to aneuploidy. As stated above, individuals have been found with one to five X chromosomes and zero to four Y chromosomes, with every possible combination totaling no more than five sex chromosomes. These include the classic XXY with Klinefelter syndrome and sterility resulting from preadolescent death of germ cells in the testes and the multiple-X females whose sexual differentiation and function is relatively normal. It is unclear why the maximum number of sex chromosomes in any complement appears to be five. The presence of multiple X chromosomes has relatively little phenotypic effect, and that of multiple Y chromosomes has even less. The mechanism usually thought to give rise to this array of numerical sex chromosome anomalies is meiotic nondisjunction. If this occurs at the first meiotic division, XX or nulli-X ova and XY or nulli-Y sperm are possible: second meiotic division nondisjunction could give rise to XX or YY sperm. Combined first and second meiotic nondisjunctional events could lead to XXX or XXXX ova and XXY, XYY, or XXYY sperm. This array of gametes is sufficient to account for all the observed karyotypes except those with three or four Y chromosomes. These could have arisen if nondisjunction occurred in a germ line mitotic division before meiosis to yield an XYY spermatocyte. Meiotic nondisjunction could then produce an XYYY gamete and MI + MII nondisjunction an XYYYY gamete. There is other evidence for the occurrence of premeiotic nondisjunction: the finding of an extra chromosome in one in 200 of the first meiotic division figures in male mice. Of course, the presence of three or four Y chromosomes might also be the result of postzygotic nondisjunction, with the reciprocal X0 cells making no significant contribution to the inner cell mass of the blastocyst or to the resultant embryo.

Although mitotic nondisjunction does occur, meiotic nondisjunction probably gives rise to most cases of sex chromosome aneuploidy other than X0. It can occur in both male and female meiosis, the latter shown by the presence of maternal genetic markers, eg, a fragile X site on both X chromosomes in some XXY males. Meiotic nondisjunction is also thought to be responsible for most cases of autosomal trisomy. Analyses of chromosome markers in more than 100 cases of trisomy 13, 14, 15, 16, 21, or 22 indicate that at least 90% of the extra chromosomes are maternal in origin and that nondisjunction at first meiotic division is about 20 times as common as at second meiotic division [Bond and Chandley, 1983]. A surprising and still unexplained observation, based on studies of the large numbers of trisomies present among spontaneous abortuses, is that a maternal age effect is quite pronounced for some chromosomes, including all the smaller chromosomes except 16, but very small for chromosome 16 or the larger chromosomes. For that matter, a maternal age effect on trisomy is seen in some strains of mice but not in others; the incidence of trisomy in the mouse is only about one-fifth that in the human [Bond and Chandley, 1983]. This is not a reporting error, based on early resorption of defective mouse embryos, because it is seen even in one-cell (first cleavage division) embryos. Although studies at a comparable stage are not possible in the human, it is now possible to examine the chromosomes of individual human sperm by in vitro fertilization of zona-free hamster ova. The incidence of aneuploid sperm and those with structural chromosome abnormalities is surprisingly high [Brandriff et al, 1984].

The range of structural changes seen in human and mouse is summarized in Table 2. Improved prometaphase chromosome banding techniques ("high-resolution cytogenetics") have led to the recognition of more and subtler karyotypic changes. This is particularly evident in the Prader-Willi syndrome of hyperphagia, obesity, mental retardation, and hypogonadotropic hypogonadism. In 94 cases of Prader-Willi syndrome, one or another chromosome abnormality involving the proximal long arm of chromosome 15 was found [Mattei et al, 1984]. Deletions are the most common, with 52 involving the q11–q13 region. Interestingly, in 11 of 13 tested, the deletion could be traced to a paternal meiotic origin. Reciprocal translocations (19 examples) with one breakpoint in proximal 15q and Robertsonian translocations (12 examples) are also common; presumably, one breakpoint is in the same region in each of these cases as well as in two pericentric inversions seen with Prader-Willi syndrome. In nine patients, the only aberration present was a bisatellited chromosome derived from a chromosome 15. It remains unclear whether the Prader-Willi syndrome is due to malfunctioning of a single gene or of a block of rather closely linked genes. The presence of a number of markers, such as RFLPs, in the proximal region of 15q would facilitate further analysis of this question.

Deletions, even those that appear to be quite small, can have profound clinical effects. This is not surprising; a chromosome band may contain more than a hundred genes. An interesting example is provided by a minute deletion limited to portions of the two terminal bands of chromosome 13, mainly 13q34, seen in a patient with ring 13 and a clotting defect resulting from a deficiency of factors VII and X [de Grouchy et al, 1984].

A clinically important point concerning many apparent deletions is that they are really more complex duplication/deficiencies resulting from aberrant meiotic segregation of reciprocal translocation chromosomes in a heterozygous carrier. In such a carrier, a ring of four chromosomes is produced by homologous pairing and subsequent terminalization of chiasmata; about half the time alternate chromosomes go to the same pole of the meiotic spindle, leading to balanced gametes. The rest of the time, adjacent chromosomes in the ring go to the same pole; in the much more common adjacent-1 segregation, the adjacent normal chromosome and the translocation chromosome with nonhomologous centromeres go to the same pole, producing reciprocally unbalanced gametes; in the much less common adjacent-2 segregation, the adjacent normal chromosome and the translocation chromosome with homologous centromeres go to the same pole. The important point to remember about this is that a translocation carrier can produce children with at least four different, though related, duplication/deficiencies. In one, the deficiency (deletion) of a segment may produce the major effect; in the reciprocal, either the duplication (partial trisomy) of the same segment or the deficiency of the other segment may produce the major effect.

The meiotic behavior of rearranged chromosomes is very important in the etiology of chromosomal syndromes. This is true of inversions [Kaiser et al, 1984] as well as translocations. A special class of rearrangement that illustrates this most clearly is the isochromosome, whose two arms are identical. An early explanation was that such chromosomes arise by misdivision of the centromere. This could account for the occasional isochromosome Yq. Clearly, however, there has to be a recombination between DNA on the two chromatids for such a misdivision to occur, and it seems more likely that in individuals with an isochromosome 21, ie, with a 45,XY,−21,−21,+i(21q) karyotype, the isochromosome arises by breakage and rejoining between homologous chromosomes and need not, therefore, carry indentical alleles at every locus. The presence of a single functional centromere means that any gamete produced by the isochromosome carrier will have either a double dose of 21q or no chromosome 21 at all, and the resultant pregnancies will end in Down syndrome or spontaneous abortion, except in the unlikely event that both partners are i(21q) carriers.

The presence of a structurally rearranged chromosome sometimes leads to secondary nondisjunction. One example involved a young man who was ascertained because of maternal concern about his secondary sexual development. He was found to have Klinefelter syndrome, was chromatin-positive, and even had 47 chromosomes per cell, but he was not the usual XXY. Instead, he had an X-autosome translocation t(X;14)(q11;q31) with an extra copy of the der(14)t(X;14) chromosome as a result of secondary nondisjunction at the second meiotic division in his mother, who also carried the translocation [Allderdice et al, 1978]. This example is informative because it indicates that an extra copy of the long arm of the X can lead to the Klinefelter phenotype. It also shows how complete the spread of inactivation can be into the autosomal segment of an X-autosome translocation; without such inactivation, this individual would have shown signs of the severe 14 trisomy syndrome, yet he had none of them!

Autosomal as well as sex chromosome imbalances lead to a wide range of phenotypic effects, including many characteristic clinical syndromes. The mechanisms by which duplication or deletion of a chromosome or chromosome segment produces particular developmental abnormalities is unknown. Is it due to a single gene dosage effect or multiple gene dosage effects? In this regard, it is important to determine the shortest region whose duplication or deficiency can produce a characteristic syndrome. Examples of the results of this kind of analysis are shown in Table 3. Clearly, duplication or deletion of the genes in rather short regions can lead to specific syndromes.

A different kind of karyotypic change, the occurrence of a fragile site, is receiving increasing attention. The nature of fragile sites is unclear. Some seem to occur spontaneously; others are enhanced by growth in an abnormal

TABLE 3. Shortest Chromosome Region Involved in Specific Syndromes

Syndrome	Region	Reference
Down	21q22.1	Crippa et al, 1984
Edwards	18q11	Mücke et al, 1982
Klinefelter	Xq	Allderdice et al, 1978
Wolf-Hirschhorn	4p16	Tranejbaerg et al, 1984
Langer-Giedion	8q23	Turleau et al, 1982
WAGR	11p13	Narahara et al, 1984
Holt-Oram	14q24.1	Turleau et al, 1984a
TKC	Xq28	Zuffardi et al, 1982
Beckwith-Wiedemann	11p15	Turleau et al, 1984b
Prader-Willi	15q11.2	Niikawa and Ishikiriyama, 1985

environment [Sutherland et al, 1983]; eg, a folate deficiency enhances the achromatic gap or chromosome break at Xq27. Fragile sites appear to be inherited as a chromosome polymorphism. Some have no phenotypic effect, whereas a few have significant effects. The fragile site at Xq27 is associated with mental retardation, macroorchidism, and reduced fertility in the male. The fragile X syndrome is perhaps the second most common cause of mental retardation, second only to trisomy 21. A recent analysis of 110 pedigrees segregating for FS Xq27 [Sherman et al, 1984] indicated that there were about 20% fewer mentally retarded males than expected among the progeny of carrier females, suggesting that there is only about 80% penetrance in males. About 30% of heterozygous females were mentally retarded, usually mildly, presumably reflecting the Lyonization phenomenon; with many cells having only the normal X active, the phenotype should be closer to normal than if the X carrying the fragile X was always active.

An interesting finding in these families is the absence of any sporadic male cases; all affected males had a carrier mother. Therefore, all mutations must have occurred in the male germ line. The mutation rate is very high, 7.2×10^{-4}, so that nearly one-third of carrier females carry a new mutation to the fragile Xq27 state. Thus, in the Prader-Willi and fragile X syndromes, there is some similarity to the phenomenon of hybrid dysgenesis in *Drosophila melanogaster*, in which a transposable element carried by the maternal genome can integrate into new positions in the X chromosome in the male germ line producing a high rate of mutation. Deeper insight into this condition and the nature of other fragile sites would be facilitated by the development of an animal model in which correlated meiotic studies could be initiated.

The Mouse Model

The mouse has long been the most important animal model system for mammalian cytogenetics. Although all mouse chromosomes are acrocentric or telocentric, with virtually no cytologically visible short arm and little difference in length, each chromosome can be readily identified by its characteristic Q-, G-, or R-banding pattern. Large numbers of translocations and inversions have been produced, and many are maintained for use in experimental studies. Reciprocal translocations are available with breakpoints scattered widely throughout the genome, including the X chromosome. X-autosome translocations in the mouse have played an important role in the development of our understanding of X inactivation. Robertsonian translocations, which join two acrocentric chromosomes into a biarmed chromosome with a single functional centromere, are available involving every chromosome except X and Y.

Robertsonian translocations occur rarely in the laboratory mouse, *Mus musculus musculus*, or its close relatives *M. m. molossinus* and *M. castaneus*. They occur frequently, however, in related European species such as *M. domesticus domesticus* and *M. d. brevirostris*. They were first found in mice from the Val Poschia, designated *M. d. poschiavinus*. These mice were homozygous for seven Robertsonian translocations, designated Rb1Bnr to Rb7Bnr (Bnr was the standard nomenclature designation of the discoverer, Alfred Gropp, who was then at Bonn). Many other such mice are now known; every isolated valley and island in Europe seems to have its own combination of homozygous Rb translocations. Because these mice are interfertile with each other and with laboratory mice, it is possible to obtain stocks of mice heterozygous or homozygous for a single Rb translocation chromosome or for as many as 18, leaving only the X, Y, and 19 in the usual acrocentric form. Crosses between CB and CD mice, each homozygous for eight different Rb translocations, produce heterozygous offspring whose meiotic pairing results in a ring of 16 bivalents [Capanna et al, 1976] reminiscent of the complete ring seen in the plant *Oenothera*, in which the chromosomes segregate regularly despite this exotic pairing configuration. That is not always the case in mice heterozygous for a Robertsonian (or other) translocation, which is one reason they are useful in and relevant to the problem of chromosome rearrangements in the human.

Robertsonian translocation heterozygotes provide a model system for producing specific autosomal trisomies and monosomies. The most common approach is to mate normal females to males doubly heterozygous for two Rb translocations with one chromosome in common. A high percentage of the zygotes will be monosomic or trisomic for this chromosome. This technique was first used by Beverly White and her collaborators, but has been most systematically exploited by the late Alfred Gropp [1982] and his associates to determine the developmental effects of every autosomal trisomy and monosomy in the mouse. All these are embryonic lethals, although the severity of the effects and the time of fetal resorption are quite variable. Monosomic embryos are the most severely affected, with preimplantation loss or resorption within 4 days of implantation in every case. Trisomic embryos develop fairly normally for 5–12 days and then show increasing abnormalities. Each trisomy has a characteristic effect. Trisomies 2, 5, and 15 are most severe, with fetal resorption by day 11. Trisomies of other large chromosomes, and chromosome 17, are next most severe in their effects, whereas trisomies 12, 13, 14, 16, 18, and 19 are compatible with the longest fetal survival. A few mice trisomic for 16, 18, or 10 can even be born alive, although few live much beyond this time.

Trisomy 16 has received particular attention as a possible mouse model of Down syndrome, because mouse chromosome 16 carries several of the genes that have been assigned to the segment of human chromosome 21 that must be present in triplicate for the Down syndrome phenotype to appear, and trisomy 16 mice share some similar clinical features [Epstein et al, 1984]. Trisomy 16 mice fail, however, to survive long enough for model studies on the developing brain. To get around this problem, chimeras (16 trisomy/normal) have been produced. These survive quite well but are again not so useful as a model system because of the highly variable proportion and distribution of trisomic cells. A possible way around this problem is to utilize reciprocal translocations and attempt to produce mice trisomic only for the segment of chromosome 16 homologous to the Down syndrome-producing segment of human chromosome 21. This approach may become much more widespread as attempts are made to understand the effects of duplications or deficiencies of specific chromosome segments homologous to those producing human disease.

Robertsonian translocations have also been used to determine the frequency of nondisjunction and chromosome loss in chromosomally normal mice [Cattanach et al, 1984]. This new approach is very exciting, in that it provides a precise method for measuring the effects of various agents on normal chromosome behavior in male and female meiosis.

The mouse is a useful model in understanding the behavior of inversions. Roderick [1979] exploited a male meiotic anaphase bridge technique to identify mice heterozygous for a radiation- or triethylene-melamine-induced paracentric inversion and identified a considerable number of these. Most inversions in the laboratory mice are paracentric, with the two breakpoints in the same arm, because there is virtually no short arm on any chromosome. Robertsonian translocation chromosomes have two arms, however, and Davisson and colleagues [1981], using triethylene-melamine, have produced a pericentric inversion, In(11.13LS)29Rk, from Rb(11.13)4Bnr. This inversion, which is associated with reduced fertility in heterozygous carriers, may be a useful model for similar inversions in the human.

CYTOGENETICS OF NEOPLASMS

Cancer has many causes, including ionizing radiation, viruses, chemicals, and certain genotypes. Chromosome changes constitute a final common pathway by which all these factors lead to neoplasia. This is so because the chromosomes carry genes whose activation, inactivation, and decrease or increase in copy number affect the malignant potential of a cell. In addition, carcinogens act directly on DNA to cause 1) aberrant centromeric function leading to extra or missing chromosomes, 2) suppression of cytokinesis leading to tetraploidy and, via multipolar spindles, to aneuploidy, and 3) chromosome breaks, leading to deletions and chromosome rearrangements. We will consider examples of each of these.

Abnormal Centromere Function

Premature separation of centromeres or centromere puffing is not uncommon in malignant cells; eg, it was seen in three of 11 human acute nonlymphocytic leukemias and was not due to a vitamin B12 or folate deficiency in these patients, as it is in some patients with pernicious anemia [Shiraishi et al, 1982]. Faulty centromeric function can lead to chromosome loss or nondisjunction, which are common events in tumor progression. Chromosome loss can lead to monosomy of a particular chromosome and hemizygosity of any mutant gene or deletion it carries; nondisjunction can lead to trisomy as well. Concurrent or sequential nondisjunction and chromosome loss can lead to homozygosity for a deletion or mutant gene, such as a trans-acting suppressor of an oncogene. Such phenomena appear to be involved in the development of a high proportion of retinoblastomas [Cavenee et al, 1983; Dryja et al, 1984] and Wilms tumors [reviewed by Solomon, 1984] and have been shown to play a role in the expression of the malignant phenotype of hybrid cells (see below).

Aberrant centromere function may also be involved in the production of ova lacking a maternal chromosome set, perhaps resulting from inclusion of the entire second meiotic division spindle in the second polar body. Fertilization of such an empty egg by a normal haploid sperm can lead, via diploidization, to a homozygous hydatidiform mole; these make up the majority of moles. Fertilization by diploid sperm can lead to heterozygous moles. These are only about one-fifth as common as homozygous moles but have a far higher malignant potential [Wake et al, 1984].

Suppression of Cytokinesis

Suppression of the first or second meiotic division can give rise to parthenogenetic embryos that form benign ovarian teratomas. About 20% of these are completely heterozygous, as expected from the failure of meiosis I, whereas about 60% are nearly completely homozygous, as expected from failure of meiosis II [Parrington et al, 1984].

Suppression of cytokinesis with resultant tetraploidy is a common response of cells to radiation, viral infection, and other insults. Tetraploid cells tend to be unstable, undergoing random chromosome loss as a result of multipolar spindle formation as a result of the extra centrioles in tetrapolar cells. It was the observation of multipolar spindles that led Boveri to his theory, developed from 1902 to 1914, that neoplasia is due to chromosome imbalance. Even today, it seems likely that tetraploidization, followed by extensive random chromosome loss, accounts for the highly variable hyperdiploid chromosome complements seen in the cells of many neoplasms; in others, they are clearly the result of a different process, with evolution of the karyotype toward increasing complexity associated with tumor progression [Sandberg, 1980; Mitelman, 1983].

Specific Chromosome Aberrations in Neoplasia

Although multiple and nonspecific changes in chromosome number or structure are common in neoplasms, there are also highly specific chromosome changes that are characteristic of particular neoplasms [Sandberg, 1980; Rowley and Testa, 1982; de la Chapelle and Berger, 1984]. The first of these to be recognized was the Philadelphia chromosome in chronic myelogenous leukemia (CML). In the prebanding era, only the Ph chromosome, thought to be a deleted G-group chromosome, could be identified. Following the introduction of chromosome banding, Janet Rowley identified the typical chromosome change in CML as a t(9;22) translocation [see Rowley and Testa, 1982]. With the increasing use of higher-resolution banding methods, specific chromosome aberrations were identified in a larger proportion of neoplasms [Yunis, 1983]. The general types of aberrations seen are indicated in Table 4. Only the more common types will be discussed here. A glance at the various tables indicates that leukemias and related tumors are heavily overrepresented. This reflects the greater difficulty of obtaining good banded karyotypes from solid tumors and indicates one of the directions for future research.

Translocations. At least 18 different specific translocations involving 15 different chromosomes have been described in various human tumors (Table 5), with more being reported all the time, eg, t(1;19)(q23;q13) in pre-B-cell leukemia [Carroll et al, 1984]. A general feature of many of the translocations is that the breakpoint in the rearrangements is immediately adjacent to an

TABLE 4. Types of Chromosome Change Related to Human Neoplasia

Point mutation	Duplication
Translocation	Trisomy
Inversion	Amplification: DMs, HSRs
Deletion	Breakage syndromes
Isochromosome	

TABLE 5. Translocations Seen in Human Neoplasms

Chromosomes involved	Breakpoints	Tumor type	Oncogenes near breakpoint
1;14	q23–qter;q32	Lymphoma	c-Sk
2;8	p11–p13;q24	Burkitt lymphoma	c-myc†
3;8	p25;q21	Salivary ca.	c-raf1, c-mos
4;11	q21;q23	ALL	
6;9	p21–p23;q33–q34	ANLL	c-Ki-ras1, c-abl
6;–	q11–q31;–	Melanoma	c-myb, c-Ki-ras1
6;12	q15;p15	Prolym. leukemia	c-myb, c-Ki-ras2
6;14	q21;q24	Ovarian ca.	c-myb
8;21	q22;q22	ANLL	c-mos†
8;14	q24;q32	Burkitt lymphoma	c-myc†
8;22	q24;q11	Burkitt lymphoma	c-myc†
9;11	p21;q23	ANLL-M4 and M5	
9;22	q34;q11	CML, ALL, ANLL	c-abl,† c-sis
11;14	q13–q23;q32	Lymphoma, ALL	
11;19	q23;p12	ANLL-M5	
11;22	q24;q12	Ewing sarcoma	c-sis
14;18	q32;q21	Lymphoma	
15;17	q22;q12–q21	ANLL-M3	c-fes, c-erbA-1

†Breakpoint at oncogene.

oncogene that has been translocated next to or otherwise brought into juxtaposition with a strong promoter or enhancer. For example, in Burkitt lymphoma, the c-myc oncogene from chromosome 8 is usually found translocated to a position abutting an immunoglobulin gene, usually one on chromosome 14 but sometimes one on 2 or 22. The breakpoint and c-myc site in chromosome 8 are at q24; the other breakpoints are at 2p12, 14q32, and 22q11, in the variable region or (more often) the switch region of the immunoglobulin gene [Dalla-Favera et al, 1982a; Taub et al, 1982]. In B-cell lymphomas and B-cell leukemias with the t(11;14)(q13;q32) translocation, the breakpoint in chromosome 14 also involves the immunoglobulin heavy-chain locus, leading Tsujimoto et al [1984] to postulate the existence of a new oncogene (bcl-1) in band 11q13 and its activation in these tumors. It must be stressed that tumors frequently manifest more than one activated oncogene; thus c-myc is not the only activated oncogene in Burkitt lymphomas. Transfection of NIH 3T3 cells with DNA from six Burkitt lymphoma DNAs has detected a transforming gene, Blym-1, which is homologous to the Blym-1 transforming gene isolated from chicken B-cell lymphomas, in which c-myc has also been activated. Blym-1 has been mapped to chromosome 1p32, a site not known to be involved in rearrangements in Burkitt lymphomas, and so the mechanism of its activation is unclear [Morton et al, 1984].

In chronic myelogenous leukemia, the c-abl oncogene at 9q34 has been translocated to the tiny Ph (22q−) chromosome, as shown by analysis of somatic cell hybrids in which the two reciprocal translocation products have been separated, and in this new position c-abl is expressed [de Klein et al, 1982; Heisterkamp et al, 1982]. In 17 patients with CML, the breakpoints in chromosome 22 are clustered within a short 5.8 kb region whose genetic content is still unknown but which presumably has a strong promoter or enhancer nearby [Groffen et al, 1984]. Interestingly, the c-sis oncogene is reciprocally translocated to chromosome 9 in CML [Groffen et al, 1983]. However, since c-sis has been mapped to 22q12.3–q13.1, it is rather far from the site of the breakpoint, 22q11; thus, even if it is involved in the malignant process in CML, which is not established, another mechanism of activation would have to be invoked. In acute myelogenous leukemia, the c-mos oncogene is very close to the breakpoint in 8q22 in the t(8;21) translocation so common in this neoplasm [Neel et al, 1982].

It remains to be seen whether oncogenes are adjacent to the translocation breakpoints in any of the other 15 or more translocations known to be associated with specific neoplasms; some are at least in the general vicinity of the breakpoints. The c-erbA-1 oncogene on 17q is closely proximal to the breakpoint in chromosome 17q21 in the t(15;17) translocation in acute promyelocytic leukemia [Dayton et al, 1984]. The c-erbB-2 oncogene, which also maps to 17q21, showed no rearrangement in seven cases of acute promyelocytic leukemia [Fukushige et al, 1986]. More than 30 oncogenes have been mapped to 18 different human chromosomes, most to fairly specific sites or regions (Table 6) [Berger et al, 1985]. The discovery and mapping of additional oncogenes and additional specific translocation breakpoints would expand the possibilities for relating oncogene activation to its translocation into the neighborhood of a strong promoter. One strong promoter that may play such a role is that of the alpha chain of the T-cell receptor; this is split by chromosome translocations involving chromosome band 14q11.2 in T-cell leukemias [Erikson et al, 1985b].

Inversions. Although inversions have been reported as one of the many karyotypic changes in single cells or single tumors [Sandberg, 1980; Mitelman, 1983], they only rarely occur as a specific change in a specific neoplasm. LeBeau et al [1983] found one such heterozygous inversion, inv(16)(p13q22), in 18 cases of acute myelomonocytic leukemia, M4 subtype, with abnormal eosinophils. In two of these patients, the metallothionein gene cluster in band 16q22 was split by one inversion breakpoint [LeBeau et al,

TABLE 6. Location of Oncogenes on Human and Mouse Chromosomes

McKusick No.	Human site	Protooncogene	Mouse site
19013	1 p36–p34	c-src 2	2
16485	1 p32	c-L-myc	
16479	1 p13–p11	c-N-ras	3
16478	1 q22-qter	c-ski	
16491	2	c-rel	
16484	2 p24–p23	c-N-myc	
16476	3 p25–p24	c-raf1	6
see 16471	4	c-raf2	6
16477	5 q33–q34	c-fms	
16489	6	c-yes2	
19011	6 p23–q12	c-K-ras1	
18999	6 q15–q24	c-myb	10
16486	7	c-met	
19014	7 pter–q22	c-erbB-1	11
16471	7 p14–q21	c-A-raf2	
19006	8 q11 or q22	c-mos	4
19008	8 q24	c-myc	15
18998	9 q34.1	c-abl	2
19002	11 p14.1–p15	c-H-ras1	7
15140	11 q13	bcl1	
16472	11 q23–q24	c-ets1	
16482	12 pter–q14	c-int1	
19007	12 p12–q24.2	c-K-ras2	6
16481	14 q21–q31	c-fos	12
16473	14 q32	c-akt-1	
19003	15 q25–q26	c-fes	7
19012	17 q11–q21	c-erbA-1	11
19015	17 q21	c-erbB-2	
16488	18 q21.3	c-yes-1	
19009	20 q12–qter	c-src-1	2
19004	22 q12.3–q13.1	c-sis	15
19002	X	c-H-ras2	
31101	X p21–q11	c-A-raf1	X

1985], suggesting that the strong promoter of a metallothionein gene activated an as yet unknown oncogene in chromosome band 16p13 and led to the leukemias. An inversion of chromosome 14, inv(14) (q11,q32), is frequently seen in T-cell leukemia. Band q11 is the location of the T-cell receptor alpha-subunit gene, and band q32 is the location of the immunoglobulin heavy-chain gene. A fusion of these two genes has been observed in T-cell leukemia, and the 5′ VH region may, in effect, activate the receptor part of the gene; the chimeric Ig/TCR protein may then play a role in T-cell proliferation [Baer et al, 1985].

Deletions. At least 16 different deletions have been found associated with specific neoplasms (Table 7). Best known among these are the del(13)(q14) in retinoblastoma and the del(11)(p13) in Wilms tumor and in the Wilms tumor-aniridia-genital abnormality-mental retardation (WAGR) complex. Analysis of many cases, looking for the shortest region of overlap among all the deletions, has led to rather precise estimates of the location of these two deleted segments: 13q14.11 in retinoblastoma [Ward et al, 1984] and the midportion of band 11p13 in Wilms tumor [Narahara et al, 1984].

The mechanism relating any deletion to the tumor associated with it is unclear, but it does not appear to involve the presence of an oncogene adjacent to either breakpoint. No oncogene has yet been found on chromosome 13, and retinoblastoma cells are sometimes monosomic for chromosome 13, making it unlikely that an unknown oncogene on this chromosome is involved. There is growing evidence that in many retinoblastomas the tumor cells are homozygous for the mutant gene or deletion in band 13q14 [Cavenee et al, 1983; Dryja et al, 1984]. This gene may therefore be a trans-acting suppressor of an oncogene on another chromosome. The situation for the Wilms-WAGR locus is less clear-cut. The H-ras1 oncogene is on the same chromosome arm as the deletion but appears to be at least several

TABLE 7. Deletions Seen in Human Tumors

Chromosome region	Tumor type	Oncogene nearby
1p32–pter	Neuroblastoma	c-src
1p11–p22	Melanoma	c-N-ras
3p14–p23	S.C. Ca. lung	c-raf1
5q12–q32†	Leukemia, etc	c-fms
6q11–q31	Melanoma	c-myb, c-Ki-ras1
6q15–q21	Ovarian ca.	c-myb
6q21–qter	Lymphoma, ALL†	c-myb
7q22–qter†	Leukemia, etc	c-erbB
11p13	Wilms, WAGR	c-Ha-ras1
11q13–q25	ANLL	
12q13–q15	Salivary ca.	c-Ki-ras2
13q14†	Retinoblastoma	
16p13–q22	ANLL	
20p12.2	MEN2	c-src
20q11–qter	PV, leukemias	c-src
22q11–qter†	Meningioma	c-sis

†Monosomy more common than deletions.

hundred kilobases away from the WAGR locus [de Martinville and Francke, 1983], and some deletions have even removed the region containing H-ras1. It is thus unlikely that this oncogene is involved in Wilms tumor, at least by the position-effect mechanism seen with some translocations. Furthermore, there is evidence from four different groups that Wilms tumor cells are homozygous or hemizygous for the chromosome with the deletion [reviewed by Solomon, 1984], suggesting that the WAGR locus, like that for retinoblastoma, codes for a trans-acting suppressor of an oncogene or other gene involved in tumorigenesis or tumor progression.

Gene Amplification

Evidence for gene amplification has been found in an increasingly broad range of neoplasms [Cowell, 1982]. The amplified genes are often in the form of multiple, tiny, paired acentric fragments of chromosome, called double minutes (DMs), or in the form of relatively homogeneously stained regions (HSRs), on one or more chromosomes in the complement. Some DMs or HSRs arise when tumor cells are selected that are resistant to high levels of an anticancer agent; the amplified regions then contain many copies of the specific gene whose product confers drug resistance, eg, dihydrofolate reductase in the case of methotrexate resistance [Alt et al, 1978]. Gene amplification is sometimes associated with a translocation in which one breakpoint is at or near the site of the amplified gene [Flintoff et al, 1984].

Gene amplification is also found in untreated tumors, where the amplified gene apparently plays some role in tumor origin or progression. In most cases, the nature of the amplified gene or genes is unknown. Significantly, however, amplification of one or another known oncogene has been found in 13 different types of human neoplasms (Table 8) and several mouse neoplasms (Table 9). Such amplification is an alternative to chromosomal rearrangement as a mechanism for increasing the level of a specific oncogene product. Amplification of ribosomal RNA genes (rDNA) has also been found in a variety of neoplasms, such as sarcomas and hepatocarcinomas [Miller et al, 1981], and may be one mechanism leading to the increased level of ribosomal RNA synthesis so commonly seen in neoplasms.

It has been suggested that rRNA gene amplification may be relatively common, because the tandemly repetitive rRNA genes are a preferred site for recombination [Wahl et al, 1983]. If this is so, then rDNA amplification in neoplasms might simply be an accompaniment of amplification of some other gene whose product is important for the malignant cells. In preliminary studies, we have looked for oncogene amplification in H4 hepatocarcinoma cells with amplified rDNA and have found no amplification of c-abl, c-myc, or c-sis sequences.

TABLE 8. Oncogene Amplification in Human Neoplasms

Neoplasm	Oncogene	Reference
Glioblastoma	c-myc	Trent et al, 1986
Promyelocytic leukemia HL60	c-myc	Dalla-Favera et al, 1982b
Neuroendocrine cancer COLO 320	c-myc	Alitalo et al, 1983
Small cell lung carcinoma	c-myc	Little et al, 1983
Mammary carcinoma	c-myc	Escot et al, 1986
Neuroblastoma	N-myc	Kohl et al, 1983
Retinoblastoma	N-myc	Lee et al, 1984
Erythroleukemia K562	c-abl	Selden et al, 1983
Epidermoid ca. A431	EGFR (c-erbB-1)	Ullrich et al, 1984
Squamous cell carcinoma	EGFR	Yamamoto et al, 1986
Glioblastomas	EGFR	Libermann et al, 1985
Colon carcinoma SW480	c-Ki-ras-2	McCoy et al, 1983
Colon carcinoma COLO 201	c-myb	Alitalo et al, 1984
Bladder carcinoma	c-Ha-ras-1	Hayashi et al, 1983
Mammary carcinoma	c-erbB-2	King et al, 1985
Salivary carcinoma	c-erbB-2	Semba et al, 1985
Gastric carcinoma MKN-7	c-erbB-2	Fukushige et al, 1986

Multiple Chromosome Changes in Neoplasia

There is ample evidence that carcinogenesis and tumor progression are mediated by multiple changes in the genome, including gene mutation and chromosome changes of the types just described. It is therefore not surprising that more intensive study of various tumors reveals the presence of more than one type of specific change. Chronic myelogenous leukemia provides the most studied example, in which a t(9;22) or similar translocation is present in a high proportion of all cases. This chromosome declines in frequency or disappears during remission, returns during relapse and during progression to blastic crisis, and is accompanied by secondary karyotypic changes: isochromosome (17q) and trisomy for various chromosomes [Rowley and Testa, 1982]. In various neoplasms, trisomies are associated mainly with tumor progression, and the chromosomes commonly involved include 8, 12, 17, 18, and 21.

Oat cell carcinoma of the lung, although still relatively little studied, also shows multiple chromosome aberrations. A deletion of a segment of the short arm of chromosome 3 has been observed in 12 of 12 cases [Whang-Peng et al, 1982], and most such tumors also contain DMs or an HSR with amplification of the c-myc oncogene [Little et al, 1983]. Neuroblastomas have long been known to contain HSRs or DMs, and recently these were shown to contain amplified N-myc genes in 14 of 18 tumors studied [Kohl et al, 1983].

TABLE 9. Types of Chromosome Changes in Neoplasms of Mice

Types of chromosome changes
Translocations
Plasmacytomas, BALB/c−, induced
T(12;15) IgH on 12, c-myc on 15
T (6;15) κ, λ on 6
Deletion
Plasmacytomas—homozygous deletion 15 band D
Myelocytic leukemia—radiation-induced
Loss of specific chromosome 4
Malignant hybrids
Trisomy 15
T-cell leukemias—spontaneous, AKR
—radiation-induced, C57 or RF
—DMBA- or BP-induced, SLJ
B-cell leukemias—spontaneous, BALB/c
—radiation-induced, BALB/FR (N2)
Tumor cell lines—erythroleukemia, BALB/c
—L-cells, malignant sublines
Trisomy 12
B-cell leukemia— DMBA-induced, SLJ
Amplification
Many tumors with DMs or HSRs
Plasmacytomas—retroviruslike gene
T-cell lymphomas—endogenous MMTV
Y1 adrenocortical carcinoma—c-Ki-ras gene
Choroid plexus (brain) tumor—SV40 transgenome

In addition, about half of neuroblastomas have a deletion of the tip of the short arm of chromosome 1, including 1p32-pter [Brodeur et al, 1981].

Retinoblastomas, too, can show multiple chromosome changes, not just a deletion of part of band 13q14. Isochromosome 6p occurs in about 60% of these tumors [Squire et al, 1984]. In one series of nine unilateral, nonfamilial retinoblastomas, Kustnetsova et al [1982] found no deletion of chromosome 13 but many other chromosome changes, perhaps the most consistent being an isochromosome of 6p, which was present in six of the nine. Marked amplification and enhanced expression of N-myc has also been observed in a minority (three of ten) of retinoblastomas [Lee et al, 1984]. Squamous cell carcinomas of the head and neck provide another example of multiple chromosome changes in malignancy. In cell lines from nine such tumors, Hauser-Urfer and Stauffer [1985] observed a structural abnormality involving chromosome 1 in all nine, with breakpoints in p13, p22, q21, or q32, and DMs or NORs ten to 30 times as large as the size expected, suggesting increased transcription and possibly amplification of the rRNA genes. Amplification of rRNA genes in one or more clusters has also been observed in several rat tumors [Miller et al, 1981].

Chromosome Breakage Syndromes

A series of eight or nine autosomal recessive disorders, of which only Bloom syndrome, Fanconi anemia, and ataxia telangiectasia are at all common, are associated with a high frequency of spontaneous or induced chromosome breaks. These are usually random breaks, so the likelihood of producing a deletion, translocation, or other rearrangement that could lead to neoplasia is enhanced in these individuals. In Bloom syndrome, for example, approximately 20% of the just over 100 individuals in the international registry have developed neoplasms, and usually at a very young age [German, 1983].

Studies in the Mouse

The same kinds of chromosomal changes seen in human neoplasms occur in neoplasms of the mouse [Miller and Miller, 1983] (Table 9); their relative frequencies are somewhat different, perhaps largely because of the greater difficulty of recognizing deletions or subtle chromosomal rearrangements in the less tractable mouse karyotype. Trisomy 15 is commonly found in a variety of spontaneous or radiation-, virus-, or chemical-induced murine tumors and tumor-derived cell lines. It is also characteristic of the malignant sublines arising from nonmalignant hybrids of malignant with nonmalignant parental cells [Jonasson et al, 1977]. The role of trisomy 15 in murine tumorigenesis and tumor progression is not understood, although this chromosome carries the c-myc and c-sis proto-oncogenes. Evidence of their activation in trisomy 15 tumor cells has not, to my knowledge, been reported. However, c-myc activation has been implicated in the T(12;15) and T(6;15) translocations commonly found in plasmacytomas; the breakpoint, as in analogous human tumors, is immediately adjacent to, or within, the c-myc gene, which is brought into juxtaposition with the strong promoter of the immunoglobulin heavy-chain gene on chromosome 12, or κ or λ light-chain gene on chromosomes 6 and 16, respectively [Crews et al, 1982; Keath et al, 1984; Erikson et al, 1985a]. More than a dozen proto-oncogenes have been assigned to specific mouse chromosomes (Table 5), but their role in murine oncogenesis is only beginning to be explored. A putative oncogene, *pim*-1, which maps to mouse chromosome 17, is activated in various murine T-cell lymphomas [Hilkens et al, 1986].

Studies in the mouse indicate that the same types of tumors, induced by the same inducing agent, can show either trisomy 15 or trisomy 12. It is, of course, possible that a more subtle change, such as a specific small deletion, went undetected, and that the more obvious trisomies, like those in many human neoplasms, are secondary changes involved in tumor progression. In this regard, it is interesting that three plasmacytomas that failed to show either of the usual translocations did have a deletion of band D in chromosome 15; furthermore, the deletion had achieved homozygosity, presumably by combined nondisjunction and chromosome loss, and the tumor cells had an increased level of expression of the c-myc gene [Wiener et al, 1984]. Deletions have also been observed in radiation-induced myelocytic leukemias [Azumi and Sachs, 1977]. As methods improve, we may expect that a growing number of even smaller deletions or other subtle rearrangements will be found.

Gene amplification occurs in a variety of murine tumors and is sometimes indicated by the occurrence of DMs or HSRs. In two related adrenocortical carcinoma lines, the DMs of one and HSR of the other contain amplified copies of the c-Ki-ras gene [Schwab et al, 1983]. In nine of 17 T-cell lymphomas (whether spontaneous or induced by X-ray, viruses, or mineral oil), amplification of endogenous mouse mammary tumor virus sequences has been observed. This is associated with generation of novel virus-cell DNA junction fragments, indicative of rearrangement of viral sequences [Dudley and Risser, 1984]. Enhanced expression of a specific set of intracisternal A-particle (endogenous retroviruslike) genes has been demonstrated in five of seven plasmacytomas, with two- to fourfold gene amplification in MOPC 315, with the extra copies integrated into new sites [Shen-Ong and Cole, 1984]. Insertion of such an endogenous viral LTR has resulted in activation of the c-mos oncogene in one plasmacytoma line [Kuff et al, 1983]. SV40 gene amplification has been seen in some, but not all, of the brain tumors that occur invariably in transgenic mice containing part of the SV40 genome [Brinster et al, 1984]. This system is of sufficient interest to warrant further discussion.

Brinster et al [1984] microinjected an engineered SV40-containing plasmid into one of the two pronuclei of a series of fertilized mouse eggs, implanted the embryos into uteri of pseudopregnant females, and examined the resultant transgenic mice and their progeny over several generations. SV40-T antigen was produced, at least by a few months of age, when tumors of the choroid plexus made their appearance in every transgenic mouse. In some lines, the affected males did not live long enough to reproduce, and transmission was via the female line to both male and female progeny. In the lines of descent from all the original transgenic mice, the pedigrees indicated a simple dominant mode of inheritance of the integrated viral genome and the brain tumors they caused. Amplification of the SV40 gene had occurred in some of the tumors but not in all. In the absence of cytogenetic studies, it is uncertain whether other characteristic chromosome changes were present in these SV40-induced neoplasms.

Studies in Somatic Cell Hybrids

Malignancy of mouse tumor cells is suppressed when they are hybridized with nonmalignant mouse cells; reexpression of the malignant phenotype in some cases occurs when one or both copies of chromosome 4 from the nontumor parent are lost [Jonasson et al, 1977]. Further analysis using naturally polymorphic chromosome 4 has shown that the gene on chromosome 4 responsible for the suppression of malignancy acts in a dose-dependent manner, with a decrease in copies of the chromosome 4 from the normal parent or an increase in copies of the chromosome 4 from the malignant parent leading to malignancy in hybrid cells [Evans et al, 1982]. It would be interesting to know whether the recurrence of the malignant phenotype in the hybrid cells is associated with activation or enhanced expression of a known oncogene. It is associated with a systematic change in glycosylation of a membrane glycoprotein that is involved in glucose transport [Bramwell and Harris, 1978]; this change may be responsible for the reduced Km (Michaelis constant) for hexose transport characteristic of most cancer cells [White et al, 1983]. It is interesting that fusion of malignant human HeLa cells with human diploid fibroblasts also produces hybrids with greatly reduced malignancy and that reexpression of the more highly malignant phenotype of the HeLa parent is associated with loss of a single chromosome 11 and a single chromosome 14 [Stanbridge, 1981].

The use of somatic cell hybrids to study the malignant process has led to other important insights into carcinogenesis, one of which concerns tissue specificity. The rearranged immunoglobulin gene and the translocated c-myc oncogene are transcribed at high levels in human Burkitt lymphoma cells and in hybrids between Burkitt lymphoma and mouse plasmacytoma cells but are repressed at the level of transcription in hybrids between Burkitt lymphoma and mouse 3T3 cells [Nishikura et al, 1984]. This means that the 3T3 cell assay is not suitable for detecting oncogenes active in B cells and serves to emphasize that, in order to be effective in carcinogenesis, a chromosome rearrangement must place an oncogene adjacent to a promoter that is active or can be activated in the particular cell or tissue in question.

REFERENCES

Alitalo K, Schwab M, Lin CC, Varmus HE, Bishop JM: Homogeneously staining chromosomal regions contain amplified copies of an abundantly expressed cellular oncogene (c-*myc*) in malignant neuroendocrine cells from a human colon carcinoma. Proc Natl Acad Sci USA 80:1707–1711, 1983.

Alitalo K, Winqvist R, Lin CC, de la Chapelle A, Schwab M, Bishop JM: Aberrant expression of an amplified c-*myb* oncogene in two cell lines from a colon carcinoma. Proc Natl Acad Sci USA 81:4534–4538, 1984.

Allderdice PW, Miller OJ, Miller DA, Klinger HP: Spreading of inactivation in an (X;14) translocation. Am J Med Genet 2:233–240, 1978.

Alt FW, Kellems RE, Bertino JR, Schimke RT: Selective multiplication of dihydrofolate reductase genes in methotrexate-resistant variants of cultured murine cells. J Biol Chem 253:1357–1370, 1978.

Azumi JI, Sachs L: Chromosome mapping of the genes that control differentiation and malignancy in myeloid leukemia cells. Proc Natl Acad Sci USA 74:253–257, 1977.

Baer R, Chen K-C, Smith SD, Rabbitts TH: Fusion of an immunoglobulin variable gene and a T cell receptor constant gene in the chromosome 14 inversion associated with T cell tumors. Cell 43:705–713, 1985.

Berger R, Bloomfield CD, Sutherland GR: Report of the Committee on chromosome rearrangements in neoplasia and on fragile sites. Cytogenet Cell Genet 40:490–535, 1985.

Bianchi NO: Gene controlled condensation in individual chromosomes. Experientia 38:792–794, 1982.

Blackburn EH: Telomeres: Do the ends justify the means? Cell 37:7–8, 1984.

Bloom KS, Carbon J: Yeast centromere DNA is in a unique and highly ordered structure in chromosomes and small circular minichromosomes. Cell 29:305–317, 1982.

Bludau H, Kopun M, Werner D: Cell cycle-dependent expression of nuclear matrix proteins of Ehrlich ascites cells studied by in vitro translation. Exp Cell Res 165:269–282, 1986.

Bond DJ, Chandley AC: "Aneuploidy." New York: Oxford University Press, 1983, pp. 55–91.

Bramwell ME, Harris H: Some further information about the abnormal membrane glycoprotein associated with malignancy. Proc R Soc Lond [Biol] 203:93–99, 1978.

Brandriff B, Gordon L, Ashworth L, Watchmaker G, Carrano A, Wyrobek A: Chromosomal abnormalities in human sperm: Comparisons among four healthy men. Hum Genet 66:193–201, 1984.

Brinster RL, Chen HY, Messing A, van Dyke T, Levine AJ, Palmiter RD: Transgenic mice harboring SV40 T-antigen genes develop characteristic brain tumors. Cell 37:367–379, 1984.

Brodeur GM, Green AA, Hayes FA, Williams KJ, Williams DL, Tsiatis AA: Cytogenetic features of human neuroblastomas and cell lines. Cancer Res 41:4678–4686, 1981.

Capanna E, Gropp A, Winking H, Noack G, Civitelli M-V: Robertsonian metacentric in the mouse. Chromosoma 58:341–353, 1976.

Carroll AJ, Crist WM, Parmley RT, Roper M, Cooper MD, Finley WH: Pre-B cell leukemia associated with chromosome translocation 1;19. Blood 63:721–724, 1984.

Cattanach BM, Papworth D, Kirk M: Genetic tests for autosomal nondisjunction and chromosome loss in mice. Mutat Res 126:189–204, 1984.

Cavenee WK, Dryja TP, Phillips RA, Benedict WF, Godbout R, Gallie BL, Murphree AL, Strong LC, White RL: Expression of recessive alleles by chromosomal mechanisms in retinoblastoma. Nature 305:779–784, 1983.

Cowell JK: Double minutes and homogenously staining regions: Gene amplification in mammalian cells. Annu Rev Genet 16:21–59, 1982.

Crews S, Barth R, Hood L, Prehn J, Calame K: Mouse c-*myc* oncogene is located on chromosome 15 and translocated to chromosome 12 in plasmacytomas. Science 218:1319–1321, 1982.

Crippa L, Ballaman J, Engel E: Trisomie partielle pour le segment 21(q11-qter) resultant d'une translocation de novo entre les chromosomes 5 et 21. Ann Genet 27:190–193, 1984.

Dalla-Favera R, Bergni M, Erikson J, Patterson D, Gallo RC, Croce CM: Human c-*myc* onc gene is located on the region of chromosome 8 that is translocated in Burkitt lymphoma cells. Proc Natl Acad Sci USA 79:7824–7827, 1982a.

Dalla-Favera RD, Wong-Staal F, Gallo RC: *Onc* gene amplification in promyelocytic leukemia cell line HL-60 and primary leukemia cells of the same patient. Nature 299:61–63, 1982b.

Davisson MT, Poorman PA, Roderick TH, Moses MJ: A pericentric inversion in the mouse. Cytogenet Cell Genet 30:70–76, 1981.

Dayton AI, Selden JR, Laws G, Dorney DJ, Finan J, Tripputi P, Emanuel BS, Rovera G, Nowell PC, Croce CM: A human c-*erb*A oncogene homologue is closely proximal to the chromosome 17 breakpoint in acute promyelocytic leukemia. Proc Natl Acad Sci USA 81:4495–4499, 1984.

de Grouchy J, Dautzenberg M-D, Turleau C, Beguin S, Chavin-Colin F: Regional mapping of clotting factors VII and X to 13q34. Expression of factor VII through chromosome 8. Hum Genet 66:230–233, 1984.

de Klein A, Kessel AG, van Grosveld G, Bartram CR, Hagemeijer A, Bootsma D, Spurr NK, Heisterkamp N, Groffen K, Stephenson JR: A cellular oncogene is translocated to the Philadelphia chromosome in chronic myelocytic leukaemia. Nature 300:765–767, 1982.

de la Chapelle A, Berger R: Report of the committee on chromosome rearrangement in neoplasia and on fragile sites. In "Human Gene Mapping 7," and Cytogenet Cell Genet 31:274–311, 1984.

de Martinville B, Francke U: The c-Ha-ras1, insulin and β-globin loci map outside the deletion associated with aniridia-Wilms' tumor. Nature 305:641–643, 1983.

Drets M, Therman E: Human telomeric 6;19 translocation chromosome with a tendency to break at the fusion point. Chromosoma 88:139–144, 1983.

Dryja TP, Cavanee W, White R, Rapport JM, Petersen R, Albert DM, Bruns GAP: Homozygosity of chromosome 13 in retinoblastoma. N Engl J Med 310:550–553, 1984.

Dudley J, Risser R: Amplification and novel locations of endogenous mouse mammary tumor virus genomes in mouse T-cell lymphomas. J Virol 49:92–101, 1984.

Earnshaw WC, Migeon BR: Three related centromere proteins are absent from the inactive centromere of a stable isodicentric chromosome. Chromosoma 92:290–296, 1985.

Epstein CJ, Cox DR, Epstein LB, Magnuson TR: Animal models for human chromosome disorders. In de la Cruz F, Shapiro RS (eds): "Research Perspectives in Cytogenetics." Baltimore: University Park Press, 1984, pp 75–95.

Erikson J, Miller DA, Miller OJ, Abcarian PW, Skurla RM, Mushinski JF, Croce CM: The c-*myc* oncogene is translocated to the involved chromosome 12 in mouse plasmacytoma. Proc Natl Acad Sci USA 82:4212–4216, 1985a.

Erikson J, Williams DL, Finan J, Nowell PC, Croce CM: Locus of the α-chain of the T-cell receptor is split by chromosome translocation in T-cell leukemias. Science 229:784–786, 1985b.

Escot C, Theillet C, Lidereau R, Spyratos F, Champeme M-H, Gest J, Callahan R: Genetic alteration of the c-*myc* proto-oncogene (MYC) in human primary breast carcinomas. Proc Natl Acad Sci USA 83:4834–4838, 1986.

Evans EP, Burtenshaw MD, Brown BB, Hennion R, Harris H: The analysis of malignancy by cell fusion. IX. Re-examination and clarification of the cytogenetic problem. J Cell Sci 56:113–130, 1982.

Evans EP, Phillips RJS: Inversion heterozygosity and the origin of X0 daughters of *Bpa*/+ female mice. Nature 256:40–44, 1975.

Fitzgerald PH, Archer SA, Morris CM: Evidence for the repeated primary non-disjunction of chromosome 21 as a result of premature centromeric division (PCD). Hum Genet 72:58–62, 1986.

Fitzgerald PH, Pickering AF, Mercer JM, Miethke PM: Premature centromere division: A mechanism of non-disjunction causing X chromosome aneuploidy in somatic cells of man. Ann Hum Genet 38:417–428, 1975.

Flintoff WF, Livingston E, Duff C, Worton RG: Moderate-level gene amplification in methotrexate-resistant Chinese hamster ovary cells is accompanied by chromosomal translocations at or near the site of the amplified DHFR gene. Mol Cell Biol 4:69–76, 1984.

Fukushige S, Matsubara K-I, Yoshida M, Sasaki M, Suzuki T, Semba K, Toyoshima K, Yamamoto T: Localization of a novel v-*erb*B-related gene, c-*erb*B-2, on human chromosome 17 and its amplification in a gastric cancer cell line. Mol Cell Biol 6:955–958, 1986.

Genest P, Genest FB, Gagnon-Blair D: Un remaniement chromosomique inhabituel. Une translocation teleomerique autosomique sur un Y a satellites (Yqs) multicentenaire. Ann Genet 26:86–90, 1983.

German J: Patterns of neoplasia associated with the chromosomal breakage syndrome. In German J (ed): "Chromosome Mutation and Neoplasia." New York: Alan R. Liss, Inc., 1983, pp 97–134.

Groffen J, Heisterkamp N, Stephenson JR, van Kessel G, de Klein A, Grosveld G, Bootsma D: c-*sis* is translocated from chromosome 22 to chromosome 9 in chronic myelocytic leukemia. J Exp Med 158:9–15, 1983.

Groffen J, Stephenson JR, Heisterkamp N, de Klein A, Bartram CR, Grosveld G: Philadelphia chromosomal breakpoints are clustered within a limited region, bcr, on chromosome 22. Cell 36:93–99, 1984.

Gropp A: Value of an animal model for trisomy. Virchows Arch [Pathol Anat] 395:117–131, 1982.

Hanks SK, Rodriguez LV, Rao PN: Relationship between histone phosphorylation and premature chromosome condensation. Exp Cell Res 148:293–302, 1983.

Hartley DA, Davies KE, Drayna D, White RL, Williamson R: A cytological map of the human X chromosome—Evidence for non-random recombination. Nucleic Acids Res 12:5277–5285, 1984.

Hauser-Urfer IH, Stauffer J: Comparative chromosome analysis of nine squamous cell carcinoma lines from tumors of the head and neck. Cytogenet Cell Genet 39:35–39, 1985.

Hayashi K, Kakizoi T, Sugimura T: *In vivo* amplification and rearrangement of the c-Ha-*ras*-1 sequence in a human bladder carcinoma. Gann 74:798–801, 1983.

Heisterkamp N, Groffen J, Stephenson JR, Spurr NK, Goodfellow PN, Solomon E, Carritt B, Bodmer WF: Chromosomal localization of human cellular homologues of two viral oncogenes. Nature 299:747–749, 1982.

Hieter P, Pridmore D, Hegemann JH, Thomas M, Davis RW, Philippsen P: Functional selection and analysis of yeast centromere DNA. Cell 42:913–921, 1985.

Hilkens J, Cuypers HT, Selten G, Kroezen V, Hilgers J, Berns A: Genetic mapping of PIM-1 putative oncogene to mouse chromosome 17. Somatic Cell Mol Genet 12:81–88, 1986.

Hotta Y, Stern H: Small nuclear RNA molecules that regulate nuclease accessibility in specific chromatin regions of meiotic cells. Cell 17:309–319, 1981.

Jabs EW, Meyers DA, Bias WB: Linkage studies of polymorphic, repeated DNA sequences in centromeric regions of human chromosomes. Am J Hum Genet 38:297–308, 1986.

Jabs EW, Wolf SF, Migeon BR: Characterization of a cloned DNA sequence that is present at centromeres of all human autosomes and the X chromosome and shows polymorphic variation. Proc Natl Acad Sci USA 81:4884–4888, 1984.

Jackson DA, Caton AJ, McCready SJ, Cook PR: Influenza virus RNA is synthesized at fixed sites in the nucleus. Nature 296:366–368, 1982.

Jonasson J, Povey S, Harris H: The analysis of malignancy by cell fusion VII. Cytogenetic analysis of hybrids between malignant and diploid cells and of tumours derived from them. J Cell Sci 24:217–254, 1977.

Kaelbling M, Miller DA, Miller OJ: Restriction enzyme banding of mouse metaphase chromosomes. Chromosoma 90:128–132, 1984.

Kaiser B, Förster W, Steuernagel P, Hillig U, Herberg K-P: Familial pericentric inversion (14)(p11;q24) with a rec dup (q) in one offspring. Clin Genet 26:73–76, 1984.

Keath EJ, Kelekar A, Cole MD: Transcriptional activation of the translocated c-*myc* oncogene in mouse plasmacytomas: Similar RNA levels in tumor and proliferating normal cells. Cell 37:521–528, 1984.

Ketupanya A, Crandall BF, Blanchard K, Rogers DW: Mosaic hexasomy 21. J Med Genet 21:228–230, 1984.

King CR, Kraus MH, Aaronson SA: Amplification of a novel c-erbB-related gene in a human mammary carcinoma. Science 229:974–976, 1985.

Kohl NE, Kanda N, Schreck RR, Bruns G, Latt SA, Gilbert F, Alt FW: Transposition and amplification of oncogene-related sequences in human neuroblastomas. Cell 35:359–367, 1983.

Kornberg A: "DNA Replication." San Francisco: WH Freeman and Company, 1980.

Kucerova M, Polivkova Z, Dluholucky S, Kvasnicova M: Terminal deletion with stable acentric fragment of 1q in a child with congenital malformations. Am J Hum Genet 35:91–95, 1983.

Kuff EL, Feenstra A, Lueders K, Rechavi G, Givol D, Canaani E: Homology between an endogenous viral LTR and sequences inserted in an activated cellular oncogene. Nature 302:547–548, 1983.

Kustnetsova LE, Prigogina EL, Pogosianz HE, Belkina BM: Similar chromosomal abnormalities in several retinoblastomas. Hum Genet 61:201–204, 1982.

LeBeau MM, Diaz MO, Karin M, Rowley JD: Metallothionein gene cluster is split by chromosome 16 rearrangements in myelomonocytic leukemia. Nature 313:709–711, 1985.

LeBeau MM, Larson RA, Bitter MA, Vardiman JW, Golomb HM, Rowley JD: Association of inv(16)(p13q22) with abnormal marrow eosinophila in acute myelomonocytic leukemia: A unique cytogenetic clinicopathologic association. N Engl J Med 309:630–636, 1983.

Lee W-H, Murphree AL, Benedict WF: Expression and amplification of the N-*myc* gene in primary retinoblastoma. Nature 309:458–460, 1984.

Libermann TA, Nubaum HR, Razon N, Kris R, Lax I, Soreq H, Whittle N, Waterfield MD, Ullrich A, Schlessinger J: Amplification, enhanced expression and possible rearrangement of the EGF receptor gene in primary human brain tumors of glial origin. Nature 313:144–147, 1985.

Little CD, Nau MM, Carney DN, Gazdar AF, Minna JD: Amplification and expression of the c-*myc* oncogene in human lung cancer cell lines. Nature 306:194–196, 1983.

Lyon MF: Distribution of crossing-over in mouse chromosomes. Genet Res 28:291–299, 1976.

Manuelidis L: Individual interphase chromosome domains revealed by in situ hybridization. Hum Genet 71:288–293, 1985.

Matsumoto Y, Yasuda H, Mita S, Marunochi T, Yamada M: Evidence for the involvement of H1 histone phosphorylation in chromosome condensation. Nature 284:181–183, 1980.

Mattei MG, Souiah N, Mattei JF: Chromosome 15 anomalies and the Prader-Willi syndrome: Cytogenetic analysis. Hum Genet 66:313–334, 1984.

McCoy MS, Toole JJ, Cunningham EH, Lowy DR, Weinberg RA: Characterization of a human colon/lung carcinoma oncogene. Nature 302:79–81, 1983.

McGhee JD, Nickol JM, Felsenfeld G, Rau DC: Higher order structure of chromatin orientation of nucleosomes within the 30 nm chromatin solenoid is independent of species and spacer length. Cell 33:831–841, 1983.

McGrath J, Solter D: Completion of mouse embryogenesis requires both the maternal and paternal genomes. Cell 37:179–183, 1984.

Miller DA, Choi Y-C, Miller OJ: Chromosome localization of highly repetitive human DNAs and amplified rDNA by use of restriction enzymes. Science 219:395–397, 1983.

Miller DA, Dev VG, Tantravahi R, Miller OJ: Suppression of human nucleolus organizer activity in mouse-human somatic hybrid cells. Exp Cell Res 101:235–243, 1976.

Miller DA, Miller OJ: Chromosomes and cancer in the mouse: Studies in tumors, established cell lines and cell hybrids. Adv Cancer Res 39:153–182, 1983.

Miller OJ, Tantravahi U, Katz R, Erlanger BF, Guntaka RV: Amplification of mammalian ribosomal RNA genes and their regulation by methylation. In Arrighi FE, Rao PN, Stubblefield E (eds): "Genes, Chromosomes and Neoplasia." New York: Raven Press, 1981, pp 253–270.

Mitchell AR, Gosden JR, Miller DA: p82H: A cloned sequence of the alphoid repetitive DNA family found at the centromere of all human chromosomes. Chromosoma 92:369–377, 1985.

Mitelman F: Catalogue of chromosome aberrations in cancer. Cytogenet Cell Genet 36:1–515, 1983.

Morton CC, Taub R, Diamond A, Lane MA, Cooper GM, Leder P: Mapping of the human *Blym*-1 transforming gene activated in Burkitt lymphomas to chromosome 1. Science 223:173–175, 1984.

Moses MJ: The synaptonemal complex and meiosis. In Sparkes RS, Comings D, Fox CF (eds): "Molecular Human Cytogenetics (ICN-UCLA Symposia on Molecular and Cellular Biology, Vol VIII)." New York: Academic Press, 1977, pp 101–125.

Mücke J, Trautmann U, Sanding K-R, Theile H: The crucial band for phenotype of trisomy 18. Hum Genet 60:205, 1982.

Mukherjee AB, Fabricant JD: Possible use of histone acetylation in mixed leukocyte cultures to assess histocompatibility. Nature 273:467–469, 1978.

Nakagome Y, Abe T, Misawa S, Takeshita T, Iinuma K: The "loss" of centromeres of aged women. Am J Hum Genet 36:398–404, 1984.

Narahara K, Kikkawa K, Kimura S, Kimoto H, Ogata M, Kasai R, Hamawaki M, Matsuoka K: Regional mapping of catalase and Wilms' tumor-aniridia, genitourinary abnormalities, and mental retardation triad loci to the chromosome segment 11p1305→p1306. Hum Genet 66:181–185, 1984.

Neel BG, Jhanwar SC, Chaganti RSK, Hayward WS: Two human c-*onc* genes are located on the long arm of chromosome 8. Proc Natl Acad Sci USA 79:7842–7846, 1982.

Niikawa N, Ishikiriyama S: Clinical and cytogenetic studies of the Prader-Willi syndrome: Evidence of phenotype-karyotype correlation. Hum Genet 69:22–27, 1985.

Nishikura K, Ar-Rushidi A, Erikson J, De Jesus E, Dugan D, Croce CM: Repression of rearranged μ gene and translocated c-*myc* in mouse 3T3 cells X Burkitt lymphoma cell hybrids. Science 224:399–402, 1984.

Parrington JM, West LF, Povey J: The origin of ovarian teratomas. J Med Genet 21:4–12, 1984.

Peretti D, Maraschio P, Lambiase S, Lo Curto F, Zuffardi O. Indirect immunofluorescence of inactive centromeres as indicators of centromeric function. Hum Genet 73:12–16, 1986.

Peterson JB, Ris H: Electron-microscopic study of the spindle and chromosome movement in the yeast *Saccharomyces cerevisiae*. J Cell Sci 22:219–242, 1976.

Reeves R, Candido EPM: Partial inhibition of histone deacetylase in active chromatin of HMG 14 and HMG 17. Nucleic Acids Res 8:1947–1963, 1980.

Ris H, Witt PL: Structure of the mammalian kinetochore. Chromosoma 82:153–170, 1981.

Roderick TH: Chromosomal inversions in studies of mammalian mutagenesis. Genetics 92:s121–s126, 1979.

Rouyer F, Simmler M-G, Johnsson C, Vergnaud G, Cooke HJ, Weissenbach J: A gradient of sex linkage in the pseudo-autosomal region of the human sex chromosomes. Nature 319:291–295, 1986.

Rowley JD, Testa JR: Chromosome abnormalities in malignant hematologic disease. Adv Cancer Res 36:103–148, 1982.

Russell LB: Numerical sex-chromosome anomalies in mammals: Their spontaneous occurrence and use in mutagenesis studies. In Hollaender A (ed): "Chemical Mutagens." New York: Plenum Press, 1976, Vol 4, pp 55–91.

Saadallah N, Hulten M: Chiasma distribution, genetic lengths, and recombination fractions: A comparison between chromosomes 15 and 16. J Med Genet 20:290–299, 1983.

Sahasrabuddhe CG, Adlakah RC, Rao PN: Phosphorylation of non-histone proteins associated with mitosis in HeLa cells. Exp Cell Res 153:439–450, 1984.

Sandberg AA: "The Chromosomes in Human Cancer and Leukemia." New York: Elsevier Press, 1980.

Schmiady H, Sperling K: Length of human prematurely condensed chromosomes during G0 and G1 phase. Exp Cell Res 134:461–465, 1981.

Schwab M, Alitalo K, Varmus HE, Bishop JM, George D: A cellular oncogene (*c*-Ki-*ras*) is amplified, over-expressed, and located within karyotypic abnormalities in mouse adrenocortical tumour cells. Nature 303:497–501, 1983.

Searle AG, Beachey CV, Boer P, De Rooij DG, Evans EP, Kirk M: A male-sterile insertion in the mouse. Cytogenet Cell Genet 36:617–626, 1983.

Seidman MM, Levine AJ, Weintraub H: The asymmetric segregation of parental nucleosomes during chromosome replication. Cell 18:439–449, 1979.

Selden JR, Emanuel BS, Wang E, Cannizzaro L, Palumbo A, Erikson J, Nowell PC, Rovera G, Croce CM: Amplified Cλ and c-*abl* genes are on the same marker chromosome in K562 leukemia cells. Proc Natl Acad Sci USA 80:7289–7292, 1983.

Semba K, Kamata N, Toyoshima N, Yamamoto T: A v-erbB-related protooncogene, c-*erb*B-2, is distinct from the c-*erb*B-1/EGF receptor gene and is amplified in a human salivary gland adenocarcinoma. Proc Natl Acad Sci USA 82:6497–6501, 1985.

Shen-Ong GLC, Cole MD: Amplification of a specific set of intracisternal A-particle genes in a mouse plasmacytoma. J Virol 49:171–177, 1984.

Sherman SL, Morton NE, Jacobs PA, Turner G: The marker (X) syndrome: A cytogenetic and genetic analysis. Ann Hum Genet 48:21–37, 1984.

Shiraishi Y, Taguchi H, Niiya K, Shiomi F, Kikukawa K, Kubonishi S, Ohmura T, Hamawaki M, Ueda N: Diagnostic and prognostic significance of chromosome abnormalities in marrow and mitogen response of lymphocytes of acute nonlymphocytic leukemia. Cancer Genet Cytogenet 5:1–14, 1982.

Singer MF: Highly repeated sequences in mammalian genomes. Int Rev Cytol 76:67–112, 1982.

Solari AJ: Synaptonemal complexes and associated structure in microspread human spermatocytes. Chromosoma 81:315–337, 1980.

Solomon E: Recessive mutation in aetiology of Wilms' tumor. Nature 309:111–112, 1984.

Sperling K, Rao PN: The phenomenon of premature chromosome condensation: Its relevance to basic and applied research. Humangenetick 23:235–258, 1974.

Squire J, Phillips RA, Boyce S, Godbout R, Rogers B, Galie BL: Isochromosome 6p, a unique chromosomal abnormality in retinoblastoma: Verification by standard staining techniques, new densitometric methods, and somatic cell hybridization. Hum Genet 66:46–53, 1984.

Stanbridge EJ, Flandermeyer RR, Daniels DW, Nelson-Rees WA: Specific chromosome loss associated with the expression of tumorigenicity in human hybrids. Somatic Cell Mol Genet 7:699–712, 1981.

Steinmetz M, Minard K, Horvath S, McNicholas J, Frelinger J, Wake C, Long E, Mach B, Hood L: A molecular map of the immune response region from the major histocompatibility complex of the mouse. Nature 300:35–42, 1982.

Sunkara PS, Wright DA, Rao PN: Mitotic factors from mammalian cells induce germinal vesicle breakdown and chromosome condensation in amphibian oocytes. Proc Natl Acad Sci USA 76:2799–2802, 1979.

Sutherland GR, Jacky PB, Baker E, Manuel A: Heritable fragile sites on human chromosomes. X. New folate-sensitive fragile sites: 6p23, 9p21, 9p32, and 11q23. Am J Hum Genet 35:432–437, 1983.

Takahashi R, Mihara K, Maeda S, Yamaguchi T, Chen HL, Aoyama N, Murao S, Hatanaka M, Sugiyama T: Secondary activation of C-ABL may be related to translocation to the nucleolar organizer region in an in vitro cultured rat leukemia cell line. Proc Natl Acad Sci USA 83:1079–1083, 1986.

Taub R, Kirsch I, Morton C, Lenoir G, Swan D, Tronick S, Aaronson S, Leder P: Translocation of the c-*myc* gene into the immunoglobulin heavy chain locus in human Burkitt lymphoma and murine plasmacytoma cells. Proc Natl Acad Sci USA 79:7837–7841, 1982.

Tomkins D, Hunter A, Roberts M: Cytogenetic findings in Roberts-SC phocomelia syndrome(s). Am J Med Genet 4:17–26, 1979.

Tranejbaerg L, Petersen A, Hove K, Rehdler H, Mikkelsen M: Clinical and cytogenetic studies in large (4;8) translocation family with pre- and post-natal Wolf syndrome. Ann Genet 27:224–229, 1984.

Trent J, Meltzer P, Rosenblum M, Horsh G, Kinzler K, Mashel R, Feinberg A, Vogelstein B: Evidence for rearrangement, amplification, and expression of c-*myc* in a human glioblastoma. Proc Natl Acad Sci USA 83:470–473, 1986.

Tsujimoto Y, Yunis J, Onorato-Showe L, Erikson J, Nowell PC, Croce CM: Molecular cloning of the chromosomal breakpoint of B-cell lymphomas and leukemias with the t(11;14) chromosome translocation. Science 224:1403–1406, 1984.

Turleau C, Chavin-Colin F, de Grouchy J, Maroteaux P, Rivera H: Langer-Giedion syndrome with and without del 8q. Assignment of critical segments to 8q23. Hum Genet 62:183–187, 1982.

Turleau C, de Grouchy J, Chavin-Colin F, Dore F, Seger J, Dautzenberg MD, Arthuis M, Jeanson C: Two patients with interstitial del (14q), one with features of Holt-Oram syndrome. Exclusion mapping of PI (alpha-1-antitrypsin). Ann Genet 27:237–240, 1984a.

Turleau C, de Grouchy J, Chavin-Colin F, Martelli H, Voyer M, Charlas R: Trisomy 11p15 and Beckwith-Wiedemann syndrome. A report of two cases. Hum Genet 67:219–221, 1984b.

Ullrich A, Coussens L, Hayflick JS, Dull TJ, Gray A, Tam AW, Lee J, Yarden Y, Libermann TA, Schlessinger J, Downward J, Mayes ELV, Whittle N, Waterfield MD, Seeburg PH: Human epidermal growth factor receptor cDNA sequence and aberrant expression of the amplified gene in A431 epidermoid carcinoma cells. Nature 309:418–425, 1984.

Verellen-Dumoulin C, Freund M, De Meyer R, Laterre C, Frederic J, Thompson MW, Markovic VD: Expression of an X-linked muscular dystrophy in a female due to translocation involving Xp21 and non-random inactivation of the normal X chromosome. Hum Genet 67:115–119, 1984.

Villasante A, Corces VG, Manso-Martinez R, Avila J: Binding of microtubule protein to DNA and chromatin: Possibility of simultaneous linkage of microtubule to nucleic acid and assembly of the microtubule structure. Nucleic Acids Res 9:895–908, 1981.

Vogelstein B, Pardoll DM, Coffey DS: Supercoiled loops and eukaryotic DNA replication. Cell 22:79–85, 1980.

Wahl GM, Vitto L, Rubnitz J: Co-amplification of rRNA genes with CAD genes in N-(phosphonacetyl)-L-aspartate-resistant Syrian hamster cells. Mol Cell Genet 3:2066–2075, 1983.

Wake N, Seki T, Fujita H, Okubo H, Sakai K, Okuyama K, Hayashi H, Shiina Y, Sato H, Kuroda M, Ichinoe K: Malignant potential of homozygous and heterozygous complete moles. Cancer Res 44:1226–1230, 1984.

Ward P, Packman S, Longhman W, Sparkes M, Sparkes R, McMahon A, Gregory T, Ablin A: Location of the retinoblastoma susceptibility gene(s) and the human esterase D locus. J Med Genet 21:92–95, 1984.

Weisbrod S, Weintraub H: Isolation of a subclass of nuclear proteins responsible for conferring a DNase I-sensitive structure on globin chromatin. Proc Natl Acad Sci USA 76:630–634, 1979.

Whang-Peng J, Bunn PA Jr, Kao-Shan CS, Lee EC, Carney DN, Gazdar A, Minna JD: A non-random chromosomal abnormality, del3p(14–23), in human small cell lung cancer. Cancer Genet Cytogenet 6:119–134, 1982.

White MK, Bramwell ME, Harris H: Kinetic parameters of hexose transport in hybrids between malignant and non-malignant cells. J Cell Sci 62:49–80, 1983.

Wiener F, Ohno S, Babonits M, Sumegi J, Wirschubsky Z, Klein G, Mushinski JF, Potter M: Hemizygous interstitial deletion of chromosome 15 (band D) in three translocation-negative murine plasmacytomas. Proc Natl Acad Sci USA 81:1159–1168, 1984.

Willard HF: Chromosome-specific organization of human alphoid repeated DNA. Cytogenet Cell Genet 37:608–609, 1984.

Willard HF, Waye JS, Skolnick MH, Schwartz CE, Powers VE, England SB: Detection of restriction fragment length polymorphisms at the centromeres of human chromosomes by using chromosome-specific α-satellite DNA probes: Implications for development of centromere-based genetic linkage maps. Proc Natl Acad Sci USA 83:5611–5615, 1986.

Woodcock CLS, Frado L-LY, Rattner JB: The higher-order structure of chromatin: Evidence for a helical ribbon arrangement. J Cell Biol 99:42–52, 1984.

Yamamoto T, Kamata N, Kawano H, Shimizu S, Kuroki T, Toyoshima N, Rikimaru K, Nomura N, Ishizaki R, Pastan I, Gamow S, Shimizu N: High incidence of amplification of the epidermal growth factor receptor gene in human squamous cell carcinoma cell lines. Cancer Res 46:414–416, 1986.

Yunis JJ: The chromosomal basis of human neoplasia. Science 221:217–236, 1983.

Zuffardi O, Fraccaro M: Gene mapping and serendipity. The locus for torticollis, keloids, cryptorchidism and renal dysplasia (31430, McKusick) is at Xq28, distal to the G6PD locus. Hum Genet 62:280–281, 1982.

Chromosome Mapping Using DNA Probes: A Case Study (Chromosome 11) and Future Possibilities

Stylianos E. Antonarakis, MD

The Johns Hopkins University School of Medicine, Baltimore, MD 21205

In the last 5 years, recombinant DNA technology has greatly enhanced our understanding of the structure and function of eukaryotic genes and has facilitated the mapping of human genes to specific human chromosomes. In this chapter some of the methods of gene mapping will be discussed, using human chromosome 11 as an example. The usefulness of creating gene maps is briefly summarized.

GENE MAPPING USING SOMATIC CELL HYBRIDS AND RESTRICTION ENDONUCLEASE ANALYSIS

Somatic cell hybrids are produced by fusing cells of different species, usually rodent and human. Since the 1960s, they have provided an important tool for gene mapping [Puck and Kao, 1982]. Human chromosomes are selectively lost from rodent-human hybrid cells, producing a hybrid cell line containing the full set of mouse chromosomes but only one or few human chromosomes. It is then possible to study the gene products from the few human chromosomes of the hybrid cell. Using this approach, many genes have been mapped to human chromosomes [McKusick and Ruddle, 1977; Ruddle, 1981]. Obviously, however, when the presence of a particular human gene is recognized by the gene product, only genes that are expressed in the hybrid cells can be mapped. On the other hand, by analysis of restriction

Birth Defects: Original Article Series, Volume 23, Number 3, pages 65-76

endonuclease digests of genomic DNA, any gene or DNA sequence can be mapped in hybrid cells if the necessary probe is available. DNA analysis determines the presence or absence of the gene regardless of its expression in the hybrid cell.

Restriction Endonuclease Analysis

Restriction endonuclease analysis (REA) of DNA from somatic cell hybrids has enabled several genes to be assigned to particular human chromosomes. In this discussion, the human chromosome 11 will be used as a case study. The β-globin gene cluster, including the genes for embryonic ϵ, fetal $^{G}\gamma$ and $^{A}\gamma$, adult δ- and β-globin genes, as well as the pseudogene $\psi\beta_1$, were localized to the short arm of chromosome 11 [Scott et al, 1979]. Shortly thereafter, as probes for other important human genes became available, the genes for insulin [de Martinville et al, 1982], c-Ha-ras-1 oncogene [McBride et al, 1982], parathyroid hormone (PTH) [Naylor et al, 1982], calcitonin [Hoppener et al, 1984; Przepiorka et al, 1984], catalase [Korneluk et al, 1984], and insulin growth factor-2 [Brissenden et al, 1984] were localized to the short arm of the same chromosome. Other genes, namely, the apoproteins A1 and C3 genes [Cheung et al, 1984], have been localized to the long arm (11q).

This method of mapping is able to localize genes to one particular chromosome or arm of the chromosome, but only in some cases to a particular chromosomal band or subband. For example, let us suppose that two hybrid cell lines contain parts of the human chromosome 11 as the only human chromosome. The first contains chromosome 11 from qter−p13 and the second from p12 to pter, and in both cell lines gene A is present. The region 11p12−p13 is the smallest region of overlap (SRO) [Sparkes et al, 1984] between the two cell lines and is the location of gene A. Hybrid cell lines that contain chromosomal rearrangements (translocations) or other chromosomal abnormalities (deletions, duplications) [The Human Genetic Mutant Repository, 1983] are also useful in gene mapping.

In some cases, the exact breakpoints of such chromosomal events (translocations or deletions) are not well defined, and this uncertainty makes mapping difficult. For example, considerable controversy has arisen in the last year concerning the localization of the β-globin gene cluster. This cluster has been localized in bands 11p1105–1108 using the so-called Puck lines, ie, hybrid cells between Chinese hamster ovary (CHO) and human cells retaining different portions of chromosome 11 after treatment of the hybrid cells with the mutagens, X-irradiation, or N-methyl-N-nitro-N-nitrosoguanidine [Puck, 1983; Gusella et al, 1979]. Another laboratory using hybrid cells with different deletions on the human chromosome 11 had localized the β-globin gene cluster on 11p15 [de Martinville and Francke, 1983]. Therefore, the need for other additional gene mapping techniques is obvious.

GENE MAPPING BY IN SITU HYBRIDIZATION

This is a powerful technique for microscopic identification of gene location by hybridization between the probe (cloned radioactive DNA sequence) and its homologous DNA sequence on a particular chromosome [Harper et al, 1981; Gerhard et al, 1981]. Briefly, in situ hybridization requires incubation of the radioactive probe with chromosomal preparations that have been previously treated with RNase and denatured to single strands. The hybridized probe releases silver grains in an overlaid autoradiographic emulsion. The chromosome preparations can be banded (G bands) to determine the precise location of the grains. Typically, not all grains appear to be in the same position on the chromosome, and a histogram of the number of grains per chromosome can be constructed for each chromosome. This allows the sequence of interest to be localized to a chromosomal band with a certain degree of confidence. Longer, ie, extended, chromosomes allow more accurate localization. Chaganti et al [1985] have developed an in situ hybridization assay using long prometaphase meiotic chromosomes from testicular tissue to better localize genes to specific chromosomes.

In recent years insulin [Harper et al, 1981], PTH [Chaganti et al, 1985], c-Ha-ras-1 [Jhanwar et al, 1983], and the β-globin gene cluster [Chaganti et al, 1985] have all been localized to the short arm of chromosome 11 using in situ hybridization (see Fig. 1). As with the somatic cell hybrid technique, considerable controversy exists concerning the results of in situ hybridization in different laboratories. For example, the β-globin gene cluster has been localized to 11p11 [Chaganti et al, 1985] or 11p15 [Morton et al, 1984], respectively, by two different groups of investigators. Considerable evidence now exists that the latter result is more accurate. The presence of pseudogenes or gene families that can be recognized by a given probe may make precise localization of genes difficult.

GENE MAPPING USING LINKAGE ANALYSIS

Another mapping approach is the linkage method. The distance between genes is determined by analyzing their genetic recombination. For reasons that will be discussed later in this chapter, these genetic maps are the most useful in clinical medicine.

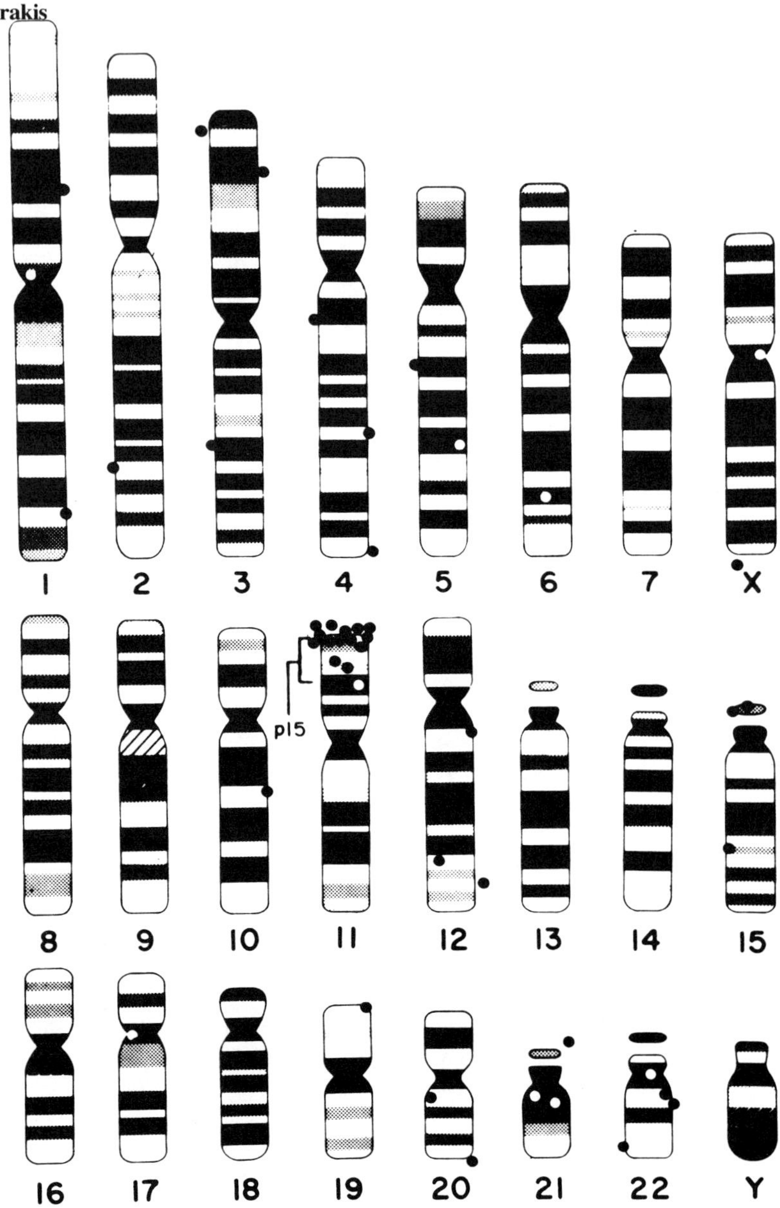

Fig. 1. Schematic representation of G-banded metaphase human chromosomes hybridized with human insulin probe. (Data compiled from 35 cells.) Note that most of the silver grains were located on the distal end of the short arm of human chromosome 11. (Reproduced from Harper et al, 1981, with permission of the publisher.)

The estimate of the genetic distance between two loci can be achieved using different methods [Conneally and Rivas, 1980]. The one most in use is the maximum likelihood estimation first introduced by N.E. Morton [1955].

Two alleles at a particular locus in an individual can be recognized at the DNA level because of the variability of the DNA sequences surrounding them. It has been estimated that one in every 500 nucleotides is different between different alleles [Jeffreys, 1979; Antonarakis et al, 1982a]. Some of these differences can be recognized using REA if the nucleotide differences happen to be in the recognition sequence of a particular endonuclease. This phenomenon is termed *restriction site DNA polymorphism* (from the Greek words πολυς: many + μορφη: face, shape) or simply *DNA polymorphism*.

There are three general categories of DNA polymorphisms.

1. Simple nucleotide changes. Most of the known polymorphisms represent changes of either one or a few nucleotides, which can be recognized by restriction analysis [Kan and Dozy, 1978].

2. Multiple insertions of small sequences into an area of DNA. For example, a sequence of 14 nucleotides can be present in a variable number of copies 5′ to the insulin gene [Bell et al, 1982], and the actual number of copies can be estimated by restriction analysis using an enzyme that cleaves DNA in sequences flanking the insertion area. In this case, not two (presence or absence of a site) but many alternative fragments can be resolved on an agarose gel after restriction analysis. Examples similar to the insulin insertion polymorphism exist in the α-globin gene cluster [Goodbourn et al, 1983] and the c-Ha-ras-1 oncogene [Capon et al, 1983].

3. The presence or absence of a pseudogene. The only known example of this rare polymorphism is the presence or absence of a processed pseudogene for the dihydrofolate reductase gene (DHFR) on chromosome 18 [Anagnou et al, 1984].

Because DNA restriction site polymorphisms are frequent, more than one polymorphic site can be found adjacent to any particular sequence. For example, in the β-globin gene cluster, 14 different polymorphic sites have been observed and studied in the last four years [Antonarakis et al, 1985a]. Let us suppose that there are two polymorphic sites adjacent to probe A. Each site can be present or absent (+ or −), and, therefore, there are four possible ways that each chromosome can be marked, namely, ++, +−, −+, −− for the two polymorphisms, respectively. The pattern of presence or absence of polymorphic sites in a single chromosome is called a *haplotype*. The number of possible haplotypes for n restriction site polymorphisms is 2^n. For example, in the β-globin gene cluster more than 15 haplotypes have been observed in studies of chromosomes 11 bearing β^{normal}, β^S, β^E, β^C, and

$\beta^{thalassemia}$ genes. Haplotypes for DNA polymorphisms are extremely useful in defining the different alleles of a given locus.

Not all polymorphic sites have the same frequency of possible alleles. Furthermore, some DNA polymorphisms are present in every ethnic group, whereas others are present in only certain groups. The DNA polymorphisms that are most useful in linkage analysis are those with a relatively high frequency of the two alleles, ie, high degree of heterozygosity in the population. A measure of the usefulness of a given polymorphism is the PIC (polymorphism information content) [Botstein et al, 1980]. The PIC value can be computed using the formula

$$PIC = 1 - P_1^2 - P_2^2 - 2\,P_1^2P_2^2.$$

For a DNA polymorphism with two alternatives (presence or absence of a given site) P_1 is the frequency of one allele and $P_2 = 1 - P_1$. (For a more detailed discussion, see Botstein et al [1980] and Bishop and Skolnick [1983].) The PIC can have values between 0 and 1. The PIC value for a marker with allele frequencies 50/50 is about 0.48 and for allele frequencies 30/70 is about 0.33. The PIC value for a multiallelic system of markers, such as the HLA-A locus, is about 0.98. In conclusion, the larger the PIC value, the more useful the DNA polymorphism is for linkage analysis studies.

Linkage Analysis

The genetic distance between two loci can be estimated by several methods. The sequential multiple likelihood analysis first described by N.E. Morton in 1955 is the most widely used. One can use DNA polymorphisms adjacent to the different genes as markers for linkage analysis. Informative families are needed in order to observe recombinants between two sets of markers. Finally, the odds for linkage vs nonlinkage at a given recombination fraction are computed, and the recombination fraction θ, where the maximum likelihood of linkage is estimated, represents the best estimate of the genetic distance between two loci. Values of lod score (logarithm of the odds of linkage vs nonlinkage for a given θ) more than 3 (which means that the likelihood of linkage is 1,000 times greater than the likelihood of nonlinkage) are considered to be very strong evidence for linkage.

Linkage map distance can be correlated approximately with physical distance on chromosomes using the assumption that cross-over events are homogeneously distributed throughout the human genome. We presently have evidence that this assumption is not correct, at least for the human β-globin gene cluster where there is an 11 kb stretch of DNA with a 30-fold increase in recombination frequency compared to neighboring sequences [Antonarakis et al, 1982b; Chakravarti et al, 1984].

It has been estimated that the total length of the haploid human genome is about 3×10^9 bp and that the total recombination distance is about 3,000 cM (centimorgans) [Renwick, 1969]. If the assumption of the previous paragraph is correct for a relatively large portion of the genome, the physical distance of 1 cM is about 1,000 kb.

The availability of DNA probes and adjacent polymorphisms has enabled several laboratories to create initial linkage maps for some human chromosomes. One of the best examples is the short arm of chromosome 11 [Antonarakis et al, 1983; Lebo et al, 1983; Fearon et al, 1984a; Gerhard et al, 1984]. We and others have conducted linkage analyses for PTH, catalase, calcitonin, β-globin, c-Ha-*ras*-1, insulin, and insulin-like growth factor-2 on 11p. DNA polymorphisms for each locus have been used in informative families, and the recombination fractions between each pair of loci have been computed. Multipoint mapping analysis has been applied for these loci [Meyers et al, 1979], and the best arrangement of all loci is shown in Figure 2 [Antonarakis et al, 1983; Fearon et al, 1984a; Kittur et al, 1985b].

Linkage maps for chromosomes X [Drayna et al, 1984], 21 [Kittur et al, 1985a], and 13 [Cavenee et al, 1984] have also been created by several

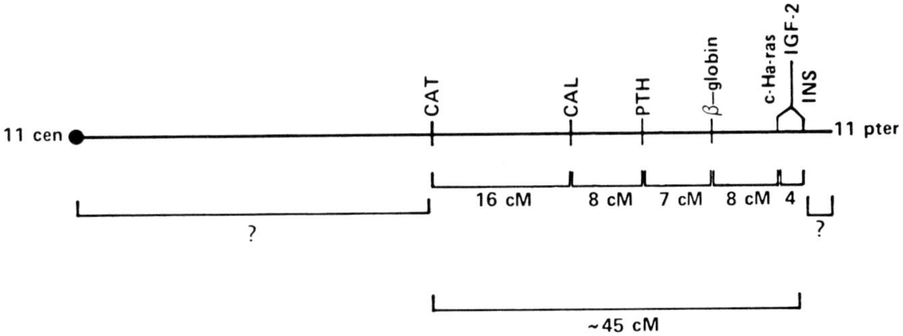

Fig. 2. Linkage map for human chromosome 11p. Genetic distances in centimorgans (cM) found between catalase, calcitonin, PTH, HBBC, HRAS1, and IGF-II loci are shown. (HBBC is designated β-globin and HRAS1 is designated C-Ha-*ras*). One centimorgan is defined as the genetic equivalent to a recombination fraction θ of 0.01. The distances between PTH, HBBC, HRAS1, and insulin genes given are those reported by us. White et al [1985] have found 11.6 cM between HBBC and insulin, 12.5 cM between HBBC and HRAS1, and 3.2 cM between insulin and HRAS1. Gerhard et al [1981] have found 9 cM between HBBC and insulin and 10 cM between HBBC and HRAS1. (Reproduced from Kittur et al, 1985b, with permission of the publisher.)

investigators, and we anticipate that in the next 3–5 years linkage maps for most of the human genome will be available. This prediction is based on the current rate of probe isolation and the rapid methods of chromosomal localization of these DNA probes. (During the preparation of this manuscript, Lebo et al [1984] published a new method of chromosomal mapping of DNA sequences. Briefly, each human chromosome can be differentially sorted using a fluorescence activated cell sorter, and the DNA from each chromosome is covalently bound to a nitrocellulose filter. A radioactive DNA probe can then be hybridized to the filter and, according to the localization of the positive signal of hybridization, the DNA sequence can be assigned to one of the human chromosomes. Using this general approach, Lebo et al first showed that the gene that is mutant in McArdle disease—muscle myophosphorylase—maps to the long arm of chromosome 11.)

Usefulness of Linkage Maps

Today the value of linkage maps to science and to clinical medicine, in particular, is quite clear. Such maps can be useful in the isolation of disease-related genes, heterozygote detection of carriers of deleterious genes, and prenatal diagnosis of single gene disorders. It has been estimated that only about 300–500 markers spaced about 10–15 cM apart from each other are needed [Botstein et al, 1980] to provide the entire genome with useful markers (DNA polymorphisms). Using such markers, the unknown gene responsible for any genetic disease could be localized to one of the human chromosomes. This endeavor simply requires a laborious linkage analysis using all the available markers in appropriate families with the disease of interest.

Three successful examples of this approach are the Huntington chorea locus, the cystic fibrosis locus, and the locus for polycystic kidney disease. A polymorphic DNA probe called *G8* from the short arm of chromosome 4 has been found to be closely linked to the phenotype of Huntington disease, a dominantly inherited disease [Gusella et al, 1983]. The genetic distance between the G8 probe and the Huntington locus is about 5 cM. Probe G8 can be used for the direct search for either the gene of interest, which is unknown to date, or adjacent DNA sequences closer (0 cM) to the Huntington gene. After such a DNA fragment is identified, carriers of the gene can be detected at any age even though the gene product or the mechanism by which it produces the disease is unknown. Similarly, the cystic fibrosis locus has been localized to the long arm of chromosome 7 [White et al, 1985; Wainwright et al, 1985; Knowlton et al, 1985] and the locus for polycystic kidney disease to chromosome 16 [Reeders et al, 1985].

Linkage maps are useful in several studies in which the identity of one chromosome has to be followed. For example, DNA polymorphisms for several genes or orphan DNA sequences (DNA sequences without known function) on chromosome 13 and chromosome 11p were used to investigate retinoblastoma [Cavenee et al, 1985], and Wilms tumor [Fearon et al, 1984b; Orkin et al, 1984], respectively. The data suggest that, for some tumors, homozygosity for one particular chromosome is essential for tumor formation. In a recent study of haplotypes of DNA polymorphisms on chromosome 21q we found that certain haplotypes had an increased probability of nondisjunction in families with trisomy 21 [Antonarakis et al, 1985b].

SUMMARY

Recombinant DNA technology has greatly facilitated the mapping of human chromosomes. Somatic cell hybrid analysis, in situ hybridization, and linkage analysis are the major methods of gene mapping. Restriction endonuclease analysis and DNA hybridization techniques create new potentials for old methods, and the human gene map is becoming more dense with mapped loci at an ever quickening pace. The major usefulness of the saturation of the gene map with DNA polymorphic markers is that one can localize and study unknown genes responsible for common genetic disorders.

REFERENCES

Anagnou NP, O'Brien SJ, Shizumura T, Nach WG, Chen MJ, Nienhuis AW: Chromosomal organization of the human DHFR genes: Dispersion, selective amplification and a novel form of polymorphism. Proc Natl Acad Sci USA 81:5170–5174, 1984.

Antonarakis SE, Boehm CD, Giardina PVJ, Kazazian HH: Non-random association of polymorphic restriction sites in the β-globin gene cluster. Proc Natl Acad Sci USA 79:137–141, 1982a.

Antonarakis SE, Kazazian HH, Orkin SH: DNA polymorphisms and molecular pathology of the human globin gene clusters. Hum Genet 69:1–14, 1985a.

Antonarakis SE, Kittur SD, Metaxotou C, Watkins PC, Patel AS: Analysis of DNA haplotypes suggests a genetic predisposition to trisomy 21 associated with DNA sequence on chromosome 21. Proc Natl Acad Sci USA 82:3360–3364, 1985b.

Antonarakis SE, Phillips JA, Kazazian HH: Genetic diseases: Diagnosis by restriction endonuclease analysis. J Pediatr 100:845–856, 1982b.

Antonarakis SE, Phillips JA, Mallonee RL, Kazazian HH, Fearon ER, Waber PG, Kronenberg HM, Ullrich A, Meyers DA: β-Globin locus is linked to the parathyroid hormone locus and lies between the insulin and PTH loci in man. Proc Natl Acad Sci USA 80:6615–6619, 1983.

Bell GI, Selby MJ, Rutter WJ: The highly polymorphic region near the human insulin gene is composed of simple tandemly repeating sequences. Nature 295:31–35, 1982.

Bishop DT, Skolnick MH: Genetic markers and linkage analysis. In Caskey CT, White RL (eds): "Recombinant DNA Applications to Human Disease." Cold Spring Harbor, NY: Cold Spring Harbor Laboratory, 1983, pp 235–250.

Botstein D, White RL, Skolnick M, Davis RW: Construction of a genetic linkage map in man using restriction fragment length polymorphisms. Am J Hum Genet 32:314–331, 1980.

Brissenden JE, Ullrich A, Francke U: Chromosomal mapping of loci for insulin-like growth factors I and II and for epidermal growth factor in man. Nature 310:781–784, 1984.

Capon DJ, Chen EY, Levison AD, Seeburg PH, Goeddel DV: Complete nucleotide sequences of the T24 human bladder carcinoma oncogene and its normal homologue. Nature 302:33–37, 1983.

Cavenee WK, Dryja TP, Phillips RA, Benedict WF, Godbout R, Gallie BL, Mulphree AL, Strong LC, White RI: Expression of recessive alleles by chromosomal mechanisms in retinoblastoma. Nature 305:779–784, 1985.

Cavenee W, Leach R, Mohandas T, Pearson P, White R: Isolation and regional localization of DNA segments revealing polymorphic loci from human chromosome 13. Am J Hum Genet 36:10–24, 1984.

Chaganti RSK, Jhanwar SC, Antonarakis SE, Hayward WS: Germ line chromosomal localization of genes in chromosome 11p linkage: Parathyroid hormone gene, β-globin, c-Ha-*ras*-1, and insulin. Somat Cell Mol Genet 11:197–202, 1985.

Chakravarti A, Buetow KH, Antonarakis SE, Waber PG, Boehm CD, Kazazian HH: Nonuniform recombination within the human β-globin gene cluster. Am J Hum Genet 36:1239–1258, 1984.

Cheung P, Kao FT, Law ML, Jones C, Puck TT, Chan L: Localization of the structural gene for human apolipoprotein A-I on the long arm of human chromosome 11. Proc Natl Acad Sci USA 81:508–511, 1984.

Conneally PM, Rivas ML: Linkage analysis in man. Adv Hum Genet 10:209–266, 1980.

de Martinville B, Leary J, Ullrich A, Francke U: The human insulin gene map on the short arm of chromosome 11. Cytogenet Cell Genet 32:265–267, 1982.

de Martinville B, Francke U: The c-Ha-*ras*-1, insulin and β-globin loci map outside the deletion associated with aniridia-Wilms' tumor. Nature 305:641–643, 1983.

Drayna D, Davis K, Hartley D, Mandel JL, Camerino G, Williamson R, White RL: Genetic mapping of the human X chromosome by using restriction fragment length polymorphisms. Proc Natl Acad Sci USA 81:2836–2839, 1984.

Fearon ER, Antonarakis SE, Meyers DA, Levine MA: c-Ha-*ras*-1 oncogene lies between β-globin and insulin loci on human chromosome 11p. Am J Hum Genet 36:329–337, 1984a.

Fearon ER, Vogelstein B, Feinberg AP: Somatic deletion and duplication of genes on chromosome 11p in Wilms' tumors. Nature 309:176–178, 1984b.

Gerhard DS, Kawasaki ES, Bancroft FC, Szabo P: Localization of a unique gene by direct hybridization in situ. Proc Natl Acad Sci USA 78:3755–3759, 1981.

Gerhard DS, Kidd KK, Housman D, Gusella JF: Data on the genetic map of the short arm of chromosome 11. In "Human Gene Mapping 7 (1984): Seventh International Workshop on Human Gene Mapping." Birth Defects: Original Article Series, vol. XX, No. 2, 1984, p 478. March of Dimes Birth Defects Foundation, White Plains, NY.

Goodbourn SEV, Higgs DR, Clegg JB, Weatherall DJ: Molecular basis of length polymorphisms in the human zeta-globin gene complex. Proc Natl Acad Sci USA 80:5022–5026, 1983.

Gusella JF, Varsanyi-Breiner A, Kao FT, Jones C, Puck TT, Keys C, Orkin S, Housman D: Precise localization of human β-globin gene complex on chromosome 11. Proc Natl Acad Sci USA 76:5239–5243, 1979.

Gusella JF, Wexler NS, Conneally PM, Naylor SL, Anderson MA, Tausi RE, Watkins PC, Ottina K, Wallace MR, Sakaguchi AY, Houng AB, Shoulson J, Bonilla E, Martin JB: A polymorphic DNA marker genetically linked to Huntington's disease. Nature 306:234–238, 1983.

Harper ME, Ullrich A, Saunders GF: Localization of the human insulin gene to the distal end of the short arm of chromosome 11. Proc Natl Acad Sci USA 78:4458–4460, 1981.

Hoppener JWM, Steenbergh PH, Zandberg J, Bakker B, Pearson PL, Geurts van Kessell AHM, Jansz HS, Lips LJM: Localization of the polymorphic human calcitonin gene on chromosome 11. Hum Genet 66:309–312, 1984.

Jeffreys AJ: DNA sequence variants in the G gamma-, A gamma-, delta- and beta-globin genes of man. Cell 18:1–10, 1979.

Jhanwar SC, Neel BG, Hayward WS, Chaganti RSK: Localization of c-ras oncogene family on human germ line chromosomes. Proc Natl Acad Sci USA 80:4794–4797, 1983.

Kan YW, Dozy AM: Polymorphisms of DNA sequence adjacent to human β-globin structural gene: Relationship to sickle cell mutation. Proc Natl Acad Sci USA 75:5631–5635, 1978.

Kittur S, Antonarakis SE, Groner Y, Phillips JA, Meyers D, Kazazian HH: Linkage map on human chromosome 21. EMBO J 4:2257–2260, 1985a.

Kittur SD, Hoppener JWM, Antonarakis SE: Linkage map of the short arm of human chromosome 11: Location of the genes for catalase, calcitonin, and insulin-like growth factor II. Proc Natl Acad Sci USA 82:5064–5067, 1985b.

Knowlton RG, Cohen-Haguenauer O, VanCong N, et al: A polymorphic DNA marker linked to cystic fibrosis is located on chromosome 7. Nature 318:380–382, 1985.

Korneluk RG, Quan F, Lewis WH, Guise KS, Willard HF, Holmes MT, Gravel RA: Isolation of human fibroblast catalase cDNA clones. J Biol Chem 259:13819–13823, 1984.

Lebo RV, Chakravarti A, Buetow kH, Cheung MC, Cann H, Cordell B, Goodman H: Recombination within and between the human insulin and β-globin gene loci. Proc Natl Acad Sci USA 80:4804–4812, 1983.

Lebo RV, Gorin F, Fletterick RJ, Kao FT, Cheung MC, Bruce BD, Kan YW: High resolution chromosome sorting and DNA spot blot analysis assign McArdle's syndrome to chromosome 11. Science 225:57–59, 1984.

McBride O, Swan DC, Santos E, Barbacid M, Tronick SR, Aaronson SA: Localization of the normal allele of T24 human bladder carcinoma oncogene to chromosome 11. Nature 300:773–774, 1982.

McKusick VA, Ruddle FH: The status of the gene map of the human chromosomes. Science 196:390–405, 1977.

Meyers DA: "Multipoint Mapping of Linkage Group I." Dissertation. Indianapolis: Indiana University School of Medicine, 1976.

Meyers DA, Merrit AD, Conneally PM, Norton JA, Rivas ML, Yu PL, Palmer CG: Linkage group I—A statistically significant locus order from family studies. In "Human Gene Mapping 4 (1977): Fourth International Workshop on Human Gene Mapping." Birth Defects. XIV(4), 1978, pp 396–400.

Morton CC, Kirsch IR, Taub R, Orkin SH, Brown JA: Localization of the β-globin gene by chromosome in situ hybridization. Am J Hum Genet 36:576–585, 1984.

Morton NE: Sequential tests for the detection of linkage. Am J Hum Genet 7:277–318, 1955.

Naylor SL, Sakagushi AY, Kronenberg HM, Shen LP, Rutter W, Chin W, Goodman G: Human genomic organization of glycoprotein and polypeptide hormone genes. Am J Hum Genet 34:165A, 1982.

Neel BG, Jhanwar SC, Chaganti RSK, Hayward WS: Two human c-onc genes are located on the long arm of chromosome 8. Proc Natl Acad Sci USA 79:7842–7846, 1982.

Orkin SH, Goldman DS, Sallan SE: Development of homozygosity for chromosome 11p markers in Wilms' tumor. Nature 309:172–174, 1984.

Przepiorka D, Baylin SB, McBride DW, Testa JR, Bustros A, Nelkin BD: The human calcitonin gene is located on the short arm of chromosome 11. Biochem Biophys Res Commun 120:493–499, 1984.

Puck TT: Human gene mapping, exposure and expression. In Caskey CT, White RL (eds): "Recombinant DNA Applications to Human Disease." NY: Cold Spring Harbor Laboratory, 1983, pp 205–213.

Puck TT, Kao FT: Somatic cell genetics and its application to medicine. Annu Rev Genet 16:225–271, 1982.

Reeders ST, Breuning MH, Davies KE et al: A highly polymorphic DNA marker linked to adult polycystic kidney disease on chromosome 16. Nature 317:542–544, 1985.

Renwick JH: Progress in mapping human autosomes. Br Med Bull 25:65–73, 1969.

Ruddle FH: A new era in mammalian gene mapping: Somatic cell genetics and recombinant DNA methodologies. Nature 294:115–120, 1981.

Scott AF, Phillips JA, Migeon BR: DNA restriction endonuclease analysis for the localization of the human β-globin genes on chromosome 11. Proc Natl Acad Sci USA 76:4563–4565, 1979.

Sparkes RS, Berg K, Evans HJ, Klinger HP (eds): "Human Gene Mapping 7." Birth Defects 20(2), 1984.

The Human Genetic Mutant Repository, NIH Publication H 83-2011, 1983. Washington, DC: U.S. Government Printing Office.

Wainwright BS, Scambler PS, Schmidtke J et al: Localization of cystic fibrosis locus to human chromosome 7 cen-q22. Nature 318:384–386, 1985.

White R, Woodward S, Leppert M et al: A closely linked genetic marker for cystic fibrosis. Nature 318:382–384, 1985.

Globin Gene Structure and the Nature of Mutation

Haig H. Kazazian, Jr., MD

Department of Pediatrics, Genetics Unit, The Johns Hopkins University School of Medicine, Baltimore, MD 21205

In this chapter I will discuss very briefly some of the information on the structure of the globin genes and their organization in the human genome. I will then outline what is known about normal variation in DNA sequences. Finally, I will discuss mutations and how knowledge of specific sites of mutation has helped us learn about nucleotide sequences important in gene expression. I will use the β-thalassemias as examples of mutations that affect gene expression in man.

In the past 5 years we have learned that nearly all mammalian genes coding for a protein product have split coding regions, ie, the coding regions are discontinuous. The example shown in Figure 1 is the first mammalian gene for which this feature was described, β-globin. This gene has three coding regions, now termed *exons*, and two intervening sequences (IVS), called *introns* [Tilghman et al, 1978]. In the β-globin gene the introns divide the protein-coding region between the codons for amino acids 30 and 31 of the 146-amino acid chain and the codons for amino acids 104 and 105. Therefore, exon-1 contains 5′ untranslated sequences and the codons for amino acids 1 to 30, exon-2 contains the codons for amino acids 31 to 104, and exon-3 contains the codons for amino acids 105 to 146 and 3′ untranslated

Support for this work was provided by grants from the NIH and the March of Dimes Birth Defects Foundation. The author thanks Emily Pastefield for expert preparation of this manuscript.

Birth Defects: Original Article Series, Volume 23, Number 3, pages 77–92

Fig. 1. A typical split gene coding for a protein. Black areas denote exons or regions encoded into protein, and open areas depict introns. This gene is quite simple, with two introns and three exons. IVS-1, intron-1; IVS-2, intron-2.

sequences. Even though only about 450 nucleotides are necessary to encode the protein (three nucleotides per amino acid), because of the introns the gene contains roughly 1,500 nucleotides. Yet this is a very simple gene. Many genes have now been described with ten or more introns (the gene for the pro-α_1-collagen chain has more than 50 introns), and some genes have a total size of 180–1,000 kilobases (kb) or 180,000 to 1,000,000 base pairs (bp). On the other hand, certain genes, such as those encoding histones and interferons, do not contain introns and are unexplained exceptions to the split gene rule.

How protein-coding genes are expressed is exemplified by the globins (Fig. 2). The entire gene, including the two introns, is transcribed into a precursor RNA. This RNA is immediately modified at both of its ends (called *5′* and *3′* ends). The modification at the 5′ end is the addition of a methylated G residue in an unusual triphosphate linkage. This modification, which is specific for messenger RNA, is called *5′ capping*. The modification at the 3′ end is the addition of about 150 A residues (the poly-A tail). The 5′ cap is thought to be important in translation of the RNA into protein, while the 3′ poly-A tail may have a role in RNA stability. Next the introns are precisely spliced out of the RNA, thereby approximating the coding region sequences. The precise mechanism and the enzymes involved in RNA splicing are not understood. Recent work suggests that the 5′ end of the intron is cleaved, and it folds back and forms a "lariat" structure by a covalent bond with an A residue 38 nucleotides 5′ to the 3′ end of the intron (T. Maniatis, personal communication). Thus, it appears that the splicing mechanism may be worked out soon. We know that splicing is important in transporting the RNA from the nucleus to the cytoplasm where it will be translated into protein. We also know that nucleotide sequences near the exon-intron junctions are important in normal splicing (see below). Translation then occurs on the fully processed messenger RNA in the cytoplasm.

The genetic code, as found in the messenger RNA, is critical to translation. There are 64 possible combinations of a code composed of three nucleotides. One combination, AUG (note that U is substituted for T in

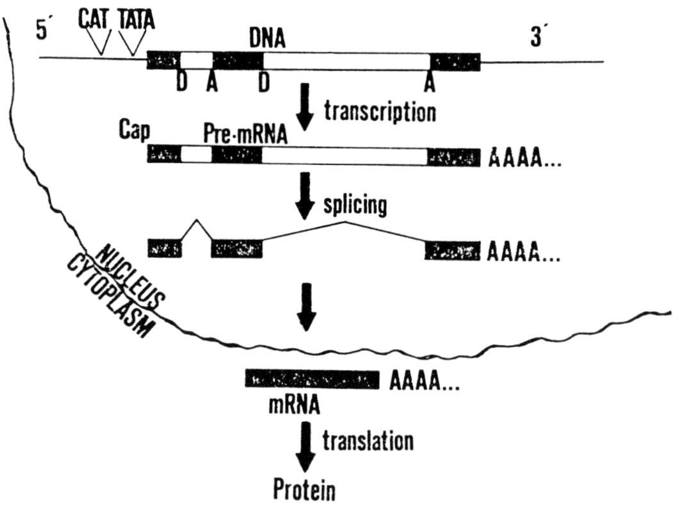

Fig. 2. Schematic presentation of gene expression. Steps include 1) transcription of precursor RNA, 2) cap and poly-A addition to the ends of the RNA, 3) splicing out of intron sequences, and 4) translation of the mature messenger RNA into protein. CAT and TATA in front of the gene refer to sequences important in transcription, while D and A refer to donor and acceptor splice junctions important in RNA splicing.

RNA), is the initiation codon and always places methionine as the initial amino acid in a protein. Methionine is usually not the first amino acid in the finished protein, because it is often removed while the nascent protein is in the early stages of production. The code also contains three terminator codons: UAG, UAA, and UGA. These codons cannot be decoded into an amino acid, and, therefore, protein synthesis terminates whenever one of these triplets is present in the reading frame. The remaining 60 codons can be decoded so that an amino acid is inserted into the growing or nascent polypeptide chain. The degeneracy of the genetic code allows for six different codons for leucine and serine, and four codons for many others of the total 20 amino acids. This degeneracy permits some variation in the coding DNA sequence without necessarily producing a change in the amino acid encoded. The relationship between the nucleotide triplet in DNA, its complement in messenger RNA, and the amino acid designated by that triplet is demonstrated for the sixth codon of the β-globin chain (Fig. 3). CTC in DNA is decoded GAG in messenger RNA and glutamate in the protein. A single nucleotide substitution in the codon (T $\rightarrow$ A) leads to sickle cell anemia, as discussed later.

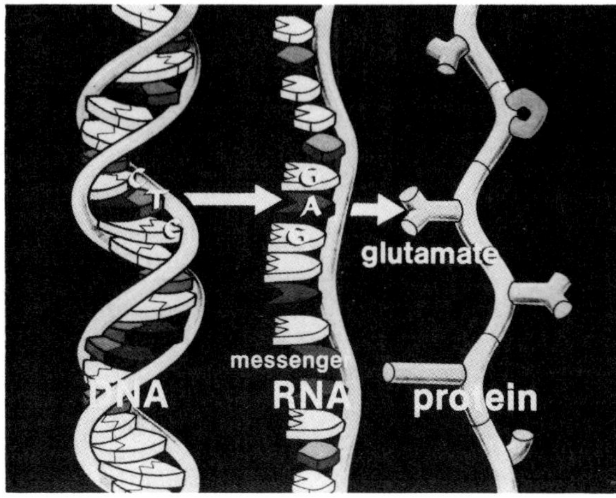

DNA and RNA Coding in Glutamate Production for Hemoglobin A

Fig. 3. Schematic presentation of the relationship between the triplet code in the strand of DNA that is transcribed, the codon in messenger RNA, and the specific amino acid inserted into the polypeptide chain. For instance, CTC in the DNA is transcribed into GAG in RNA, and the amino acid inserted is glutamate.

I now need to discuss the environment within the genome in which a gene is located. This brings us to the topic of gene families. Again my example is the β-globin gene cluster. In this family of related genes there are five functional genes, one embryonic (ϵ), two fetal ($^{G}\gamma$ and $^{A}\gamma$), a minor adult (δ), and a major adult (β) (Fig. 4). These genes account for only about 7,500 nucleotides or 15% of the roughly 50,000 nucleotides in this cluster [Maniatis et al, 1980; Fritsch et al, 1980]. It may be noteworthy that the functional genes are arranged in order of their expression in development, ie, embryonic, fetal, and adult. A single pseudogene sequence, $\psi\beta_1$, is also found in this cluster. This pseudogene sequence is very similar to the β-gene sequence, but it contains mutations that prevent gene expression. Pseudogenes may be evolutionary vestiges much like the anatomic appendix.

The function of the roughly 40 kb of flanking sequences in this cluster is not known, but it is known that interspersed between the functional globin genes are repeated sequences [Collins and Weissman, 1984]. Some of these sequences are repeated over 500,000 times in the genome. These sequences are called *Alu* sequences, because many of them contain an *Alu*I restriction endonuclease site (see below). They are about 300 nucleotides in length and when compared to one another are roughly 95% homologous. Another repeat

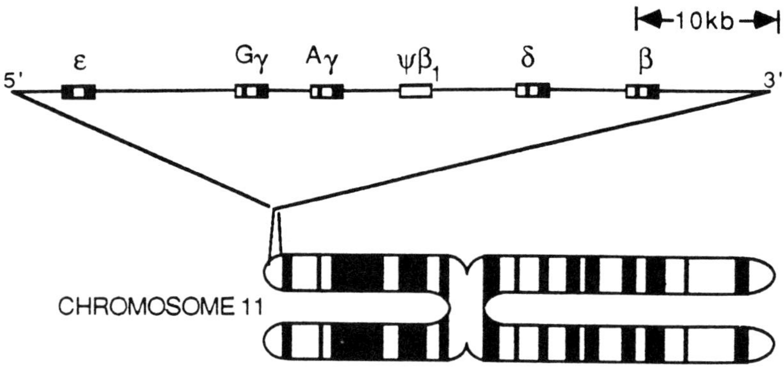

Fig. 4. β-Globin gene cluster on chromosome 11. Since kb refers to kilobases, the cluster contains about 50,000 bp. See text for details. The orientation of the β-globin gene cluster vis-á-vis the centromere is centromere-ϵ-β-telomere.

sequence is 6.4 kb in length. This sequence is located 3′ to the β gene and is also highly repeated in the genome. The appearance of repeated sequences interspersed every few kilobases by single copy or functional gene sequences is a general characteristic of the human genome. The function of these repeat sequences is still unknown.

Each gene cluster is unique in its organization. For example, the α-globin cluster, which is found on the short arm of chromosome 16, contains two pseudogenes, no fetal genes, one embryonic ζ gene, two adult α genes, and spans only about 25 kb of DNA [Collins and Weissman, 1984]. In addition, the two introns in the α genes are each only about 150 bp in length.

The switching in gene expression from embryonic genes to fetal genes to adult genes is not a consequence of DNA rearrangement during development. The DNA in both the β- and α-gene clusters appears to be static and the mechanism of hemoglobin switching remains unknown, although changes in DNA methylation have recently been implicated. In contrast, DNA rearrangement in the immunoglobin gene clusters plays an important role in expression of these important genes.

NORMAL VARIATION

To discuss mutations and their consequences, a brief introduction to normal variation is required. We have all heard a great deal about normal variation at the protein or enzyme level. Now we are learning that variation in the DNA is even more extensive than we imagined from protein data. Common normal variation in DNA has been termed *DNA polymorphism*

[Kan and Dozy, 1978]. This polymorphism can be detected if it affects a restriction endonuclease site or if it is an insertion or deletion of a number of nucleotides. An example of a restriction endonuclease polymorphism is shown in Figure 5. This polymorphism affects a *Hinc*II site near the ϵ-globin gene [Antonarakis et al, 1982]. When genomic DNA is digested with *Hinc*II, electrophoresed, and hybridized with a radioactive probe containing the ϵ gene, chromosomes that contain this site yield a 3.7 kb fragment, and those that lack the site demonstrate an 8 kb fragment. Since this polymorphism is very frequent, nearly 50% of individuals in the population are heterozygous and demonstrate both fragments by this analysis. Single gene fragments of this kind are detected by the Southern blotting technique [Southern, 1975].

Using the procedure of Southern blotting one can find a number of DNA polymorphisms in the β-gene cluster. Many of these polymorphisms show unusual associations. Even though there are some 17 DNA polymorphisms known, instead of 217 different sequences in the β-gene cluster only nine are quite commonly observed. However, because these nine sequences are quite common, there is a very good chance that any individual in the population is heterozygous for some region of the β-gene cluster.

This variation also holds for the β-gene itself [Orkin et al, 1982a]. In most populations there are three common β-globin gene sequences, which differ from each other by a small number of nucleotide substitutions in IVS-2 and a substitution in codon 2 of the β chain. This latter substitution does not alter the amino acid inserted at that position (Fig. 6). Since all populations have three normal β genes, all of which are quite common, most individuals in any population group are heterozygous for two normal β genes. Thus, normal variation in DNA is extensive. The reader should realize that the 50,000

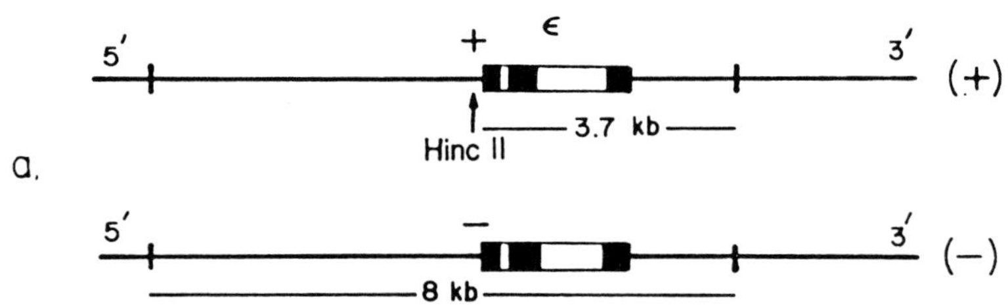

Fig. 5. A polymorphism of a restriction endonuclease site near the globin gene. Near many normal ϵ-globin genes is a *Hinc*II site, but this site is often missing in normal individuals. When the site is present, the 8 kb fragment is cut into 3.6 and 4.4 kb fragments. The 3.6 kb fragment is readily detected by use of a piece of DNA containing the gene. When the site is absent, the 8 kb fragment is not cleaved. The presence or absence of this *Hinc*II site can be discovered by Southern blotting.

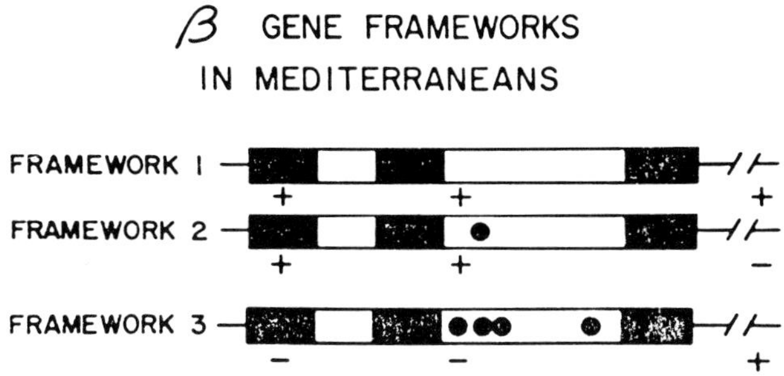

Fig. 6. Three β genes occur commonly in various populations. These are called *frameworks*, and in Mediterraneans 50% of β genes have framework 1, 30% have framework 2, and 20% have framework 3. Each black dot represents a single nucleotide substitution in that framework.

nucleotides of the β-gene cluster derived from the father contain 100 or more nucleotide differences from the 50,000 nucleotides of this cluster derived from the mother.

MUTATION

For many years we have known about nucleotide substitutions that produce an amino acid substitution in a protein. The best example of this is the T-A substitution in the sixth codon of the β gene, which produces a glutamate-to-valine substitution in the chain. This is the mutation that produces the β-globin of sickle cell anemia. A large number of other nucleotide substitutions that produce abnormal hemoglobins (over 350) have been described. Only recently have we learned about the mutations that alter gene expression. These mutations lead to reduced production of the encoded protein, and they tell us what sequences within and adjacent to the gene are important in expression. The best example of such mutations is still the point mutations that produce the β-thalassemias, inherited disorders of β-globin gene expression.

MOLECULAR DEFECTS IN THE β-THALASSEMIAS

The genetic defects in β-thalassemia are best reviewed in the context of the normal sequence of events in gene expression (Fig. 2). To review, in erythroid cells the β-globin gene is transcribed by RNA polymerase into a globin RNA precursor that contains two intron sequences. Within the nucleus these introns are excised during RNA processing, and the coding blocks (or exons) are ligated together precisely to form the processed mRNA that is transported to the cytoplasm and translated into β-globin. Genetic lesions that interrupt the normal sequence of events lead to the decreased β-globin production that defines β-thalassemia. When no β-globin protein is synthesized from an affected gene, a β°-thalassemia defect is present. When some β-globin is made, the defect is of the β^+ variety. We can consider the defined mutations by the phase of the gene expression pathway that is affected. All β-thalassemia defects that are known to date involve DNA sequences of the β-globin gene rather than distant nucleotides in the genome. We can consider the following classes:

1. Sizeable deletion of the β-globin gene
2. Transcription mutations
3. RNA processing mutations
4. mRNA function mutations
5. RNA cleavage mutations

The known molecular defects in β-thalassemia are displayed in Table 1.

Sizeable Deletion of the β-Globin Gene

To a considerable extent the definition of a sizeable deletion of the β-globin gene depends on the sensitivity of the methods that can be employed to search for missing DNA. In general, β-thalassemia is not associated with obvious deletion of the β-globin gene. Only one common variety of simple β-thalassemia has been identified in which gross deletion of β-gene DNA is detectable by routine restriction enzyme mapping. In this form of β°-thalassemia, found in about 30% of Asian Indian β-thalassemic chromosomes, a 619 bp deletion involves the terminal portion of the second intron, the entire third exon, and a short stretch of DNA 3′ to the β gene. Although other small deletions have been identified in the β-globin gene leading to β-thalassemia, they are generally too small to be detected by routine restriction mapping (see below).

Transcription Mutants

In vitro studies with several gene systems and comparisons of DNA sequences flanking several normal genes have defined 5′ or upstream DNA segments that are necessary for proper RNA synthesis from a gene [Breathnach and Chambon, 1981]. These include sequences located approximately −75 to −90 bp (relative to the position of the cap site or start of RNA transcription), surrounding a DNA sequence known commonly as the CAT box [Efstratiadis et al, 1980]. These sequences are particularly important in setting the level of RNA expression. Sequences more proximal at about −30 bp in the TATA box homology (CATAAAA) are important for RNA transcription as well, but also strongly influence the actual start site of RNA synthesis (the cap site). Nucleotide substitutions in these two controlling regions have been found in association with β-thalassemia.

The first and most fully defined transcription mutant involves a C–G change at position −87. Position −87 is just 11 bp upstream from the beginning of the CAT box. That the sequence of the β-globin gene itself was otherwise normal led to the prediction that this allele was a transcriptional mutant. Treisman et al [1983] have studied the expression of this gene and demonstrated that the C–G change does lead to a five- to tenfold reduction in the level of β-globin RNA that is produced [Treisman et al, 1983]. This mutation is particularly instructive in that it demonstrates the in vivo significance of a DNA segment thought to be important in promoter function on the basis of studies of various other genes in artificial cell environments. The precise means by which this position −87 substitution alters gene transcription is as yet unknown. However, recently a C–T mutation at position −88 was described in an American black, and this mutation further suggests the importance of this distal promoter region.

Three presumptive transcription mutants have been identified within the TATA box located more proximal to the gene. Two curiously involve position −28. The first was found in a Kurdish patient and was an A–C change in the DNA [Poncz et al, 1982]. Subsequently, we identified an A–G change at the same position in a Chinese gene [Cheng et al, 1984]. An A–G change has also been found at position −29 in a black American [Antonarakis et al, 1984] and in Chinese. All three mutations appear to be of the β^+ type, with about 10% to 25% of normal expression in the in vitro expression system and in erythroid cells in vivo.

RNA Processing Mutants

When it first became apparent that introns were a common feature of globin and other eukaryotic genes, it was proposed that some forms of thalassemia might result from abnormalities in RNA processing. Indeed, this is the case. We may subdivide these mutants into four classes: 1) splice junction mutants, 2) consensus region mutants, 3) internal intron mutants, and 4) coding region mutants that affect RNA processing.

Comparisons of numerous normal gene sequences led to formulation of consensus sequences for exon-intron junctions [Breathnach and Chambon, 1981]. Although flexible in precise sequences, the donor or 5′-splice site

TABLE 1. Mutations in β-Thalassemia (Total number = 38, June 1986)

Mutant class	Type	Origin	Direct detection
Nonfunctional mRNA	$0 = \beta^{o}$		
Nonsense mutants	$+ = \beta^{\dagger}$		
Codon 17 (A-T)	0	Chinese	Oligonucleotide
Codon 39 (C-T)	0	Mediterranean	Oligonucleotide
Codon 15 (G-A)	0	Asian Indian	Oligonucleotide
Codon 121 (A-T)	0	Polish	*Eco*RI
Codon 37 (G-A)	0	Saudi Arabian	*Ava*II
Frameshift mutants			
−2 codon 8	0	Turkish	Oligonucleotide
−1 codon 16	0	Asian Indian	Oligonucleotide
−1 codon 44	0	Kurdish	Oligonucleotide
+1 codons 8/9	0	Asian Indian	Oligonucleotide
−4 codons 41/42	0	Asian Indian	Oligonucleotide
−1 codon 6	0	Mediterranean	*Mst*II
+1 codons 71/72	0	Chinese	Oligonucleotide
RNA processing mutants			
Splice junction changes			
IVS-1 position 1 (G-A)	0	Mediterranean	Oligonucleotide
IVS-1 position 1 (G-T)	0	Asian Indian	Oligonucleotide
IVS-2 position 1 (G-A)	0	Mediterranean	*Hph*I
IVS-1 3′-end −17 bp	0	Kuwaiti	*Fnu*4H
IVS-1 3′-end −25 bp	0	Asian Indian	*Fnu*4H
IVS-2 3′-end (A-G)	0	American black	Oligonucleotide
IVS-2 3′-end (A-C)	0	American black	Oligonucleotide
Consensus changes			
IVS-1 position 5 (G-C)	+	Indian	Oligonucleotide
IVS-1 position 5 (G-T)	+	Mediterranean	Oligonucleotide
IVS-1 position 6 (T-C)	+	Mediterranean	Oligonucleotide
IVS-2 3′ -end CAG-AAG	+	Iranian	Oligonucleotide
Internal IVS changes			
IVS-1 position 110 (G-A)	+	Mediterranean	Oligonucleotide
IVS-1 position 116 (T-G)	?	Mediterranean	*Mae*I
IVS-2 position 705 (T-G)	+	Mediterranean	Oligonucleotide
IVS-2 position 745 (C-G)	+	Mediterranean	*Rsa*I
IVS-2 position 654 (C-T)	0	Chinese	Oligonucleotide
Coding region substitutions affecting processing			
Codon 26 (G-A)	E	S.E. Asian	HbE
Codon 24 (T-A)	+	American black	Oligonucleotide
Codon 27 (G-T)	Knossos	Mediterranean	Hb Knossos

continued

TABLE 1. Mutations in β-Thalassemia (Total number = 38, June, 1986)† (continued)

Mutant class	Type‡	Origin	Direct detection
Transcriptional mutants			
−88 C-T	+	American black	Oligonucleotide
−87 C-G	+	Mediterranean	*Avr*II
−31 A-G	+	Japanese	Oligonucleotide
−29 A-G	+	American black	Oligonucleotide
−28 A-C	+	Kurdish	Oligonucleotide
−28 A-G	+	Chinese	Oligonucleotide
RNA cleavage + polyadenylation mutants			
AATAAA-AACAAA	+	American black	Oligonucleotide

†This table has been updated since the manuscript was prepared in August 1984. It includes four mutations not known at that time. These are nonsense codon 121; IVS-1, 3′-end−17 bp; IVS-1 position 5 (G-T); and −31 A-G. Presumably, this list will be slightly incomplete at the time of publication.

always has a GT dinucleotide and the 3′ (or acceptor) an AG. Neighboring these invariant dinucleotides are preferred sequences that vary among the different normal genes that have been sequenced.

One class of RNA processing mutant is defined by mutations that affect the invariant nucleotides of these splice sites. Two β-thalassemic genes in Mediterraneans have been identified in which the G of the GT at the donor site is mutated to A at either intron-1 or intron-2 (see Table 1) [Orkin et al, 1982a; Treisman et al, 1982]. A third gene in an Asian Indian has a G–T substitution in the GT of intron-1. In all of these genes the nucleotide alteration prevents normal RNA splicing at the donor site and leads to alternative processing of RNA at other donor-like sequences located within either coding or intervening sequences. These mutations are of the $\beta°$ type. Another, which is not a point mutation but a small deletion, involves a 25 bp deletion that removes the IVS-1 acceptor including the invariant AG. A fifth is an A–G substitution in the invariant AG in the American black. Splicing at the normal acceptor site is completely prevented in these situations.

A second class of processing mutant includes nucleotide changes in the consensus region of the intron junctions, but not within the invariant nucleotides. Two such mutations have been identified within the consensus sequence of the intron-1 donor site [Orkin et al, 1982a]. Both mutations lead to decreased normal splicing at the donor site and low level alternative RNA processing into donor-like sequences in both exon-1 and intron-1 [Treisman et al, 1983]. The splicing patterns of the RNA transcripts from one of these two genes is compared with those for the normal gene in Figure 7. In general,

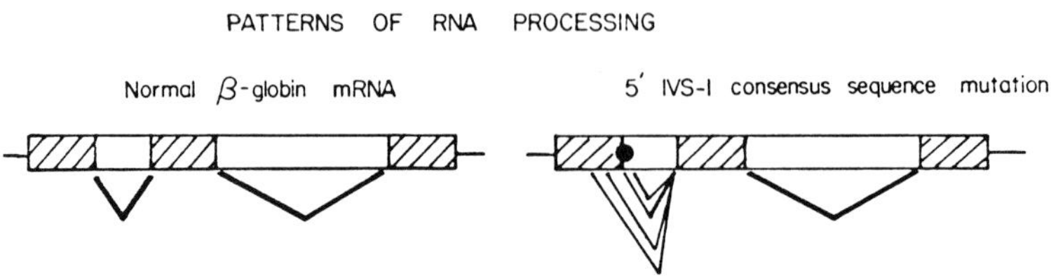

Fig. 7. Patterns of RNA splicing in the normal β-globin messenger RNA and the messenger RNA containing a mutation near one end of intron-1. Note that in addition to some normal splicing, three alternative splices occur between the normal acceptor site and donor sites not normally utilized.

the degree of impairment in globin production documented in vitro mirrors the relative severity of these thalassemic genes in vivo. The consensus region mutations are both of the β^+ variety, since some normal globin chain is produced from each. These splicing mutants emphasize the complexity of RNA processing. Seemingly trivial nucleotide sequence changes within the consensus sequences produce significant effects on both the efficiency and accuracy of RNA processing.

Nucleotide substitutions within an intron that generate new splicing signals constitute the third class of processing mutants. Five mutations of this kind are known, occurring at intron-1 position 110 [Spritz et al, 1981], intron-1 position 116 (F. Collins, personal communication), and intron-2 positions 654 [Cheng et al, 1984], 705 [Orkin et al, 1982a], and 745 [Dobkin et al, 1983]. In the first characterized gene of this type, a common β^+-thalassemia gene in Mediterraneans, a G–A change within intron-1, produces an internal acceptor-like sequence that is used in RNA processing in preference to the normal site at the end of intron-1. The majority of RNA processed from this gene is, therefore, abnormal and contains an extra 19 nucleotides derived from the intron, as shown in Figure 8. Since some normal RNA is also formed, this mutant gene is of the β^+ type. The three intron-2 mutants (at positions 654, 705, and 745) generate donor-like splicing sites within the intron that ultimately lead to abnormal RNA processing as well. In these instances, a considerable fraction of the RNA transcripts are processed incorrectly.

The fourth class of RNA processing mutants includes those genes in which nucleotide substitutions within a *coding* region rather than an intron alter RNA processing. The β-globin chain variant of HbE is the first defined member of this group [Orkin et al, 1982b]. This β-chain mutant has an amino acid replacement (lys-glu) in position 26 of the β-polypeptide. As is characteristic of β-thalassemia rather than a hemoglobinopathy, the β^E chain is

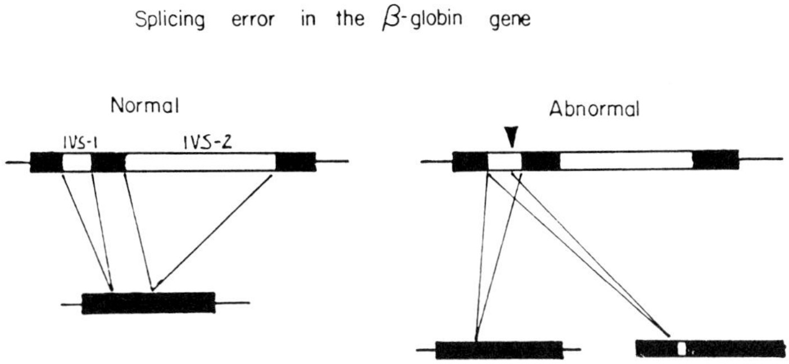

Fig. 8. Normal and abnormal splicing in intron-1 (IVS-1) associated with mutation (located in intron-1 at arrowhead). Normal splicing is shown on the left, and abnormal splicing associated with the mutant gene is on the right. With this mutation some β-messenger RNAs are normally spliced, while others are spliced into the intron at the mutation site. These latter messenger RNAs contain 19 extra nucleotides of intron-1 in the protein coding region (open area) and cannot produce normal protein.

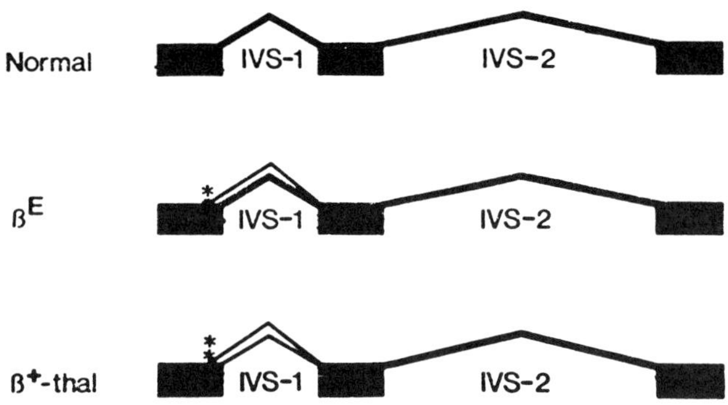

Fig. 9. Mutations in exons (coding regions) can affect RNA splicing. A potential, but normally unused, splice site in exon-1 is activated by the two mutations shown. In these genes both normal and abnormal splicing occurs.

underproduced in erythroid cells and its mRNA is deficient relative to β-mRNA. Cloning, sequencing, and expression studies of β^E-globin gene revealed the following results: The normal gene sequence in exon-1 that includes codon 26 resembles a donor splice site (Fig. 9). This region, however, is not normally used in RNA processing, hence its designation as a cryptic splice site. It is used, nevertheless, in processing of the abnormal RNA species produced from the β^E gene. This mutation alters the preexisting cryptic donor site and renders it functional at a low level during processing.

Therefore, the exon mutation affects RNA processing by enhancing the affinity of a cryptic splicing site for the splicing apparatus. One might also envision a mutation that is silent with respect to amino acid reading information producing a similar effect, and Goldsmith et al [1983] have provided evidence for this in their analysis of a β^+-thalassemia gene isolated from an American black individual. In this instance, a translationally silent substitution in codon 24 (see Fig. 9), located within the same exon-1 cryptic donor site activated by the β^E mutation, leads to alternative RNA processing and to a more severe thalassemic state than seen in the HbE situation.

Mutations Affecting mRNA Function

Some forms of β-thalassemia are due to mutations within the β-gene coding sequences that render the final mRNA nonfunctional. Chang and Kan [1979] first demonstrated the existence of this mutant class in their studies of β°-thalassemia in a Chinese individual. In this instance, they found that a point mutation within codon 17 changed a lysine codon to a stop codon for translation. This type of defect is generally referred to as a *nonsense mutation*. More recently several groups have independently identified a nonsense mutation in codon 39 in the Mediterranean population. Premature termination of translation of mRNA can also result from small deletions or insertions that lead to frameshifts and the generation of downstream chain terminator codons. The first example of this type of mutation was a dinucleotide deletion in codon 8 found in a Turkish patient [Orkin and Goff, 1981]. A series of frameshift mutations are now known, and they are summarized in Table 1.

RNA Cleavage Mutations

Recently a T–C mutation in the AATAAA signal for cleavage of the primary RNA transcript was found in an American black. A similar RNA cleavage mutation has been observed in an α-thalassemia gene in Saudi Arabia [Higgs et al, 1983]. The β-thalassemia mutation is associated with in vivo production of an RNA transcript that is not cleaved in the normal position just 3′ to the AATAAA signal and continues 900 nucleotides downstream to the next AATAAA signal. At this point cleavage of the transcript and polyadenylation occur. This transcript, which is 1,600 nucleotides long, is presumed to be unstable, leading to the β-thalassemia condition [Orkin et al, 1985].

REDUCED β-GENE EXPRESSION BY POSITION EFFECT

Two examples now exist of markedly reduced or absent β-gene expression in the presence of a normal β gene and a deletion of part of the β-gene cluster

5′ to the β gene that brings into the cluster new upstream DNA sequences. One deletion ends 2.5 kb 5′ to the β gene, while the other ends about 20 kb 5′ to the β gene and leaves not only the β gene, but also the $^{A}\gamma$ and δ genes intact. In the former example, which has been studied extensively, the β gene has a normal nucleotide sequence and normal function in vitro but has lost its DNAse hypersensitivity and its hypomethylation [Kioussis et al, 1983]. These two deletions are important examples of position effects, which were previously well worked out in *Drosophila*, and they indicate that DNA sequences surrounding single copy genes can have an important role in the expression of those genes.

SUMMARY

In this chapter I have tried to relate the salient features of globin gene structure, genomic organization, normal variation, and mutations affecting gene expression. The lessons learned from the normal β-globin gene cluster and mutations producing β-thalassemia should be highly applicable to studies of other inherited diseases. As more and more gene probes become available, which have relevance to the study of human disease, the striking extent of genetic heterogeneity producing single gene disorders of man will be illuminated.

REFERENCES

Antonarakis SE, Boehm CD, Giardina PJV, Kazazian HH Jr: Nonrandom association of polymorphic restriction sites in the β-globin gene cluster. Proc Natl Acad Sci USA 79:137–141, 1982.

Antonarakis SE, Orkin SH, Cheng T-C, Scott AF, Sexton JP, Trusko S, Charache S, Kazazian HH Jr: β-Thalassemia in American blacks: Novel mutations in the TATA box and IVS-2 acceptor splice site. Proc Natl Acad Sci USA 81:1154–1158, 1984.

Breathnach R, Chambon P: Organization and expression of eucaryotic split genes coding for proteins. Annu Rev Biochem 50:349–383, 1981.

Chang JC, Kan YW: $\beta°$-Thalassemia, a nonsense mutation in man. Proc Natl Acad Sci USA 76:2886–2889, 1979.

Cheng T-C, Orkin SH, Antonarakis SE, Potter MJ, Sexton P, Markham AF, Giardina PVJ, Li A, Kazazian HH Jr: β-Thalassemia in Chinese: Use of in vivo RNA analysis and oligonucleotide hybridization in systematic characterization of molecular defects. Proc Natl Acad Sci USA 81:2821–2825, 1984.

Collins FS, Weissman SM: The molecular genetics of human hemoglobins. Prog Nucleic Acid Res Mol Biol 31:315–462, 1984.

Dobkin C, Pergolizzi RG, Bahre P, Bank A: Abnormal splice in a mutant human β-globin gene not at the site of a mutation. Proc Natl Acad Sci USA 80:1184–1188, 1983.

Efstratiadis A, Posakanoy JW, Maniatis T, Lawn RM, O'Connell C, Spritz RA, deRiel JK, Forget BG, Weissman SM, Slightom JL, Blechl AE, Smithies O, Baralle RE, Shoulders

CC, Proudfoot NJ: The structure and evolution of the human β-globin gene family. Cell 21:653–668, 1980.

Fritsch EF, Lawn RM, Maniatis T: Molecular cloning and characterization of the human β-like globin gene cluster. Cell 19:959–972, 1980.

Goldsmith ME, Humphries RK, Ley TJ, Cline A, Kantor JA, Nienhuis AW: Silent nucleotide substitution in β^+-thalassemia gene activates a cryptic splice site in globin RNA coding nucleotide. Proc Natl Acad Sci USA 80:2318–2322, 1983.

Higgs DR, Goodbourn SEY, Lamb J, Clegg JB, Weatherall DJ: α-Thalassemia caused by a polyadenylation signal mutation. Nature 306:398–400, 1983.

Kan YW, Dozy AM: Polymorphism of DNA sequence adjacent to the human β-globin structural gene: Relation to sickle mutation. Proc Natl Acad Sci USA 75:5631–5635, 1978.

Kioussis D, Vanin E, deLange T, Flavell RA, Grosveld FG: β-Globin gene inactivation by DNA translocation in $\gamma\beta$-thalassemia. Nature 306:662–666, 1983.

Maniatis T, Fritsch EF, Lauer J, Lawn RM: The molecular genetics of human hemoglobin. Annu Rev Genet 14:145–178, 1980.

Orkin SH, Goff SC: Nonsense and frameshift mutations in β°-thalassemia detected in cloned β-globin genes. J Biol Chem 256:9782–9784, 1981.

Orkin SH, Kazazian HH Jr, Antonarakis SE, Goff SC, Boehm CD, Sexton JP, Waber PG, Giardina PJV: Linkage of β-thalassemia mutations and β-globin gene polymorphisms with DNA polymorphisms in the human β-globin gene cluster. Nature 296:627–631, 1982a.

Orkin SH, Kazazian HH Jr, Antonarakis SE, Ostrer H, Goff SC, Sexton JP: Abnormal RNA processing due to the exon mutation of the β^B-globin gene. Nature 300:768–769, 1982b.

Orkin SH, Cheng TC, Antonarakis SE, Kazazian HH Jr: Thalassemia due to a mutation in the cleavage-polyadenylation signal of the human beta-globin gene. EMBO J 4:453–456, 1985.

Poncz M, Ballantine M, Soloweijczyk D, Barak I, Schwartz E, Surrey S: β-thalassemia in a Kurdish Jew. J Biol Chem 257:5994–5996, 1982.

Southern EM: Detection of specific sequences among DNA fragments separated by gel electrophoresis. J Mol Biol 98:503–517, 1975.

Spritz RA, Jagadeeswaran P, Choudary PV, Biro PA, Eider JT, deRiel JK, Manley JL, Gefter ML, Forget BG, Weissman SM: Base substitution in an intervening sequence of a β^+ thalassemic human globin gene. Proc Natl Acad Sci USA 78:2455–2459, 1981.

Tilghman SM, Tiemeier DC, Seidman JG, Peterlin BM, Sullivan M, Maizel JV, Leder P: Intervening sequence of DNA identified in the structural portion of a mouse β-globin gene. Proc Natl Acad Sci USA 75:725–729, 1978.

Treisman RA, Proudfoot NJ, Shander M, Maniatis T: A single base change at a splice site in a β^o-thalassemia gene causes abnormal RNA splicing. Cell 29:903–911, 1982.

Treisman RA, Orkin SH, Maniatis T: Specific transcription and RNA splicing defects in 5 cloned β-thalassemia genes. Nature 302:591–596, 1983.

Identification of Recessive Mutations at Human Cancer Loci

Webster K. Cavenee, PhD

University of Cincinnati College of Medicine, Cincinnati, OH 45267

The etiologic involvement of genetic components in the development of many types of cancer has been postulated since pioneering studies by Boveri [1], which showed that abnormalities in the mitotic cycle of sea urchin eggs frequently preceded abnormal growth and chromosomal content. Later studies using inbred mouse strains showed that tumors arising in one strain could be transplanted to the autologous strain but could not be efficiently transplanted to various other heterologous strains [2]. Indeed, epidemiologic analysis of human populations has shown strong familial components in subgroups of cancers of the breast [3], kidney [4], eye [5], of neural crest origin [6], endocrine glands [7], bone or connective tissue [8], esophagus [9], stomach [10], gastrointestinal tract [11], skin [12], gonads [13], lung [14], and leukemias and lymphomas [8,15].

This work was supported by grants from the National Eye Institute, NIH (EY 05510), the National Cancer Institute, NIH (CA 38583), and the March of Dimes Birth Defects Foundation (Basil O'Connor 5-496).

I gratefully acknowledge my many colleagues and collaborators who have contributed enormously to this work: R. White, T. Dryja, R. Phillips, B. Gallie, W. Benedict, A.L. Murphree, L. Strong, M. Nordenskjold, A. Koufos, M. Hansen, M. Shull, B. Lampkin, K. Bove, N. Copeland, N. Jenkins, P. Pearson, T. Mohandas, and P. Dignan.

Present address: Ludwig Institute for Cancer Research, Montreal, Quebec, Canada H3A 1A1.

Birth Defects: Original Article Series, Volume 23, Number 3, pages 93–107

Relatively recent advances in techniques for differential and subregional staining of human chromosomes have resulted in a wealth of information about karyotypic abnormalities specifically associated with different tumor types. In general, these aberrancies take the form of deletion of a specific segment of a chromosome or reciprocal translocations between two different chromosomes [16,17]. In the latter case, it is common that the chromosomal breakpoints of these interchanges are specific at the karyotypic level, suggesting that genes on one chromosome are either inactivated during the translocation process or, conversely, are moved next to a gene on the other chromosome with subsequent alteration of its expression. Although the nature of the genes affected in these karyotypically altered cells is unknown for the most part, the number of specific chromosomal changes in neoplastic cells of various types is impressive. Several examples of these changes and their associated cancers are deletion of 13q14 in retinoblastoma [18,19]; deletion of 11p13 in Wilms tumor [20,21]; deletion of 1p31 in neuroblastoma [22]; deletion of 3p21 in small cell lung carcinoma [23]; deletions of 5q22–q33 and 7q33–q36 in acute nonlymphocytic leukemia [24,25]; translocation between chromosomes 9 and 22 in chronic myelogenous leukemia [26]; between chromosomes 9 and 22, or 8 and 21, or 15 and 17, in acute nonlymphocytic leukemia [27–29]; between chromosomes 11 and 14 in chronic lymphocytic leukemia [30]; between chromosomes 8 and 14, or 4 and 11, or 9 and 22 in acute lymphocytic leukemia [31–33]; and between chromosomes 8 and 14 in Burkitt lymphoma [34].

RECESSIVE ALLELES IN RETINOBLASTOMA

Retinoblastoma is a neoplasm of embryonic retina that occurs in heritable and sporadic forms. It is a disease that occurs at a frequency of about 1:10,000 live births, affects children between birth and 7 years of age, and shows no discernible geographic, racial, or sex-specific clustering. The heritable form of the disease has been extensively documented [35–37] and is generally attributed to a single autosomal dominant mutation with high penetrance. Thus, this tumor is often considered to be a member of a family of embryonal tumors, including retinoblastoma, neuroblastoma, Wilms tumor, and medulloblastoma, with forms that display a clear tendency to cluster in families, and thus may have a genetic component in their etiology.

Retrospective examination of the constitutional karyotypes of retinoblastoma patients has revealed occasional deletions of various sized parts of chromosome 13 [18,19]. The common overlap of these deletions (so-called SRO [smallest region of overlap]) was band 13q14. It has been postulated that this chromosomal region contains a locus (retinoblastoma-1 [RB1]) in which alteration results in predisposition to tumor formation. This regional map location is supported by several studies. First, karyologic examination of the tumor from individuals with a normal constitutional karyotype has revealed deletions or rearrangements involving 13q14 in some tumors [18]. Second, quantitative analysis of the activity levels of the enzyme esterase D (ESD) in cells from patients with congenital deletions of chromosome 13 has allowed the mapping of the structural gene for this enzyme to band q14 [38]. Subsequent examination of the segregation of alleles at the ESD locus in families who are also genetically susceptible to retinoblastoma indicates tight linkage between the two loci [39]. Finally, at least one family has been reported [40] that shows a balanced insertional translocation of a piece of 13q into chromosome 3 and with segregation for deletion of the translocated 13q segment in many members who developed retinoblastoma. The region of the chromosome that was deleted in the predisposed children was 13q13.1–13q14.5, thus including the proposed RB1 locus.

Detailed statistical analyses of heritable and sporadic retinoblastoma have shown a relationship between these two forms of the disease [3–5]. The incidence of affected progeny in bilateral cases approaches 50%, as would be expected for an autosomal dominant trait, whereas the corresponding incidence in unilateral cases is about 10%. However, the two forms of the disease do not appear to be due to two different mutations, since a bilaterally affected parent may have a unilaterally affected child, and children of a unilaterally affected parent may also be bilaterally affected [37]. A comparison of the number of tumors, age of onset, and laterality has shown [5] that the hereditary cases generally have more tumors (an average of three versus one for nonhereditary cases), occur about a year earlier, and are most often affected bilaterally rather than unilaterally. It was estimated from epidemiologic data [5] that about 60% of retinoblastoma cases are nonhereditary and unilateral, 15% are hereditary and unilateral, and 25% are hereditary and bilateral.

The model that was proposed to explain these observations [5] was that as few as two mutations were required for malignant transformation. The hereditary cases would have inherited a germinal mutation that does not, in itself, cause the tumor but rather predisposes each retinal cell to a further transforming event. In this model, the nonhereditary cases would also result from two mutations except that these events would have to occur in the same somatic cell. Thus, the two forms of the disease could be viewed as resulting from the same two-step process, at the level of the aberrant retinal cell, the difference being the inheritance or somatic occurrence of the first mutation.

We proposed that the second step in tumorigenesis in both heritable and sporadic retinoblastoma involves somatic alteration of the normal allele at the RB1 locus in a way that unmasks the mutant allele [41]. Thus, the first mutation in this process, although it may be inherited as an autosomal dominant trait at the organismal level, would be expected to have the properties of a recessive defect at the level of the individual retinal cell.

The model that we have constructed to take account of these considerations is described in Figure 1. In this model, the heritable form of the disease arises as a germinal mutation of the RB1 locus (Fig. 1a) and is inherited by an individual who, therefore, is an obligate heterozygote (rb/+) at the RB1 locus in each of his somatic and germ cells (Fig. 1b). A subsequent event in any of his retinal cells that results in homozygosity or hemizygosity for the mutant allele (that is, mutant at the RB1 locus on both chromosome 13 homologs) will result in a tumor clone. The chromosomal mechanisms that could accomplish this loss of constitutional heterozygosity include mitotic

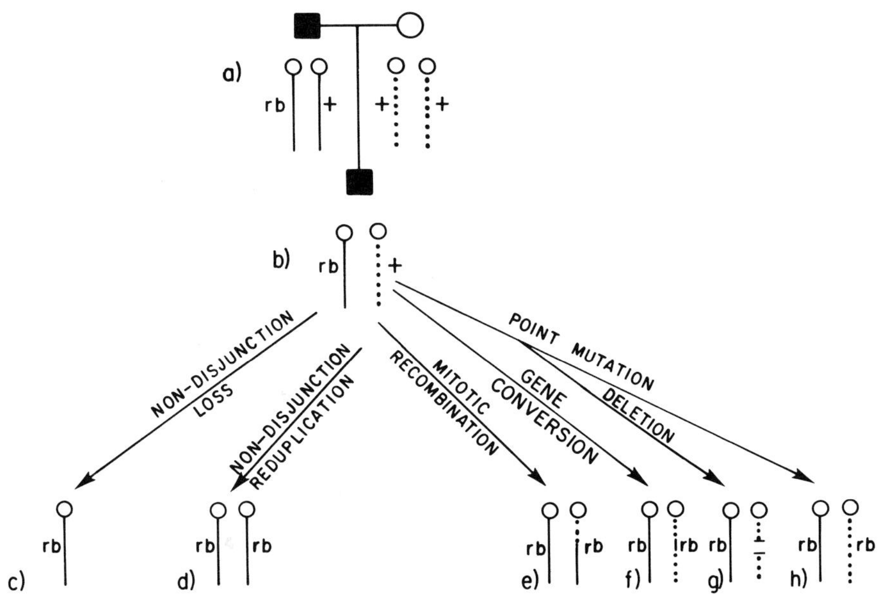

Fig. 1. Chromosomal mechanisms that could reveal recessive mutations. A child (**b**) inherits a chromosome 13 from his affected parent that carries a recessive defect at the RB1 locus, designated "rb", and so is genotypically rb/+ in all of his cells. A retinoblastoma may develop by eliminating the dominant wild-type allele at the RB1 locus by one of the mechanisms shown in **c–h**. (Reprinted with permission from Nature 305:779–784; copyright © 1983, Macmillan Journals, Ltd.)

nondisjunction with loss of the wild-type chromosome, which would result in hemizygosity at all loci on chromosome 13 (Fig. 1c); mitotic nondisjunction with reduplication of the mutant chromosome, which would result in homozygosity at all loci on the chromosome (Fig. 1d); or mitotic recombination between the chromosomal homologs with a breakpoint between the RB1 loci and the centromeres, which would result in heterozygosity at loci in the proximal region of the chromosome and homozygosity throughout the rest of the chromosome including the RB1 locus (Fig. 1e). Regional events such as gene conversion (Fig. 1f), deletion (Fig. 1g), or mutation (Fig. 1h) must also be considered. Heritable and sporadic retinoblastoma could each arise through the appearance of homozygosity at the RB1 locus, the difference being two somatic events in the sporadic case compared to one germinal and one somatic event in the heritable case.

The approach we have taken to examine these hypotheses relies on the variability of DNA sequences among humans [42,43], which results in heritable differences in recognition sequences for bacterial restriction endonucleases. Segments of the human genome are isolated in recombinant DNA form, and the loci homologous to these probe segments are tested for their inclusion of restriction endonuclease recognition sequences that vary between unrelated individuals (RFLPs). Two types of such variation have been defined. The first, and most abundant, results from simple base pair changes within the recognition site sequence for a particular restriction endonuclease and yields alleles of longer length (when the effect of the mutation is loss of a site) or shorter length (when the effect of the mutation is the gain of a site) [42]. The second type results from the insertion or deletion of varying numbers of blocks of like DNA sequence into or out of the genomic locus [43]. Practically, the net result is the observation of two alleles at the locus encompassing a site change (presence or absence of the site) in the first case, or numerous alleles at a locus subject to insertion or deletion of larger segments of DNA. In either case, however, any given individual will reveal only two alleles at the locus, one from the paternally derived chromosome and one from the maternally derived homolog. In all cases examined to date these types of markers have been shown to behave in family studies as would be predicted for simple mendelian codominant alleles.

Recombinant DNA segments have been isolated from human chromosome 13 [44] and used as probes to examine loci scattered along the chromosome to detect changes in the somatic genotypes of tumors as compared to the germ line genotypes of the individuals harboring the tumors [41]. One example of the results obtained in this study is shown in Figure 2. A somatic, or mitotic, recombination between the mutant and wild-type chromosome

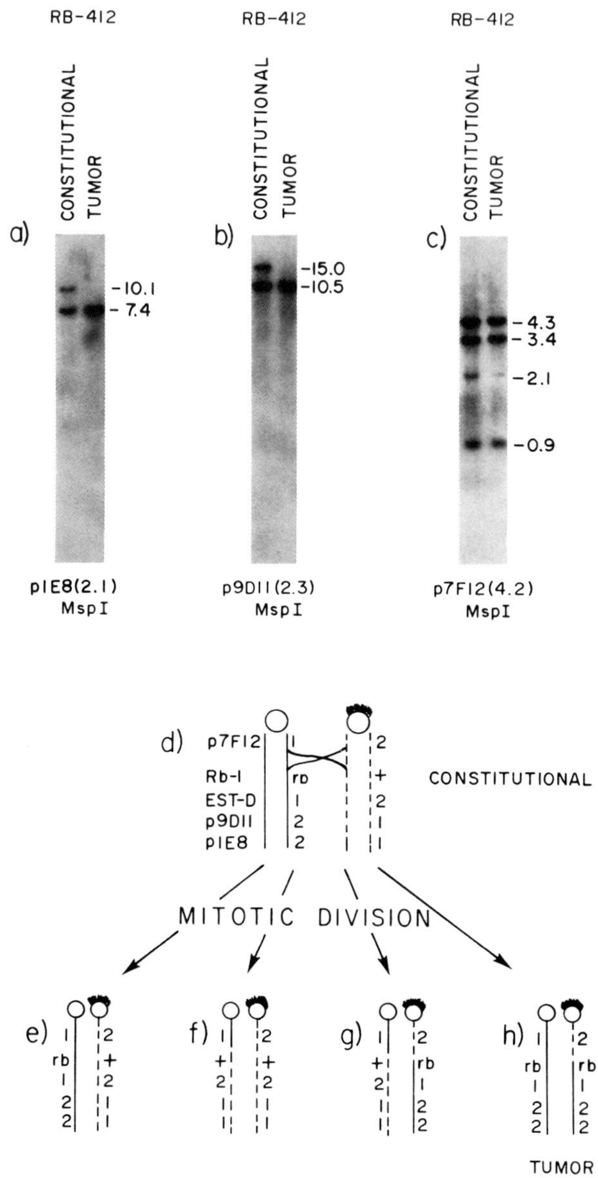
RB-412
CONSTITUTIONAL
TUMOR
a)
-10.1
- 7.4
pIE8(2.1)
MspI
RB-412
CONSTITUTIONAL
TUMOR
b)
-15.0
-10.5
p9D11(2.3)
MspI
RB-412
CONSTITUTIONAL
TUMOR
c)
- 4.3
- 3.4
-2.1
-0.9
p7F12(4.2)
MspI
d)
p7F12
Rb-1
EST-D
p9D11
pIE8
CONSTITUTIONAL
MITOTIC DIVISION
e)
f)
g)
h)
TUMOR

homologs, with subsequent segregation, can result in a cell that maintains heterozygosity at loci proximal to the breakpoint of the recombinational event but that shows homozygosity at loci distal to such a breakpoint. The first example of such an event in mammals was shown to have occurred in the tumor that arose in patient Rb-412 [45]. The constitutional cells from this person were heterozygous at the ESD locus and for a quinacrine-staining satellite heteromorphism on the short arm of the chromosomes 13. An examination of the tumor cells derived from this patient showed the presence of both types of satellite staining, but only one isozymic form of esterase D. A reasonable previous interpretation of these data [45] was that both chromosomes 13 were present, in their entirety, in the Rb-412 tumor and that a somatic inactivation of the ESD locus on one homolog resulted in hemizygosity and subsequent detection of only one isozymic form of the enzyme. Alternatively, a mitotic recombination event occurring between the centromere and the ESD locus, as described above, could also generate chromosomes consistent with these results. Constitutional and tumor genotypes of this patient were examined at chromosome 13 loci defined by seven DNA probes, and the results shown in Figure 2 were obtained. Three of the markers showed both codominant allelic forms in skin fibroblasts from Rb-412: p1E8, which maps distal to q22; p9D11, which maps to q21; and p7F12, which maps to q13. Only one allele of the loci homologous to p1E8 and p9D11 was apparent in DNA from the tumor tissue. Both p1E8 and p9D11 reduced to homozygosity in the tumor tissue. However, p7F12, the marker locus that maps between the RB1 locus and the centromere, remained heterozygous. These results, together with the heterochromatic staining and ester-

---

Fig. 2. Homozygosity for chromosome 13 allows expression of recessive mutant alleles at the retinoblastoma locus. **a)** Pattern obtained when the hybridization probe was p1E8 (homologous to a locus on chromosome 13 mapping between band q22 and the terminus of the long arm); **b)** with the hybridization probe p9D11 (homolgous to a locus on chromosome 13 mapping to band q22); **c)** with hybridization probe p7F12 (homologous to a locus on chromosome 13 mapping between bands q12 and q14). **d)** Schematic diagram showing inferred haplotypes on each chromosome. The cap on the chromosome homolog represented by the dashed lines denotes a fluorescent-staining heterochromatin region. The point of cross over must lie between the RB1 and 7F12 loci and is shown occurring between chromatids of the chromosome 13 homologs at the four-strand stage of mitotic chromosome replication. **e–h)** Schematic diagram of the four possible combinations of wild-type and recombinant homologs. Possibilities e–g result in a phenotypically wild-type cell and are inconsistent with the allelic data shown in a–c, whereas the combination shown in h corresponds to the experimental data and results in homozygosity for the mutant (rb) allele at the RB1 locus. (Reprinted with permission from Nature 305: 779–784; copyright © 1983, Macmillan Journals, Ltd.)

ase D typing, are consistent with the occurrence of somatic recombination between the mutant and wild-type chromosomes 13 in the cell that gave rise to the tumor. The cross-over point was between the 7F12 locus and the RB1 and ESD loci such that all loci distal to the cross-over point, including RB1, have become homozygous. Between RB1 and the terminus of the short arm, however, two markers maintain both the maternal and paternal haplotypes.

Including patient Rb-412, 23 retinoblastoma cases have been examined using this approach, and loss of heterozygosity has been observed with at least one probe in 17 of the 23 cases [41,46] (also, our unpublished results). We have observed examples of nondisjunctional loss of the wild-type chromosome, nondisjunctional loss of one chromosome 13 with reduplication of the remaining homolog, and mitotic recombination between the two chromosome 13 homologs [41,46]. These results emphasize the feasibility of this approach and indicate that, unexpectedly, greater than 75% of the cases examined allow expression of the first recessive defect at the RB1 locus not by mutation at the homologous locus but rather by its elimination through chromosomal segregation or recombination mechanisms. Molecular genetic support for this idea has been obtained recently by comparing constitutional and tumor genotypes from heritable cases and their families. In all cases examined to date, the chromosome 13 retained in these tumors derived from the affected parent and the chromosome 13 lost during tumorigenesis derived from the normal parent [47].

These studies suggest chromosomal mechanisms that can affect the expression of a predisposed mutant allele at the RB1 locus. To determine the chromosomal specificity of these losses of constitutional heterozygosity, we also compared constitutional and tumor genotypes at loci on other chromosomes. Constitutional heterozygosity was retained in the tumors in every case. These data support the 13q14 map location of the RB1 locus and lend credence to the specific corollaries we have made to Knudson's multistep hypothesis [3–5] for the pathway to tumorigenesis. The genetic inference of recessive mutations predisposing to the development of cancer may be of wider significance if similar mechanisms can be shown to occur during the development of other, more common, tumors. These studies suggest the possibility of identifying other chromosomal loci that predispose to other types of tumors by examining the incidence of loss of heterozygosity at loci on each arm of each chromosome.

RECESSIVE ALLELES IN WILMS TUMOR

Our first attempt at testing the proposal that loci predisposing to tumorigenesis could be mapped by searching for somatic loss of heterozygosity was performed with the embryonal nephroblastoma Wilms tumor [48]. This tumor, the most common abdominal neoplasm in children, exhibits both heritable and sporadic forms [4]. The autosomal dominant heritable form is often associated with other specific congenital malformations such as aniridia, genitourinary abnormalities, and mental retardation [20]. A chromosomal locus that plays a role in the development of this tumor has been mapped to chromosome 11 at band p13 by the observation of interstitial deletions of 11p13 in constitutional karyotypes of patients [49,50], by direct cytogenetic determination of 11p13 deletions in Wilms tumor cells in patients with normal constitutional karyotypes [51], and by the description of a family in whom cells from unaffected members carried an interstitial deletion including 11p13 but also contained this part of the chromosome as an interstitial translocation while cells from affected members carried only the deleted chromosome [21]. These cases imply that mutant forms of one or more loci contained within the 11p13 band predispose to the formation of Wilms tumor. They are not, however, sufficient to elicit the cancer, since discrete tumor foci are seen amid the background of normal kidney, even in cases in which the 11p13 band of one homolog is missing in the germ line. Thus, a second, postzygotic event must be required as well.

To determine whether mechanisms similar to those described for retinoblastoma play a role in the development of Wilms tumor, we isolated high molecular weight DNA from primary tumor biopsies and corresponding normal tissue (either from peripheral blood leukocytes or from unaffected kidney where possible), digested these samples with appropriate restriction endonucleases, separated the resultant fragments by electrophoresis through agarose gels, and transferred the DNA fragments to nylon membranes [41,44]. Each of these paired samples was then independently hybridized to four polymorphic recombinant DNA probes that have been mapped to subregion p15 of the short arm of chromosome 11 [52]. The probe pADJ-762 reveals polymorphic alleles, in *Bcl*I-digested DNA, of 11.6 and 4.2 kb, as well as a constant band of 6.9 kb [52,53]. It also showed an independent polymorphism, in *Msp*I-digested DNA, with alleles of 2.3 and 1.8 kb [52,53]. The probe p*Hins*-310 derives from the 5′ flanking region of the human insulin gene [54] and shows polymorphic alleles of various lengths caused by the insertion or deletion of blocks of DNA [55]. The probe JW151 derives from the human β-globin region, or gene cluster [56], and reveals two independent RFLPs, one with alleles of 8.0 and 7.2 kb representing the γ^G locus and one with alleles of 3.5 and 2.7 kb representing the γ^A locus. The probe pTBB2 is homologous to the c-Ha-ras-2 protooncogene [57] and reveals polymorphic alleles, in *Taq*I-digested DNA, of various lengths caused by insertion or deletion of blocks of DNA [52,57]. Each of these probes has been independently mapped to the chromosomal region 11p15–11pter [52,58,59].

Table 1 displays the constitutional and tumor genotypes obtained in the patients we examined. Patient 1 was constitutionally heterozygous (ie, revealed both codominant alleles) at four loci on chromosome 11p, those defined by pADJ-762 in *Bcl*I-digested DNA, p*Hins*-310 in *Pvu*II-digested DNA, and the γ^G- and γ^A-globin loci defined by JW151 in *Hind*III-digested DNA. One of the alleles at each of these loci was lost in primary tumor tissue from this patient. It is unlikely that these allele losses resulted from a deletion of the terminal portion of the short arm of chromosome 11, since the hybridization intensity of each of the alleles remaining in the tumor tissue was greater than that obtained from corresponding alleles in normal tissue from this patient. These results suggest that this loss of alleles at multiple loci during Wilms tumor oncogenesis was not due to a simple deletion event and may, by analogy with the previous analysis of retinoblastoma tumors [41], indicate that loss of one chromosome 11 homolog occurred with subsequent reduplication of the chromosome carrying the initial predisposing recessive mutation.

TABLE 1. Alleles at Chromosome 11 Loci in Normal (N) and Tumor (T) Tissue From Six Wilms Tumor Patients

			Patient No.					
Locus	Enzyme	Tissue	1	2	3	4	5	6
	*Msp*I	N	2,2	2,2	2,2	1,2	2,2	1,2
		T	2,2	2,2	2,2	2,2	2,2	2,2
pADJ-762								
	*Bcl*I	N	1,2	2,2	2,2	1,1	1,2	2,2
		T	2,2	2,2	2,2	1,1	1,1	2,2
p*Hins*-310	*Pvu*II	N	1,2	1,2	1,3	1,3	2,3	1,2
		T	2,2	2,2	1,3	3,3	3,3	2,2
	*Hind*III (γ^G)	N	1,2	1,2	2,2	1,2	1,2	1,2
		T	1,1	2,2	2,2	1,1	2,2	1,1
JW151								
	*Hind*III (γ^A)	N	1,2	1,2	1,1	1,1	1,1	1,1
		T	1,1	1,1	1,1	1,1	1,1	1,1
PTBB2	T*aq*I	N	2,2	2,2	—	2,2	2,2	2,2
		T	2,2	2,2	—	2,2	2,2	2,2

Alleles for each locus are named with 1 as the longer allele and 2 or 3 as the shorter. Constitutional tissue from patients 1, 2, 3, 4, and 5 was peripheral blood lymphocytes; in patient 6 normal kidney was obtained. —, Not examined.

Similarly, patient 2 was constitutionally heterozygous at the locus homologous to p*Hins*-310 and at the γ^G- and γ^A-globin loci defined by JW151. Each of these loci had become homozygous in primary tumor tissue from this patient, and the hybridization intensities of the alleles remaining in the tumor were again greater than that of the corresponding constitutional alleles. Karyotypic analysis of tumor cells from this patient showed no discernible deletion of an 11p homolog. Patient 5 was also constitutionally heterozygous at the loci homologous to pADJ-762 and p*Hins*-310 as well as to the γ^G-globin loci homologous to JW151. Each of these loci became homozygous in tumor tissue from this patient. Although technical considerations precluded analysis of the differential hybridization intensities in this case, karyotypic analysis of these tumor cells did not show any obvious deletion of either 11p homolog.

Based on our previous work with retinoblastoma [41], a likely explanation for these results is that a mitotic segregation event has occurred in each predisposed kidney cell such that one chromosome homolog was lost and the remaining homolog was reduplicated during the process of tumorigenesis. We infer that the remaining chromosomes are each defective at the Wilms tumor locus contained within band 11p13. It is also possible that mitotic recombination events occurred between the most proximal marker locus (defined by JW151) and the centromere such that one parental and one recombinant chromosome remained in these tumors. We would hypothesize that such events would have breakpoints between the Wilms locus and the centromere, resulting in homozygosity for recessive mutant alleles at the Wilms locus similar to the retinoblastoma case Rb-412 [41]. Distinguishing between these mechanisms will require the development of probes homologous to loci between band p13 and the centromere.

THE FUTURE

These results describe mechanisms whereby abnormal mitotic events allow expression of recessive mutations predisposing to cancer [41] and are operative in more than one tumor type and affect more than one chromosome. These results suggest that the strategies previously proposed [41] for mapping recessive genes involved in the oncogenesis of various tumor types are feasible. The utility of this approach in identifying which parental chromosome carries the predisposing defect and subsequent genetic counseling in hereditary cases is now apparent.

Several questions arise from these studies. How many loci are there in the human genome in which recessive mutations can predispose to cancer? Two possible alternatives are that there is one such gene for each tumor type or that there are a limited number of such genes each with a broad tissue specificity. Perhaps an examination of tumor types that are clinically associated at significant frequencies will allow a resolution of this question. Can these approaches be adapted to be clinically useful? The inferences of tumor-specific chromosomal mechanisms described here require that we be able to distinguish a mutant from a normal chromosome. Since the inheritance of the mutant chromosome results in a virtual certainty of tumor formation, this ability to discriminate chromosomes in families would seem highly suitable to premorbid predictions of cancer. Do these genes play a role in normal development? It is indeed curious and provocative that the existence of recessive oncogenes has been demonstrated first in embryonal cancers. These tumors have been postulated for many years to represent as much a problem of embryonic development as of oncology. It is tantalizing to hypothesize that some developmental anomaly syndromes, called hamartomatoses, may represent a less expressive recessive mutant gene or even, perhaps, the heterozygote phenotype. What might be the normal function of the product encoded by these loci? Might they encode proteins whose function is to repress other tumor-causing loci or, rather, might they encode proteins necessary in the logical, step-wise, ordered pathway to normal organismal development? The answers to these questions promise to provide exciting insights into basic processes in biology and will rely on our ability to isolate, analyze, and manipulate these genes and then to reintroduce them into appropriate target cells. The studies described here indicate a new approach to the somatic cell genetics of cancer and provide the first step in the elucidation of their function: their identification and localization.

REFERENCES

1. Boveri T: "The Origin of Malignant Tumors." Baltimore: Williams & Wilkins, 1929.
2. Cloudman AM: A comparative study of transplantability of eight mammary gland tumors arising in inbred mice. Am J Cancer 16:568–630, 1932.
3. Knudson AG, Strong LC, Anderson DE: Heredity and cancer in man. Prog Med Genet 9:113–158, 1973.
4. Knudson AG, Strong LC: Mutation and cancer: A model for Wilms' tumor of the kidney. JNCI 48:313–324, 1972.
5. Knudson AG: Mutation and cancer: Statistical study of retinoblastoma. Proc Natl Acad Sci USA 68:820–823, 1971.
6. Warthin AS: Heredity with reference to carcinoma. Arch Intern Med 12:546–558, 1913.
7. Pollock RS: Carotid body tumors—Idiosyncracies. Oncology 27:81–87, 1973.
8. Miller RW: Deaths from childhood leukemia and solid tumors among twins and other sibs in the United States, 1960–67. JNCI 46:203–209, 1971.

9. Harper PS, Harper RMJ, Howel-Evans AW: Carcinoma of the oesophagus with tylosis. Q J Med 39:317–333, 1970.
10. Creagan ET, Fraumeni JF: Familial gastric cancer and immunologic abnormalities. Cancer 32:1325–1331, 1973.
11. Erbe RW: Inherited gastrointestinal-polyposis syndromes. N Engl J Med 294:1101–1104, 1976.
12. Ramsay CA, Coltart TM, Blunt S, Pawsey SA, et al: Prenatal diagnosis of xeroderma pigmentosum. Lancet 2:1109–1112, 1974.
13. Simpson JL, Photopoulos G: Hereditary aspects of ovarian and testicular neoplasia. Birth Defects 12(1):51–72, 1976.
14. Tokhuta GK, Lilienfeld AM: Familial aggregation of lung cancer in humans. JNCI 30:289–312, 1963.
15. Kurita S, Kamei Y, Ota K: Genetic studies on familial leukemia. Cancer 34:1098–1106, 1974.
16. Sandberg AA: "The Chromosomes and Human Cancer." New York: Elsevier Press, 1980.
17. Yunis JJ: The chromosomal basis of human neoplasia. Science 221:227–236, 1983.
18. Balaban-Malenbaum G, Gilbert F, Nichols W, Hill R, Shields J, Meadows AT: A deleted No. 13 in human retinoblastoma cells: Relevance to tumorigenesis. Cancer Genet Cytogenet 3:243–250, 1981.
19. Yunis JJ, Ramsay N: Retinoblastoma and subband deletion of chromosome 13. Am J Dis Child 132:161–163, 1978.
20. Riccardi VM, Sujansky E, Smith AC, Francke U: Chromosomal imbalance in the aniridia-Wilms' tumor association: 11p interstitial deletion. Pediatrics 61:604–610, 1978.
21. Yunis JJ, Ramsay N: Familial occurrence of the aniridia-Wilms' tumor syndrome with deletion 11p13–14.1. J Pediatr 96:1027–1030, 1980.
22. Gilbert F, Balaban G, Moorhead P, Bianchi D, Schlesinger H: Abnormalities of chromosome 1p in human neuroblastoma tumors and cell lines. Cancer Genet Cytogenet 7:33–42, 1982.
23. Whang-Pen J, Kao-Shan CS, Lee EC, Bunn PA, Carney DN, Gazdar AF, Minna JD: Specific chromosome defect associated with human small-cell lung cancer: Deletion 3p (14–23). Science 215:181–182, 1982.
24. Borgstrom GH, Teerenhovi L, Vuopio P, de la Chapelle A, van den Berghe H, Brandt L, Golomb HM, Louwagie A, Mitelman F, Rowley JD, Sandberg AA: Clinical implications of monosomy 7 in acute nonlymphocytic leukemia. Cancer Genet Cytogenet 2:115–130, 1980.
25. van den Berghe H, David G, Michaux JL, Sokal G, Verwilgen R: 5q− Acute myelogenous leukemia. Blood 48:624–625, 1976.
26. Rowley JD: A new consistent chromosomal abnormality in chronic myelogenous leukemia identified by quinacrine fluorescence and Giemsa staining. Nature 243:290–292, 1973.
27. Bloomfield CD, Peterson LC, Yunis JJ, Brunning RD: The Philadelphia chromosome (Ph′) in adults presenting with acute leukemia: A comparison of Ph'^{+} and Ph'^{-} patients. Br J Haemataol 36:347–358, 1977.
28. Rowley JD: Identification of a translocation with quinacrine fluorescence in a patient with acute leukemia. Ann Génét 16:109–112, 1973.
29. Rowley JD, Golomb HM, Vardiman J, Fukuhara S, Dougherty C, Potter D: Further evidence for a non-random chromosomal abnormality in acute promyelocytic leukemia. Int J Cancer 20:869–872, 1977.

30. Autio K, Turunen O, Penttila O, Eramaa E, de la Chapelle A, Schroder J: Human chronic lymphocytic leukemia: Karyotypes in different lymphocyte populations. Cancer Genet Cytogenet 1:147–155, 1979.
31. Berger R, Bernheim A, Brouet JC, Daniel MT, Flandrin G: t(8;14) Translocation in a Burkitt's type of lymphoblastic leukemia (L3). Br J Haematol 43:87–90, 1979.
32. Bloomfield CD, Lindquist LL, Brunning RD, Yunis JJ, Coccia PF: The Philadelphia chromosome in acute leukemia. Virchows Arch Cell Pathol 29:81–91, 1978.
33. van den Berghe H, David G, Broechaet-van Orshoven A, Louwagie A, Uerwilghen R, Casteels-van Daele M, Eggermont E, Ecckels R: A new chromosome anomaly in acute lymphoblastic leukemia (ALL). Hum Genet 46:173–180, 1979.
34. Yunis JJ, Oken MM, Kaplan ME, Ensrud KM, Howe RR, Theologides A: Distinctive chromosomal abnormalities in histological subtypes of non-Hodgkin's lymphoma. N Engl J Med 307:1231–1236, 1982.
35. Falls HF, Neel JV: Genetics of retinoblastoma. Arch Ophthalmol 46:367–389, 1951.
36. Macklin MT: A study of retinoblastoma in Ohio. Am J Hum Genet 12:1–43, 1960.
37. Schappert-Kimmijser J, Hemmes GD, Nijland R: The heredity of retinoblastoma. Ophthalmologica 151:197–213, 1966.
38. Sparkes RS, Sparkes MC, Wilson MG, Towner JW, Benedict WF, Murphree AL, Yunis JJ: Regional assignment of genes for human esterase D and retinoblastoma to chromosome band 13q14. Science 208:1042–1044, 1980.
39. Sparkes RS, Murphree AL, Lingua RW, Sparkes MC, Field LL, Funderburk SJ, Benedict WF: Gene for hereditary retinoblastoma assigned to human chromosome 13 by linkage to esterase D. Science 219:971–973, 1983.
40. Strong LC, Riccardi VM, Ferrell RE, Sparkes RS: Familial retinoblastoma and chromosome 13 deletion transmitted via an insertional translocation. Science 213:1501–1503, 1981.
41. Cavenee WK, Dryja TP, Phillips RA, Benedict WF, Godbout R, Gallie BL, Murphree AL, Strong LC, White RL: Expression of recessive alleles by chromosomal mechanisms in retinoblastoma. Nature 305:779–784, 1983.
42. Barker DF, Schaefer M, White RL: Restriction sites containing CpG show a higher frequency of polymorphism in human DNA. Cell 36:131–138, 1984.
43. Wyman AR, White RL: A highly polymorphic locus in human DNA. Proc Natl Acad Sci USA 77:6754–6759, 1980.
44. Cavenee WK, Leach RJ, Mohandas T, Pearson P, White RL: Isolation and regional localization of DNA segments revealing polymorphic loci from human chromosome 13. Am J Hum Genet 36:10–24, 1984.
45. Godbout R, Dryja TP, Squire J, Gallie B, Phillips RA: Somatic inactivation of genes on chromosome 13 is a common event in retinoblastoma. Nature 304:451–454, 1983.
46. Dryja TP, Cavenee WK, White R, Rapoport J, Petersen R, Albert D, Bruns G: Homozygosity of chromosome 13 in retinoblastoma. N Engl J Med 310:550–553, 1984.
47. Cavenee WK, Hansen MF, Nordenskjold M, Kock E, Maumenee I, Squire JA, Phillips RA, Gallie BL: Genetic origin of mutations predisposing to retinoblastoma. Science 228:501–503, 1985.
48. Koufos A, Hansen MF, Lampkin BC, Workman ML, Copeland NG, Jenkins NA, Cavenee WK: Loss of alleles at loci on human chromosome 11 during genesis of Wilms' tumor. Nature 309:170–172, 1984.
49. Francke U, Riccardi VM: Aniridia-Wilms' tumor association: Evidence for specific deletion of 11p13. Cytogenet Cell Genet 24:185–193, 1979.

50. Riccardi VM, Hittner HM, Francke U, Yunis JJ, Ledbetter D, Borges W: The aniridia-Wilms' tumor association: The critical role of chromosome band 11p13. Cancer Genet Cytogenet 2:131–137, 1980.
51. Slater RM, Kraker J: Chromosome number 11 and Wilms' tumor. Cancer Genet Cytogenet 5:237–245, 1982.
52. Gerald PS, Grezeschik KH: Report of the committee on the genetic constitution of chromosome 10, 11, and 12. Cytogenet Cell Genet 37:103–126, 1984.
53. Barker DF, White RL: A locus on chromosome 11p with multiple restriction site polymorphisms. Am J Hum Genet 36:1159–1171, 1984.
54. Bell GI, Karam J, Rutter WJ: Polymorphic DNA region adjacent to the 5′ end of the human insulin gene. Proc Natl Acad Sci USA 78:5759–5763, 1981.
55. Bell GI, Selby MJ, Rutter WJ: The highly polymorphic region near the human insulin gene is composed of simple tandemly repeated sequences. Nature 295:31–35, 1981.
56. Antonarakis SE, Boehm CD, Giardina PJV, Kazazian HH: Nonrandom association of polymorphic restriction sites in the β-globin gene cluster. Proc Natl Acad Sci USA 79:137–141, 1982.
57. Goldfarb M, Shimizu K, Perucho M, Wigler M: Isolation and preliminary characterization of human transforming gene from T24 bladder carcinoma cells. Nature 296:404–409, 1982.
58. deMartinville B, Francke U: The c-Ha-*ras*1, insulin and β-globin loci map outside the deletion associated with aniridia-Wilms' tumour. Nature 305:641–643, 1983.
59. Huerre C, Despoisse S, Gilgenkrantz S, Lenoir G, Junien C: c-ha-*ras*1 is not deleted in aniridia-Wilms' tumour association. Nature 305:638–641, 1983.

Retroviral DNA Content of the Mouse Genome

Nancy A. Jenkins, PhD, and Neal G. Copeland, PhD

Mammalian Genetics Laboratory, BRI-Basic Research Program, NCI-Frederick Cancer Research Facility, Frederick, MD 21701

INTRODUCTION

Retroviruses, RNA tumor viruses, have been studied intensively for many years primarily because of their ability to induce disease in a wide variety of vertebrate species. The single feature that distinguishes retroviruses from all other animal viruses is their unusual strategy for replication. Members of this family of viruses carry a single-stranded RNA genome. The RNA genomes of all replication-competent retroviruses contain three coding domains: *gag*, for the synthesis of core proteins; *pol*, for the synthesis of virion-associated RNA-dependent DNA polymerase (reverse transcriptase); and *env*, for the synthesis of envelope glycoproteins.

Following infection of a sensitive cell, which is mediated by an interaction between the virus particle and a host-encoded cellular receptor, the viral RNA genome is transcribed into a double-stranded DNA molecule through a series of highly ordered processes [for review see Varmus, 1982]. An obligate step in the life cycle of a retrovirus is the integration of this DNA molecule into the host genome. The integrated DNA copy of the retroviral

This research was sponsored by the National Cancer Institute, DHHS, under Contract No. N01-CO-23909 with Litton Bionetics, Inc. The contents of this publication do not necessarily reflect the views or policies of the DHHS, nor does mention of trade names, commercial products, or organizations imply endorsement by the U.S. Government.

Birth Defects: Original Article Series, Volume 23, Number 3, pages 109–122

genome, which is called a *provirus,* contains all information encoded by the RNA genome as well as an additional copy of sequences derived from the 3′ and 5′ termini of the viral RNA. These sequences, which are generated during synthesis of viral DNA, are arranged as direct repeats at the ends of the provirus and are referred to as *long terminal repeats* (LTRs). Within the LTRs there are regulatory sequences including promoters, enhancers, and polyadenylation signals [for review see Varmus, 1982].

Integration of the viral DNA into host chromosomes is specific with respect to the virus. Integration occurs at the ends of the viral LTRs, precisely two base pairs (bp) are removed from each LTR upon integration, and a short duplication of host DNA occurs at the site of viral integration [for review, see Varmus, 1982]. However, there appears to be little, if any, specificity with respect to the site of virus integration in the host genome. The structure of a nondefective provirus is illustrated in Figure 1.

The provirus is a stable component of host DNA. It is replicated and inherited along with flanking cellular DNA. Transcription of viral RNA and translation into protein occurs via normal cellular mechanisms. Viral particle assembly then occurs, and the new virion buds from the cell membrane thus completing the virus life cycle.

Retroviruses are unique among the animal viruses because they are carried in the germ line of a wide variety of vetebrate species.These "endogenous" retroviruses are transmitted vertically as normal mendelian genes and thus stand in contrast to other animal viruses that are exogenous viruses, being transmitted horizontally from one animal to another. Retroviruses can also be transmitted as exogenous viruses, and comparisons between similar families of endogenous and exogenous retroviruses of a single species via heteroduplex analysis, T_1 oligonucleotide fingerprinting, and/or nucleic acid hybridization studies indicate that these two classes of viruses are closely related [for review see Coffin, 1982].

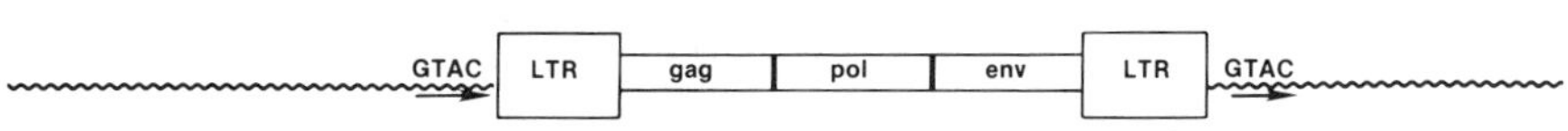

Fig. 1. The structure of a nondefective provirus. The provirus of a replication-competent retrovirus contains the DNA sequence coding for three genes, *gag, pol,* and *env,* and is flanked on each end by long terminal (direct) repeats (LTRs). Upon integration of the provirus in the host genome, a duplication of cell DNA is generated. In the case of ecotropic murine leukemia viruses, this duplication (indicated by GTAC) is 4 bp in length. As retroviruses integrate at multiple sites in mouse chromosomes, the actual sequence duplicated varies between different proviruses.

In fact, though the origin of endogenous retroviruses is not well understood, it is likely that at least some of these viral loci actually arose from the rare exogenous infection of the host germ line. Following infection of a germ cell by a replication-competent retrovirus, a provirus would be formed. As this event occurs in a germ cell rather than a somatic cell, the new proviral locus would then be replicated and transmitted as part of the normal host genetic complement. The consequences to the host carrying endogenous proviruses depend on many factors, including the nature or type of retrovirus, whether the virus is expressed, and the particular site in the chromosomes at which the resident provirus is located. Many endogenous proviruses have no detectable effect on the host. However, certain murine endogenous viruses have been shown to be both pathogenic and mutagenic.

ENDOGENOUS MURINE RETROVIRUSES

It has been estimated that more than 0.3% of the mouse genome [Jaenisch, 1983] and 3% of the mouse Y chromosome [Phillips et al, 1982] is retrovirus-related. These sequences have been classified into several different families, including A-type, B-type, and C-type particles according to their morphology as visualized under the electron microscope. One of the most intensively studied families of endogenous murine retroviruses is the C-type murine leukemia viruses (MuLVs). It is this family of viruses that are classically associated with high leukemic inbred mouse strains. Endogenous MuLVs can be further classified into three groups based on host range, a property conferred by the product of the *env* gene (Fig. 1). Uninfected mouse cells of certain strains will spontaneously begin producing virus that is either 1) ecotropic and will infect only murine cells or 2) xenotropic and will infect cells of many species other than mouse. Mink cell focus-forming viruses (MCF), which will infect both murine cells and cells of heterologous species, have also been isolated. Several studies suggest that MCF viruses are *env* gene recombinants between ecotropic and xenotropic MuLVs. *Env* gene sequences homologous to those carried by MCF virus have also been identified in normal mouse cell DNA, indicating that they are carried in the mouse genome and may represent the progenitor of the MCF virus *env* gene [Chattopadhyay et al, 1981,1982].

Endogenous MuLVs are structurally related. Ecotropic and xenotropic MuLVs share approximately 60% sequence homology [Callahan et al, 1974]. Nuclei acid hybridization studies have shown that all inbred mouse strains carry multiple copies of xenotropic and MCF virus-related sequences. In contrast, the number of ecotropic proviral loci carried by inbred strains is low. This low copy number has made it possible to study this group of endogenous proviruses in some detail.

ENDOGENOUS ECOTROPIC MuLV PROVIRUSES

The prototypic endogenous ecotropic virus was isolated from AKR mice. Mice of this inbred strain contain at least two highly expressed endogenous ecotropic proviral loci and produce virus throughout life. With the advent of recombinant DNA technology the AKR ecotropic MuLV was molecularly cloned [Chattopadhyay et al, 1980]. From this molecular clone it was possible to isolate a small 400 bp DNA fragment of the *env* gene (Fig. 1) that was specific for ecotropic MuLVs and did not hybridize, following restriction enzyme analysis and Southern blotting, to any other endogenous MuLV sequences [Chattopadhyay et al, 1980]. Using restriction enzyme analysis, Southern blotting, and hybridization with the ecotropic MuLV DNA-specific probe, it was then possible to evaluate the organization and distribution of these proviruses in DNAs of various inbred mouse strains [Chattopadhyay et al, 1980; Jenkins et al, 1982] (also, Jenkins and Copeland, unpublished observations). These studies indicate the following: 1) the number of ecotropic proviral genomes in inbred strains is indeed low, 0 to ten copies per haploid mouse genome; 2) most endogenous ecotropic MuLVs appeared structurally nondefective by this analysis and resembled the prototype AKR provirus; 3) at least 45 distinct ecotropic proviral integration sites in chromosomes of inbred mouse strains were identified. 4) in general, the distribution of these proviruses followed known strain and substrain relationships, suggesting that these viral loci were acquired after speciation, but before the establishment of inbred strains, and that these integrations are relatively stable. Of the more than 45 ecotropic proviral loci identified, 17 have been mapped by various techniques to specific locations in mouse chromosomes (Fig. 2). These proviruses are distributed over ten different chromosomes, confirming that retroviruses can integrate at many different sites. These endogenous proviruses will serve as useful genetic markers for future linkage studies.

For example, *Emv-16* and *Emv-17* (endogenous ecotropic murine leukemia virus locus 16 and 17, respectively), two closely linked proviruses of RF/J mice, have been placed on chromosome 1 in the vicinity of leaden, *ln* (Buchberg, Taylor, Jenkins, and Copeland, manuscript submitted). *Emv-15* is closely associated with the lethal yellow, A^Y, coat-color mutation of the agouti, *a*, locus on chromosome 2 [Copeland et al, 1983b]. *Emv-11*, *Emv-12*, *Emv-13*, and *Emv-14* of AKR mice have been assigned to chromosomes 7,

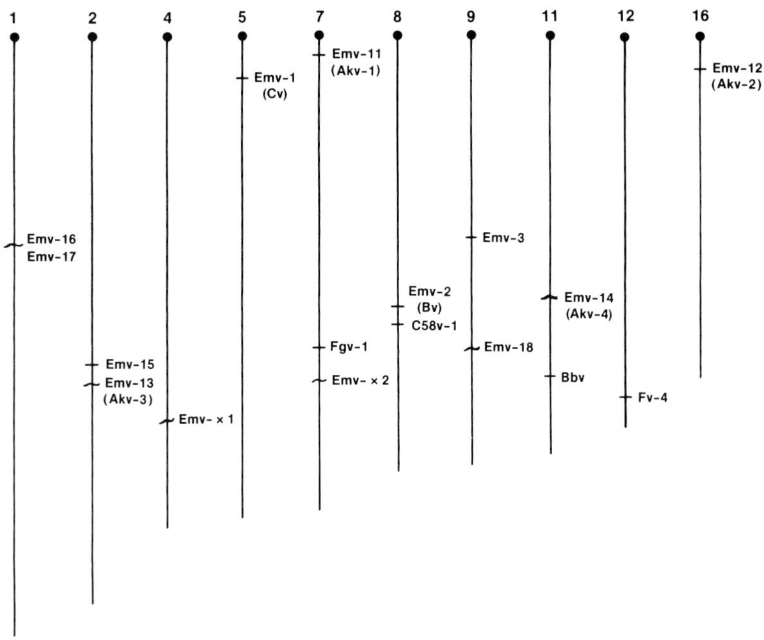

Fig. 2. Chromosomal assignments of ecotropic murine leukemia proviral sequences. A genetic map of the ecotropic proviral loci is shown. Seventeen proviruses have been assigned to ten different chromosomes. Proviruses associated with straight lines (–) represent those loci whose exact map position is known, while viral sequences shown with curved line (~) are assigned only to a particular region of a chromosome. *Emv-1*, *Emv-2*, *Emv-11*, *Emv-12*, *Emv-13*, and *Emv-14* were previously designated *Cv*, *Bv*, *Akv-1*, *Akv-2*, *Akv-3*, and *Akv-4*, respectively. (See text for references.)

16, 2, and 11, respectively [Taylor et al, 1985; Kozak and Rowe, 1980b]. Two of the six proviruses of C58/J mice, which have not been named and are referred to as *Emv-x1* and *Emv-x2,* have been linked via analysis of the NX58 recombinant inbred strains to *Gpd-1* on chromosome 4 (*Emv-x1*) and *Hbb* on chromosome 7 (*Emv-x2*) (Riblet, Copeland, and Jenkins, unpublished results). *Emv-1* mapped to chromosome 5 [Kozak and Rowe, 1979; Ihle et al, 1979] and is carried by many different strains including BALB/c, A, and C3H. *Emv-2*, carried by the C57BL and C57BR strains, is located on chromosome 8 [Kozak and Rowe, 1982]. *Emv-3* is causally related to the dilute, *d*, mutation of mouse chromosome 9 [Jenkins et al, 1981]. Kozak and Rowe [1982] have shown that an endogenous ecotropic provirus of C3H/Fg mice, *Fgv-1*, is on chromosome 7; *Bbr*, of B10.Br/SgL mice, is on chromosome 11; and *C58v-1*, of C58 mice, is on chromosome 8. Another endogenous provirus, designated *Emv-18*, has been placed near *Fv-2* (Friend virus restriction locus-2) on chromosome 9 [Mowat and Bernstein, 1983]. Finally, an ecotropic viral resistance gene, *Fv-4,* on chromosome 12, is associated with a defective ecotropic MuLV [Kozak et al, 1984].

Expression of endogenous ecotropic MuLV proviruses varies greatly with inbred mouse strain and age. Strains such as AKR and C58 spontaneously produce high titers of virus beginning early in life, and they carry multiple copies per haploid genome of ecotropic proviral loci [Quint et al, 1981; Steffen et al, 1982; Buckler et al, 1982a,b; Jenkins et al, 1982]. In contrast, several other strains, including DBA/2, BALB/c, C57BL/6, and C3H/He, express ecotropic MuLV only upon aging or chemical induction [Pincus, 1980] and carry a single copy per haploid mouse genome of ecotropic MuLV sequences [Buckler et al, 1982a,b; Jenkins et al, 1982; McCubrey and Risser, 1982]. However, the exact mechanisms regulating ecotropic provirus expression remain unclear. Regulation may occur at the transcriptional level [Cabradilla et al 1976; Thompson et al, 1980] and be controlled by linked cis-acting cellular DNA sequences [Copeland and Cooper, 1979], DNA methylation [Groudine et al, 1981; Harbers et al, 1981; Hoffmann et al, 1982], or both. The chromosomal integration site of a given proviral locus also seems to influence the time in development and the tissue of virus activation [Jaenisch et al, 1981]. Additionally, some endogenous ecotropic proviruses have been molecularly cloned and shown to carry small mutations within their genome that inhibit expression [Copeland et al, 1984].

Results from numerous studies indicate that there is a strong correlation between the presence of multiple endogenous ecotropic proviral loci that are expressed early in life and the propensity to develop leukemia. Nearly all AKR and C58 mice develop T-cell lymphomas by about 12 months of age, whereas the incidence of lymphomas in DBA/2, BALB/c, C57BL/6, and C3H/He mice is low [for review, see Pincus, 1980]. An additional consequence of viremia caused by early ecotropic MuLV expression is that the provirus copy number in these mice increases over time. For example, by comparing the endogenous ecotropic proviral DNA content of various AKR sublines [Quint et al, 1982; Steffen et al, 1982], AKR/J-derived recombinant inbred mouse strains (Jenkins et al, 1981; Steffen et al, 1982] and congenic NIH.AKV sublines of mice [Rowe and Kozak, 1980], it was shown that these mice spontaneously acquire new endogenous ecotropic proviruses. Acquisition of new ecotropic proviral loci apparently results from rare infection of the germ line by the virus these mice produce spontaneously throughout life [Rowe and Kozak, 1980]. An estimate of the rate at which new endogenous ecotropic proviruses are becoming fixed in the germ line of these viremic strains is one new provirus every 15 to 30 years or 37 to 75 generations of inbreeding [Jenkins et al, 1981; Quint et al, 1982].

Since retroviruses integrate at multiple sites in mouse chromosomes, it is possible that they may act as insertional mutagens by interrupting normal gene expression or by activating the expression of cellular sequences flanking the proviral integration site. That spontaneous mutating insertions have occurred in the mouse germ line was first demonstrated by studies of the recessive dilute, *d*, coat-color mutation of DBA/2J mice [Jenkins et al, 1981]. The *d* mutation cosegregated with an endogenous ecotropic MuLV genome (termed *Emv-3*) in all inbred, recombinant inbred, and mutant mouse strains tested. The *d* mutation is relatively unstable and is spontaneously lost from the germ line (d^+) at a frequency of 3×10^{-6}. Analysis of several independent dilute coat-color revertants showed that loss of the mutation is always associated with loss of most of the associated provirus, confirming virus integration did indeed cause the *d* mutation [Copeland et al, 1983a] (Fig. 2). Two revertant sites, as well as the preintegration site, were cloned and sequenced [Hutchison et al, 1984]. Exactly one LTR remained behind in each revertant site, indicating that the virus was excised by homologous recombination involving the viral LTRs. Preliminary sequence analysis of the proviral integration site indicated that the virus did not integrate in coding sequence. As the dilute mutation is recessive, the most likely interpretation of these results is that the provirus induced the dilute mutation by blocking normal gene expression through integration into either an intron or 5′ or 3′ noncoding sequence.

One other mouse coat-color mutation, the lethal yellow, A^y, coat-color mutation of the agouti locus on chromosome 2 (Fig. 2), is also closely linked to the site of integration of an ecotropic provirus [Copeland et al, 1983b]. Unlike the dilute mutation, this coat-color mutation is dominant and homozygous lethal. As a lethal yellow coat-color revertant has yet to be identified, it will require further molecular studies to determine if this virus caused the A^y mutation.

Jaenisch et al [1983] have found that germ line integration of Moloney murine leukemia virus at the *Mov-13* locus of mice led to a recessive lethal mutation. Integration was subsequently shown to have occurred at the 5′ end of the αl(l) collagen gene, causing a complete transcriptional block [Schnieke et al, 1983]. Developmental arrest of embryos homozygous for the mutant *Mov-13* allele coincided with the time at which high expression of αl (l) collagen is seen in normal embryogenesis.

ENDOGENOUS XENOTROPIC AND MCF MuLV PROVIRUSES

Multiple copies of endogenous xenotropic-related MuLV DNA sequences are present in the genome of all inbred strains of mice. The viruses encoded by at least some of these endogenous sequences are capable of infecting cells of many mammalian species but not mice. Xenotropic MuLVs grow rather poorly and are frequently mixtures of several related viruses. Because of both the heterogeneity of this family of viruses and the lack of specific hybridization probes, these viruses have in general been more difficult to study than ecotropic MuLVs.

The expression of xenotropic MuLVs, like ecotropic MuLVs, varies between inbred strains. Strains such as NZB spontaneously produce high titers of viruses throughout life. Several strains, including BALB/c, do not spontaneously produce virus, but fibroblast cultures can be induced to produce virus (referred to as *inducible* or *class II* xenotropic MuLV). Occasionally, xenotropic MuLV can be isolated from NIH-Swiss or C57L mice, but cells from these mice are not inducible for xenotropic virus (referred to as *noninducible* or *class III* xenotropic MuLV) [for review, see Coffin, 1982].

Several xenotropic MuLV isolates have been molecularly cloned [Chattopadhyay et al 1981; Buckler et al, 1982 a,b]. A small fragment of the *env* gene (Fig. 1) from a xenotropic MuLV, pX_{env}, has been identified that will not hybridize to ecotropic MuLV genomes [Buckler et al 1982 a,b]. This sequence does, however, hybridize to mink-cell focus-forming viruses MCFs).

Mink cell focus-forming viruses, which infect and replicate in both mouse cells and cells of other species, are isolated from preleukemic or leukemic mice. MCFs are thought to be recombinant viruses derived from at least two endogenous MuLVs, one of which is ecotropic. Ecotropic and MCF MuLVs differ largely in the *env* gene, and it appears likely that the MCF *env* gene sequence is encoded in the mouse germ line [Chattopadhyay et al, 1982].

Several MCF virus isolates have been molecularly cloned and studied [Chattopadhyay et al, 1982], but the xenotropic and MCF *env* sequences are closely related and it has been difficult to isolate a fragment from this gene that will hybridize to MCF but not xenotropic viruses. Therefore, these viruses will be considered as a single family for the purpose of molecular genetic analysis.

Restriction enzyme analysis, Southern blotting, and hybridization with the pX_{env} or closely related probes have revealed that most inbred strains carry approximately 30–50 pX_{env}-related proviruses [Blatt et al, 1983; Hoggan et al, 1983; Meuruelo et al, 1983; Wejman et al, 1984]. However, considerable variability in the location of these sequences has been seen in different inbred mouse strains. Several of these loci have been mapped [Blatt et al, 1983; Meuruelo et al, 1983; Wejman et al, 1984], primarily by molecular genetic analysis of recombinant inbred mouse strains to many different mouse chromosomes (Fig. 3). A number of these proviral loci are very tightly linked to loci encoding mouse lymphocyte cell surface antigens. These lymphocyte antigens are differentially expressed on the lymphocyte membrane and, in some cases, are expressed primarily on B cells, on T cells, on B and T cells, or in some cases on lymphoid as well as nonlymphoid cells. The reason for this close association is not yet known. Although these proviruses have not been associated with any of the known mouse mutations, their close association with genes encoding lymphocyte differentiation antigens suggests that they may be useful molecular markers for cloning and characterizing regions of the mouse genome important in lymphocyte development.

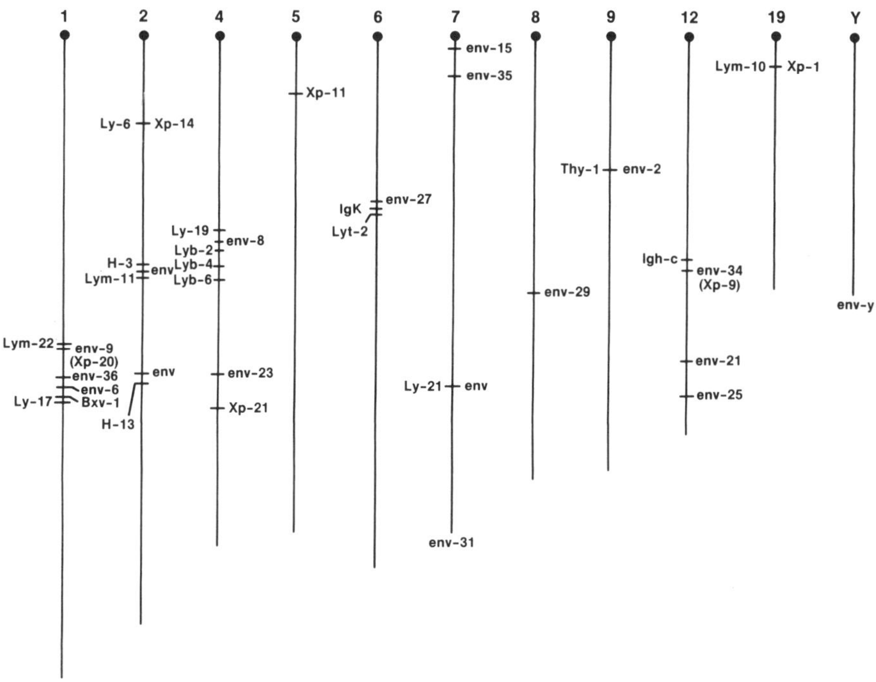

Fig. 3. Chromosomal assignments of pX_{env}-related proviral sequences. Endogenous proviruses homologous to the pX_{env} hybridization probe have been mapped to 11 different mouse chromosomes. Virus-related sequences are denoted *env* (endogenous virus) or Xp (xenotropic virus-related sequence detected following digestion of DNA with the enzyme *Pvu*II) [Blatt et al, 1983; Wejman et al, 1984] followed by a numerical designation. Sequences described only by Meuruelo et al [1983] have been labeled *env* without a numerical designation. Additionally, *Bxv-1* (the inducible xenotropic MuLV locus of BALB/c mice) has been assigned to chromosome 1 [Kozak and Rowe, 1980a]. Many of the loci are located near sequences coding for lymphocyte or minor histocompatibility antigens.

ENDOGENOUS MOUSE MAMMARY TUMOR VIRUSES

Mouse mammary tumor viruses (MMTVs) are the prototype type-B retrovirus. In general, the conclusions drawn from studies on the endogenous type-C MuLVs apply to MMTVs. 1) MMTV sequences are present in the mouse germ line. These sequences are arranged in the form of proviruses, and it is likely that these endogenous viruses were acquired by multiple, independent infection events of the mouse germ line after speciation but before inbreeding. 2) All inbred strains of mice carry from approximately three to ten endogenous MMTV proviruses that are located on several different chromosomes. There are, however, some California wild mice that lack endogenous MMTV proviruses. 3) The various MMTV proviruses carried by inbred mouse strains can be distinguished by restriction enzyme analysis, Southern blotting, and hybridization with an MMTV-specific DNA sequence, isolated from the MMTV *env* gene. 4) The endogenous MMTV proviruses differ greatly in their level of expression and oncogenic potential. A feature unique to MMTVs is the regulation of their gene expression by glucocorticoid hormones [for review, see Coffin, 1982].

ENDOGENOUS VIRUS-LIKE ELEMENTS OF MICE

Intracisternal A-Type Particles

A-type particles are frequently detected in early mouse embryos and in certain mouse tumors. These particles are unlike type-B and type-C retroviruses in that they never appear to be released from the cell as virions. That these particles are actually retrovirus-like was suggested by studies that showed that they contain an RNA genome as well as an associated reverse transcriptase.

A-type proviruses are abundant in the mouse. It has been estimated that there are at least 1,000 copies per haploid genome. Several endogenous A-type proviruses have been molecularly cloned and shown to be flanked by LTR sequences. Similar to other retroviruses, there is a short duplication of cell DNA upon integration. By heteroduplex analysis there is a great deal of variability between various A-type proviruses. Neither the mechanisms controlling the expression nor the origin of these A-type sequences has been resolved [for review, see Coffin, 1982]. Like ecotropic viruses, A-type viruses have been shown to induce mutations via integration [Kuff et al, 1983; Canaani et al, 1983].

Endogenous Retrovirus-Related Sequences

Several families of retrovirus-related sequences have been reported. These include the VL30 (virus-like 30S RNA) genes, which are structurally analogous to retroviruses [for review see Coffin, 1982], and the M720-related sequences [Phillips et al, 1982], which in addition to being present on autosomes, compose as much as 3% of the mouse Y chromosome. As there is no compelling evidence for site-specific integration of retroviruses, it is possible that this high copy number of M720 sequences on the Y chromosome is the result of an amplification event that occurred after virus acquisition [Phillips et al, 1982].

New families of retroviral-related sequences continue to be identified. Recent additions include the ETn, BglII repeats, LTR-IS, and MuRRS sequences summarized in Table 1 [Brulet et al, 1983; Propst and Vande Woude, 1984; Wirth et al, 1984; Schmidt et al, 1985, respectively]. The study of the structure and expression of these sequences will increase our knowledge about the biology of murine retroviruses. The mapping of additional proviruses (or retrovirus-related sequences) will provide more genetic markers in mouse chromosomes and, perhaps, interesting associations between retroviral integration sites and mouse mutations.

SUMMARY

Mice, like many other vertebrate species, carry multiple copies of retrovirus-related sequences as part of their normal genetic complement. These endogenous retroviral DNA sequences are, in general, structurally and functionally similar to proviruses formed via exogenous virus infection. They behave as relatively stable genetic entities, are transmitted in a normal mendelian manner, and can be expressed by normal cellular mechanisms. Chromosome mapping studies have shown that these virus-related sequences are dispersed throughout the mouse genome, in most cases were acquired

TABLE 1. Classification of Murine Retrovirus-Related Sequences

Family	Copies per haploid mouse genome
Ecotropic MuLV	0–10
Xenotropic and mink cell focus-forming viruses	~30–50
Mouse mammary tumor viruses	0–10
A-particle related sequences	~1,000
VL30 sequences	~20–50
M720-related sequences	>100
ETn	~1,000
*Bgl*II repeats	~1,500
LTR-IS	500–1,000
MuRRS	50–100

after speciation but before inbreeding, and occasionally cause mutations as a result of integrating within the mouse genome.

REFERENCES

Blatt C, Mileham K, Haas M, Nesbitt MN, Harper ME, Simon MI: Chromosomal mapping of the mink cell focus-inducing and xenotropic *env* gene family in the mouse. Proc Natl Acad Sci USA 80:6298–6302, 1983.

Brulet P, Kaghad M, Xu YS, Croissant O, Jacob F: Early differential tissue expression of transposon-like repetitive DNA sequences of the mouse. Proc Natl Acad Sci USA 80:5641–5645, 1983.

Buckler CE, Hoggan MD, Chan HW, Sears JF, Khan AS, Moore JL, Hartley JW, Rowe WP, Martin MA: Cloning and characterization of an envelope-specific probe from xenotropic murine leukemia proviral DNA. J Virol 41: 228–236, 1982a.

Buckler CE, Staal SP, Rowe WP, Martin MA: Variation in the number of copies and in the genomic organization of ecotropic murine leukemia virus proviral sequences in sublines of AKR mice. J Virol 43:629–640, 1982b.

Cabradilla CD, Robbins KC, Aaronson SA: Induction of mouse type-C virus by translational inhibitors: Evidence for transcriptional depression of a specific class of endogenous virus. Proc Natl Acad Sci USA 73:4541–4545, 1976.

Callahan R, Benveniste RE, Lieber RE, Todaro GJ: Nucleic acid homology of murine type-C viral genes. J Virol 14:1394–1403, 1974.

Canaani E, Dreazen O, Klar A, Rechavi G, Ram D, Cohen JB, Givol D: Activation of the c-mos oncogene in a mouse plasmacytoma by insertion of an endogenous intracisternal A-particle genome. Proc Natl Acad Sci USA 80:7118–7122, 1983.

Chattopadhyay SK, Cloyd MW, Linemeyer DL, Lander MR, Rands E, Lowy DR: Cellular origin and role of mink cell focus-forming viruses in murine thymic lymphomas. Nature 295:25–31, 1982.

Chattopadhyay SK, Lander MR, Gupta S, Rands E, Lowy DR: Origin of mink cytopathic focus-forming (MCF) viruses: Comparison with ecotropic and xenotropic murine leukemia virus genomes. Virology 113:465–483, 1981.

Chattopadhyay SK, Lander MR, Rands E, Lowy DR: Structure of endogenous murine leukemia virus DNA in mouse genomes. Proc Natl Acad Sci USA 77:5774–5778, 1980.

Coffin J: Endogenous viruses. In Weiss R, Teich N, Varmus H, Coffin J (eds): "RNA Tumor Viruses, Molecular Biology of Tumor Viruses," 2nd Ed. Cold Spring Harbor, NY: Cold Spring Harbor Laboratory Publications, 1982, pp 1109–1203.

Copeland NG, Bedigian HG, Thomas CY, Jenkins NA: DNAs of two molecularly cloned endogenous ecotropic proviruses are poorly infectious in DNA transfection assays. J Virol 50:437–444, 1984.

Copeland NG, Cooper GM: Transfection by exogenous and endogenous murine retrovirus DNAs. Cell 16:347–356, 1979.

Copeland NG, Hutchison KW, Jenkins NA: Excision of the DBA ecotropic provirus in dilute coat-color revertants of mice occurs by homologous recombination involving the viral LTRs. Cell 33:379–387, 1983a;

Copeland NG, Jenkins NA, Lee BK: Association of the lethal yellow (A^Y) coat-color mutation with an ecotropic murine leukemia virus genome. Proc Natl Acad Sci USA 80:247–249, 1983b.

Groudine M, Eisenman R, Weintraub H: Chromatin structure of endogenous retroviral genes and activation by an inhibitor of DNA methylation. Nature 292:311–317, 1981.

Harbers K, Schnieke A, Stuhlmann H, Jähner D, Jaenish R: DNA methylation and gene expression: Endogenous retroviral genome becomes infectious after molecular cloning. Proc Natl Acad Sci USA 78:7609–7613, 1981.

Hoffmann JW, Steffen D, Gusella J, Tabin C, Birds S, Cowing D, Weinberg RA: DNA methylation affecting the expression of murine leukemia proviruses. J Virol 44:144–157, 1982.

Hoggan MD, Buckler CE, Sears JF, Rowe WP, Martin MA: Organization and stability of endogenous xenotropic murine leukemia virus proviral DNA in mouse genomes. J Virol 45:473–477, 1983.

Hutchison KW, Copeland NG, Jenkins NA: Dilute coat-color locus of mice: Nucleotide sequence analysis of the d^{+2J} and d^{+Ha} revertant alleles. Mol Cell Biol 4:2899–2904, 1984.

Ihle JN, Joseph DR, Domotor JJ: Genetic linkage of C3H/HeJ and BALB/c endogenous ecotropic C-type viruses to phosphoglucomutase-1 on chromosome 5. Science 204:71–73, 1979.

Jaenisch R: Endogenous retroviruses. Cell 32:5–6, 1983.

Jaenisch R, Harbers K, Schnieke A, Löhler J, Chumakov I, Jähner D, Grotkopp D, Hoffman E: Germline integration of Moloney murine leukemia virus at the *Mov-13* locus leads to recessive lethal mutation and early embryonic death. Cell 32:209–216, 1983.

Jaenisch R, Jähner D, Nobis P, Simon I, Löhler J, Harbers K, Grotkopp D: Chromosomal position and activation of retroviral genomes inserted into the germline of mice. Cell 24:519–529, 1981.

Jenkins NA, Copeland NG, Taylor BA, Lee BK: Dilute (*d*) coat-colour mutation of DBA/2J mice is associated with the site of integration of an ecotropic MuLV genome. Nature 293:370–374, 1981.

Jenkins NA, Copeland NG, Taylor BA, Lee BK: Organization, distribution, and stability of endogenous ecotropic murine leukemia virus DNA sequences in chromosomes of *Mus musculus*. J Virol 43:26–36, 1982.

Kozak CA, Gromet NJ, Ikeda H, Buckler CE: A Unique sequence related to the ecotropic murine leukemia virus is associated with the *Fv-4* resistance gene. Proc Natl Acad Sci USA 81:834–837, 1984.

Kozak CA, Rowe WP: Genetic mapping of the ecotropic murine leukemia virus-inducing locus of the BALB/c mouse to chromosome 5. Science 204:69–71, 1979.

Kozak CA, Rowe WP: Genetic mapping of xenotropic murine leukemia virus-inducing loci in five mouse strains. J Exp Med 152:219–228, 1980a.

Kozak CA, Rowe WP: Genetic mapping of ecotropic virus-inducing locus *Akv-2* of the AKR mouse. J Exp Med 152:1419–1423, 1980b.

Kozak CA, Rowe WP: Genetic mapping of ecotropic murine leukemia virus-inducing loci in six inbred strains. J Exp Med 155:524–534, 1982.

Kuff EL, Feenstra A, Lueders K, Smith L, Hawley R, Hozumi N, Sullivan M: Intracisternal A-particle genes as movable elements in the mouse genome. Proc Natl Acad Sci USA 80:1992–1996, 1983.

McCubrey J, Risser R: Genetic interactions in induction of endogenous murine leukemia virus from low leukemic mice. Cell 28:881–888, 1982.

Meuruelo D, Rossomando A, Offer M, Bulbaum J, Pellicer A: Association of endogenous viral loci with genes encoding murine histocompatibility and lymphocyte differentiation antigens. Proc Natl Acad Sci USA 80:5032–5036, 1983.

Mowat M, Bernstein A: Linkage of the *Fv-2* gene to a newly reinserted ecotropic retrovirus in *Fv-2* congenic mice. J Virol 47:471–477, 1983.

Phillips SJ, Birkenmeier EH, Callahan R, Eicher EM: Male and female mouse DNAs can be discriminated using retroviral probes. Nature 297:241–243, 1982.

Pincus T: The endogenous murine type C viruses. In Stephenson JR (ed): "Molecular Biology of RNA Tumor Viruses." New York: Academic Press, 1980, pp 77–130.

Propst F, Vande Woude GF: A novel transposon-like repeat interrupted by an LTR element occurs in a cluster of B1 repeats in the mouse c-*mos* locus. Nucleic Acids Res 12:8381–8392, 1984.

Quint W, Quax W, Van Der Putten H, Berns A: Characterization of AKR murine leukemia virus sequences in AKR mouse substrains and structure of integrated recombinant genomes in tumor tissues. J Virol 39:1–10, 1981.

Quint W, Van Der Putten H, Janssen F, Berns A: Mobility of endogenous ecotropic murine leukemia viral genomes within mouse chromosomal DNA and integration of a mink cell focus-forming virus-type recombinant provirus in the germline. J Virol 41:901–908, 1982.

Rowe WP, Kozak CA: Germline reinsertions of AKR murine leukemia virus genomes in *Akv-1* congenic mice. Proc Natl Acad Sci USA 77:4871–4874, 1980.

Schmidt M, Wirth T, Kroger B, Horak I: Structure and genomic organization of a new family of murine retrovirus-related DNA sequences (MuRRS). Nucleic Acids Res 13:3461–3470, 1985.

Schnieke A, Harbers K, Jaenisch R: Embryonic lethal mutation in mice induced by retrovirus insertion into the $\alpha 1$ (I) collagen gene. Nature 304:315–320, 1983.

Steffen DL, Taylor BA, Weinberg RA: Continuing germline integration of AKV proviruses during the breeding of AKR mice and derivative recombinant inbred strains. J Virol 42:165–175, 1982.

Taylor BA, Rowe L, Jenkins NA, Copeland NG: Chromosomal assignment of two endogenous ecotropic murine leukemia proviruses of the AKR/J mouse strain. J Virol 56:172–175, 1985.

Thomson JA, Laipis PJ, Stein GS, Stein JL, Lander MR, Chattopadhyay SK: Regulation of endogenous type C viruses: Evidence for transcriptional control of AKR viral expression. Virology 101:529–533, 1980.

Varmus HE: Form and function of retroviral proviruses. Science 216:812–820, 1982.

Wejman JC, Taylor BA, Jenkins NA, Copeland NG: Endogenous xenotropic murine leukemia virus-related sequences map to chromosomal regions encoding mouse lymphocyte antigens. J Virol 50:237–247, 1984.

Wirth T, Schmidt M, Baumruker T, Horak I: Evidence for mobility of a new family of mouse middle repetitive DNA elements (LTR-IS). Nucleic Acids Res 12:3603–3610, 1984.

Eukaryotic Transposable Elements: Probes for Identifying and Studying Genes Important in Mammalian Development

Neal G. Copeland, PhD, and Nancy A. Jenkins, PhD

Mammalian Genetics Laboratory, BRI–Basic Research Program, NCI–Frederick Cancer Research Facility, Frederick, MD 21701

INTRODUCTION

Until recently, the genomes of both prokaryotes and eukaryotes were regarded as generally static sequences subject to change only on the leisurely time scale of evolution. This stability is dramatically indicated by the conservation of genetic linkage relationships that have been observed between distantly related species, for example, between mouse and man [Nadeau and Taylor, 1984]. Even in prokaryotes, where generation times are quite fast, the overall organization of the genome has been found to change only relatively slowly. For example, very similar genetic linkage maps have been described for two different bacterial species, *Escherichia coli* and *Salmonella typhimurium* [O'Brien, 1984]. This "conventional" evolution of the genome relies on mechanisms that have been familiar for many years. These include

This research was sponsored by the National Cancer Institute, DHHS, under contract No. N01-CO-23909 with Litton Bionetics, Inc. The contents of this publication do not necessarily reflect the views or policies of the DHHS, nor does mention of trade names, commercial products, or organizations imply endorsement by the U.S. Government.

Birth Defects: Original Article Series, Volume 23, Number 3, pages 123–135

conjugation in bacteria as well as gene duplication or rearrangement in eukaryotes resulting from factors such as unequal crossing-over or gene conversion. Within this framework, however, are mechanisms that have the potential of rapidly scrambling the genome and consequently of greatly increasing evolutionary rates. Such forces ultimately may even be responsible for generating new species. This fluidity of the genome has been recognized only relatively recently and is invariably due to the action of transposable elements.

DISCOVERY OF TRANSPOSABLE ELEMENTS

Transposable elements were first discovered in the corn plant maize in the 1940s by a series of ingenious genetic studies done by Barbara McClintock for which she was recently awarded the Nobel prize [for a review of this work as well as transposable elements in general see Shapiro, 1983]. The maize transposable elements were identified in part because of their ability to induce unstable mutations that could be visualized on the surface of the corn kernel. For the most part, these mutations were induced via the insertion of transposable elements into genes affecting pigment synthesis in such a way that the gene was partially or totally inactive. Subsequent excision of the transposable element from the genome resulting in the restoration of wild-type gene expression could then be seen by a change in pigmentation. The ability of transposable elements to excise from the host genome is a general property of transposable elements (see below), and, as a result, transposable element-induced mutations are often recognized because they are phenotypically unstable. If excision occurs in the germ line, then all kernels would be pigmented. However, if excision occurs infrequently and only in somatic tissue, then kernels would be identified that contain spots of wild-type pigmentation. The size of the spots reflects the time in development at which excision occurs (ie, the earlier in development excision occurs, the larger the spots that result).

McClintock [1945, 1946] also demonstrated that transposable elements can induce chromosome breaks. In fact, the well-known maize transposable element termed *dissociation* (Ds) was first identified by its ability to induce chromosome breaks. These breaks could be visualized directly under the microscope, or, in strains of the appropriate genetic constitution, they could be seen phenotypically as variegated colored kernels. The exact physical mechanism by which transposable elements induce chromosome breaks is not known. However, transposable elements of other species have also been reported to induce chromosome breaks [Engels, 1983].

In addition to the unstable mutations of maize, transposable element-induced mutations have also been identified in a number of other species [Shapiro, 1983]. In fact, in *Drosophila* it has been estimated that as many as 50% of all spontaneous mutations are transposable element-induced. It is apparent therefore that the presence of transposable elements can have profound effects on the genome and on evolutionary rates.

TRANSPOSABLE ELEMENTS OF *DROSOPHILA MELANOGASTER*

Among the first eukaryotic transposable elements to be molecularly cloned and studied in detail were those of the fruit fly *D. melanogaster.* Therefore, much of what we know about the molecular nature of transposable elements has come from the study of these elements. Interestingly, the gross structure of eukaryotic transposable elements has been found to be remarkably similar from species to species; therefore, a brief description of the transposable elements of *D. melanogaster* will serve as a useful introduction to eukaryotic transposable elements in general.

Three classes of *Drosophila* transposable elements have been identified, which include the *copia*-like elements, *FB* elements, and *P* elements (Fig. 1) [Shapiro, 1983]. The *copia*-like elements comprise as much as 1% of the genome of *D. melanogaster.* At least seven (and probably many more yet to be identified) *copia*-like families have been identified with individual elements ranging in size from 5 to 8.5 kb. Although not related by sequence, each family is present at about 10 to 100 copies per genome and is present at widely scattered and highly polymorphic locations. These elements terminate in short direct repeats 276 to 500 bp in length that in turn are bounded by smaller 3 to 18 bp inverted repeats. Several long open reading frames have been identified in the body of these elements that are thought to encode functions necessary for transposition. The overall structure of the *copia*-like elements is strikingly similar to that of the yeast *Ty1* elements and to retrovirus proviruses. In fact, recent evidence has been accumulating to suggest that both *copia*-like and *Ty1* elements transpose via a retroviral-like intermediate [Flavell, 1984; Saigo et al, 1984; Boeke et al, 1985]. A recent report by Johns et al [1985] demonstrating the presence of *copia*-like elements in maize suggests that species as distantly related as plants also contain retrovirus-related sequences.

The *foldback* or *FB* elements are structurally dissimilar to the *copia*-like elements [Potter, 1982]. They are 500 to 5,000 bp in length and terminate in long inverted repeats of 250 to 1,250 bp, which are themselves composed of short 10 to 30 bp repeats. Some elements contain a central region of nonre-

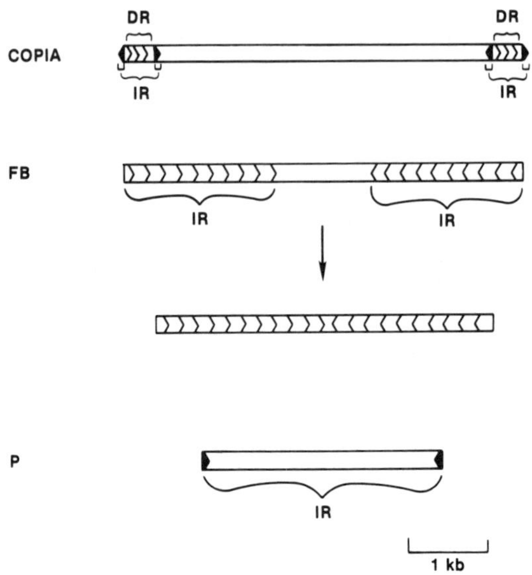

Fig. 1. Structure of three classes of *Drosophila melanogaster* transposable elements. The *copia*-like elements are bounded by short direct repeats (DR), which in turn are each bounded by inverted repeats (IR). The *FB* elements terminate in large inverted repeats and in some cases contain an internal segment of unique sequence DNA. *P* elements are bounded by short inverted repeats.

peated sequence, whereas others do not (Fig. 1) [Shapiro, 1983]. The mechanism of transposition of *FB* elements is unknown. However, the average *Drosophila* genome contains approximately 50 to 100 *FB* elements.

P elements represent the third structurally distinct family of *Drosophila* transposable elements. These elements were first identified because of their key role in hybrid dysgenesis. The phenomenon of hybrid dysgenesis is a "syndrome of correlated genetic traits that is spontaneously induced in hybrids between certain mutually interacting strains, usually in one direction only" [Kidwell et al, 1977]. These traits can include reduced fertility as well as increased rates of mutation, male recombination, and chromosome rearrangement. Two independent systems of hybrid dysgenesis have been described, called I-R and P-M [Kidwell, 1979]. In the case of the P-M system, dysgenesis occurs when males of a P strain are mated with females of an M strain, but not in reciprocal crosses, or when P males are crossed to P females or M males to M females. Molecular genetic data have indicated that dysgenesis is the result of the action of a family of transposable elements termed the *P* elements. *P* elements are carried by P strains but not M strains.

P element transposition is repressed in P strains but not when placed in the context of M cytotype. Therefore, in hybrids, transposition of *P* elements is derepressed, and P elements transpose at high rates, leading to a variety of *P* element-induced chromosomal aberrations.

A typical P strain contains about 30 to 50 dispersed *P* element copies [O'Hare and Rubin, 1983]. These elements are heterogeneous in size, ranging from 0.5 to 2.9 kb. This heterogeneity is thought to result from the presence of defective *P* elements containing deletions of varying sizes within the central region of the element. This region is thought to encode functions required for *P* element transposition. However, transposition is not limited to nondefective elements since some defective *P* elements have been shown to transpose when complemented in *trans* with factors encoded by nondefective *P* elements [Engels, 1984]. Unlike *copia* or *FB* elements, *P* elements are bounded by short 31 bp inverted repeats.

The finding that many of the spontaneous mutations occurring in dysgenic crosses are the result of *P* element insertion has made it possible to generate, almost at will, *P* element-induced mutations at any locus where one has a convenient selection scheme. These have included not only pigment mutations but also mutations that affect normal development. Element-induced mutations have the additional advantage in that they are particularly amenable to further experimental intervention, since the element serves as a molecular "tag" that can be used to clone molecularly sequences from the mutant and ultimately the wild-type gene. In fact, *P* element-induced mutations have been instrumental in identifying and cloning genes that are important in development [Engels, 1983].

MECHANISMS OF TRANSPOSITION

As discussed previously, *Drosophila copia*-like elements and yeast *Ty1* elements most likely transpose via an RNA intermediate involving the formation of a retrovirus-like intermediate. However, in no case has anyone found evidence for the presence of extracellular *copia,* or *Ty1* particles (only intracellular particles have been observed), suggesting that they are more analogous to mouse intracisternal type-A particles rather than to the classic infectious type-C retroviruses. It is also important to point out that retroviruses, which have been isolated from a variety of different species including humans, are in every sense of the word transposons. In fact, retroviruses represent the largest group of eukaryotic transposable elements identified to date.

In the case of transposable elements such as *P* or *FB* elements that have no known homology with retroviruses, transposition is thought to occur through a DNA intermediate. Although the exact mechanism(s) of transposition of these elements is unclear, it is thought that transposition involves two different types of reactions. In the first, replication of the transposon occurs without replication of adjacent chromosomal sequences. Second, breakage of the target (chromosomal) DNA occurs to generate a site into which the transposon is inserted (Fig. 2). The term "transposition" is therefore almost a misnomer in that a transposon does not leave its former site in the course of transposition. In most cases studied, breakage of the target DNA appears to occur by staggered single-stranded nicks with the net result that small sequences of cellular DNA are duplicated and flank the transposon as direct repeats (Fig. 2). The size of the direct repeat varies from transposon to transposon but usually ranges from 4 to 9 bp in length. Finally, retrovirus proviruses are also bounded by short (4 to 6 bp) direct repeats of cellular DNA, suggesting that viral DNA integration also involves the formation of staggered single-stranded breaks.

TYPES OF TRANSPOSABLE ELEMENT-MEDIATED DNA REARRANGEMENTS

Rearrangements of host DNA can take place when two homologous transposons occupy chromosomal sites near each other. Subsequent reciprocal recombination between the two elements can then occur, the products of which are determined by whether the elements are in the same or inverted orientation.

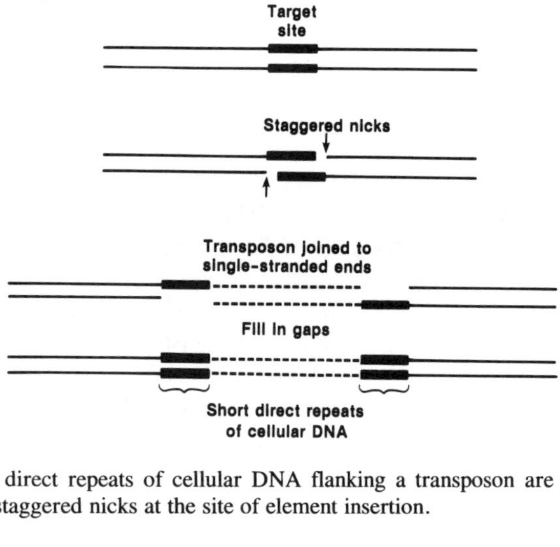

Fig. 2. Short direct repeats of cellular DNA flanking a transposon are generated by the introduction of staggered nicks at the site of element insertion.

Reciprocal recombination between a pair of direct repeats will result in the deletion of chromosomal DNA between the elements (Fig. 3). The intervening cellular DNA is excised as a circle containing one copy of the element, the chromosome retaining a single copy of the element.

Reciprocal recombination between inverted repeats results in the inversion of the chromosomal DNA between the elements (Fig. 3); the repeats themselves remain in place and are available for further inversions. Thus reciprocal recombination between adjacent transposons can lead to both inversions and deletions of cellular DNA.

Reciprocal recombination can also take place between the repeats that are present at the termini of transposable elements. If a transposable element (ie,

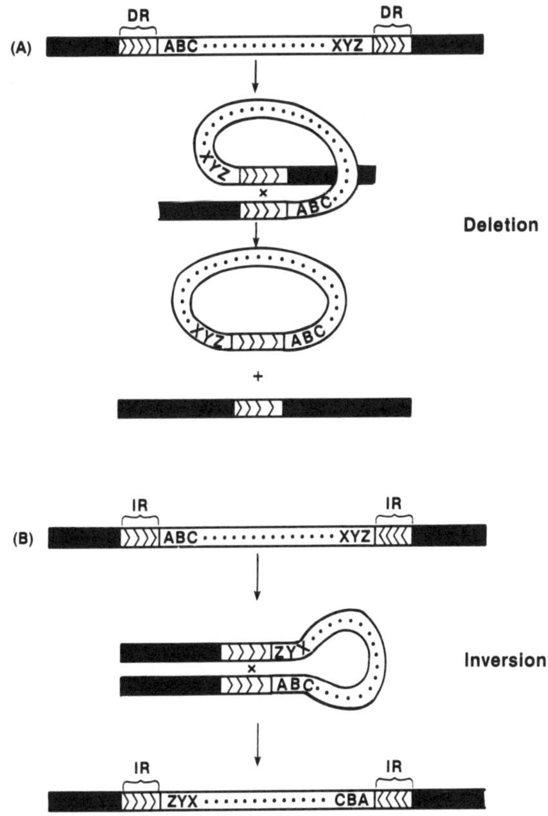

Fig. 3. Deletion and inversion of cellular DNA produced by reciprocal recombinant between homologous transposable elements that are present as direct repeats (DR) or indirect repeats (IR).

P element) is terminated by inverted repeats, then recombination would simply lead to an inversion of the DNA between the repeats. However, if the element is terminated by direct repeats (ie, *copia*-like elements and retroviruses), then recombination would result in excision of most of the transposable element. One copy of the direct repeat would remain behind in the host genome. In fact, the instability of many transposable element-induced mutations is often attributable to these types of excision events [Shapiro, 1983; Copeland et al, 1983]. In these cases, the transposable element is usually integrated into noncoding sequence (either intron or 5′ or 3′ noncoding sequence) in such a way that the presence of the complete element prevents wild-type gene expression, whereas a single direct repeat has no obvious phenotypic effect. It is also possible for transposable elements that end in inverted repeats to excise from the host genome and consequently restore wild-type gene expression. However, in these cases, reciprocal recombination usually takes place between the small direct repeats of cellular DNA that flank the transposon. When this occurs, the complete element is excised along with one copy of the small direct repeat generating a completely wild-type chromosome.

As pointed out earlier, transposable elements can also induce chromosomal breaks with one end point of the break usually occurring next to or within the transposon. In addition, reciprocal recombination between homologous transposable elements located on different chromosomes would produce a chromosome translocation. Thus it is apparent that transposons (and retroviruses) have the potential of rapidly inducing a variety of chromosomal aberrations, and they may in fact be instrumental in evolutionary change. This seems especially true when one considers that as much as 0.05% or more of the genome of many organisms (ie, mouse, Drosophila, yeast) is composed of transposon-like sequence (including retrovirus-like transposons). This suggests that the cell has some mechanism for suppressing transposon-induced chromosomal rearrangements. Some support for this suggestion, at least in yeast, has recently been reported [Jackson and Fink, 1985].

RETROVIRUSES AS INSERTIONAL MUTAGENS OF MICE

That retroviruses can spontaneously act as insertional mutations in mice via replicative transposition was first suggested by our studies of the dilute (*d*) coat-color mutation of DBA/2J mice [Jenkins et al, 1981]. This mutation, which is a very old mutation of the "mouse fancy," is recessive and in homozygotes produces mice that are lighter in color than their wild-type counterparts. In a series of molecular genetic studies [Jenkins et al, 1981; Copeland et al, 1983] it was possible to show that this mutation was caused by the integration of an ecotropic (retrovirus that can infect and replicate in mouse cells) murine leukemia virus (MuLV) into the mouse genome, presumably within or near the wild-type *d* gene. Using the provirus as a "molecular tag," it was possible to clone molecularly sequences from the *d* locus, which, in turn, allowed us to characterize molecularly many of the mutagen-induced (X-rays, γ-rays, chemicals) mutations that have been generated at this locus [Rinchik et al, 1986].

Jaenisch et al [1983] have also identified a retrovirus-induced mutation that they experimentally induced in mice by infecting postimplantation mouse embryos with Moloney MuLV. This mutation is recessive lethal, induces a dominant coat-color change in older animals, and has been shown to result from virus integration within the αl (I) collagen gene leading to a complete transcriptional block [Schnieke et al, 1983]. Again, this mutation will be very useful for analyzing the effects of abnormal collagen gene expression during early mouse embryogenesis.

HIGH-FREQUENCY GERM LINE ACQUISITION (TRANSPOSITION) OF ECOTROPIC MuLV PROVIRUSES IN SWR/J-RF/J HYBRID MICE

Throughout this review, we have tried to emphasize the special importance of transposable elements or retrovirus-induced mutations for identifying, cloning, and studying genes that affect development. The advantages of a murine system analogous to hybrid dysgenesis of Drosophila are obvious. Mutations that affect mammalian development could be induced via transposon mutagenesis that subsequently could be characterized exhaustively at both the genetic and molecular genetic levels. Since these mutations would be induced in a mammal, their relationship to similar mutations identified in humans could be evaluated more readily. Once the genes themselves are cloned, it should ultimately be possible, by using standard recombinant DNA techniques, to determine if related genes are present in the human genome and subsequently address their functions in human development. Recently, we have identified such a system that, at least in principle, meets this goal.

In attempting to derive SWR/J lines (SWR/J does not carry any endogenous ecotropic proviruses and is permissible for the replication of most endogenous ecotropic viruses) congenic for two endogenous RF/J ecotropic proviruses, *Emv-16* and *Emv-17* (endogenous ecotropic proviral locus 16 and locus 17, respectively), we found that both proviruses are tightly linked (no recombinants in more than 3,000 animals analyzed) and segregate with the high viremic phenotype in hybrid mice [Jenkins and Copeland, 1985].

Importantly, many progeny derived from female but not male virus carriers were found to have acquired new ecotropic proviruses (detected by Southern analysis of spleen or tail DNAs) that were not detected in the genome of either parent. In most cases, these new ecotropic proviruses appeared to be carried in the germ line and could be transmitted genetically to progeny mice. In addition, these proviruses segregated as unlinked genes, and no new proviruses were identified that were either linked or allelic with the parental (*Emv-16* and *Emv-17*) proviruses. These results are consistent with an infection mechanism for virus amplification. Additional genetic crosses also have supported this hypothesis [Jenkins and Copeland, 1985], suggesting that new proviruses result from infection of germ cell or of embryos by viremic females. Approximately 20% of the progeny from these crosses acquired new proviruses, and, in many cases, animals were identified that had acquired more than one (two to ten) new ecotropic provirus. On average, about 0.5 new proviruses per animal were acquired. This transposition frequency is considerably higher than previously reported for ecotropic proviruses [Rowe and Kozak, 1980; Quint et al, 1982; Steffen et al, 1982] and is only ten- to 20-fold lower than that reported for *P* element transposition in dysgenic crosses of *D. melanogaster* [Bingham et al, 1982].

Why new proviruses should be acquired at such high frequencies in SWR/J-RF/J hybrid mice is unclear, but preliminary experiments suggest that 1) SWR/J is a highly permissive strain for virus acquisition and 2) *Emv-16* and *Emv-17* are expressed at the right time in development and/or in the right tissues for virus acquisition to occur. At present, we are evaluating the frequency of virus acquisition in other hybrid strain combinations in the hope of identifying hybrids that acquire new proviruses at frequencies even greater than SWR/J-RF/J hybrid mice.

To date, no phenotypic dominant mutations have been identified in SWR/J-RF/J hybrid mice. This was not unexpected; dominant mutations are generally much rarer events than recessive mutations [Swaroop et al, 1985]. At present, we are screening for the presence of phenotypically visible recessive mutations by intercrossing related male and female progeny carrying the same new ecotropic viral insertion. Recessive lethal mutations will be identified by the presence of new proviruses that cannot be made homozygous. This approach actually represents one of the few and most powerful approaches for generating mutations that affect early development. In addition, these types of recessive lethal mutations have the advantage that they are "marked" by the presence of an ecotropic provirus and the inheritance of the mutation can readily be assayed in the heterozygous state by scoring (using Southern analysis) for the presence or absence of the associated proviruses.

Finally, by breeding these "dysgenic" animals to tester stocks that carry multiple recessive mutations, it may be possible also to select for specific recessive mutations. For example, more than 50 loci and many hundred mutant alleles have been identified that affect the mouse coat color [Silvers, 1979]. Approximately one-half of these loci have alleles with known neurologic or developmental effects. With this approach it may be possible to generate insertion mutations that can be scored phenotypically by coat color, have alleles with known effects on development, and are amenable to further experimental intervention.

HUMAN TRANSPOSABLE ELEMENTS

Although very little is known about the content and nature of transposable elements in human populations, it has recently become evident that humans, like most other eukaryotes studied to date, carry within their genome endogenous retrovirus-related sequences [Martin et al, 1981; Bonner et al, 1982; Callahan et al, 1982, 1985; O'Connell et al, 1984]. Several distinct families of endogenous human retroviruses have been identified that, in turn, appear to each be present in multiple copies. It would not be unexpected if further work identified additional new families of endogenous human retroviruses as well as other transposable element families that are more analogous to the *P* and *FB* elements of *Drosophila*. Therefore, the potential impact of these elements on the human genome seems large, and it would not be surprising to uncover mutations in humans that are the result of transposition.

CONCLUSIONS

Transposable elements represent discrete genetic and physical entities that are capable of moving from one position in the genome to another. They are ubiquitous in nature and have been identified in many prokaryotic and eukaryotic species. Because these elements can insert themselves at many different sites in the host genome, they have the potential of causing mutations via the physical interruption of normal gene expression. This mutagenic property of transposable elements is being exploited by many laboratories to induce mutations that are particularly amenable to study at the molecular level. These elements should prove to be invaluable experimental reagents for identifying and studying genes important in development.

REFERENCES

Bingham PM, Kidwell MG, Rubin GM: The molecular basis of P-M hybrid dysgenesis: The role of the P element, a P-strain-specific transposon family. Cell 29:995–1004, 1982.

Boeke JD, Garfinkel DJ, Styles CA, Fink GR: *TY* elements transpose through an RNA intermediate. Cell 40:491–500, 1985.

Bonner TI, O'Connell C, Cohen M: Cloned endogenous retroviral sequences from human DNA. Proc Natl Acad Sci USA 79:4709–4713, 1982.

Callahan R, Chiu IM, Wong JFH, Tronick SR, Roe BA, Aaronson SA, Schlom J: A new class of endogenous human retroviral genomes. Science 228:1208–1211, 1985.

Callahan R, Drohan W, Tronick S, Schlom J: Detection and cloning of human DNA sequences related to the mouse mammary tumor virus genome. Proc Natl Acad Sci USA 79:5503–5507, 1982.

Copeland NG, Hutchison KW, Jenkins NA: Excision of the DBA ecotropic provirus in dilute coat-color revertants of mice occurs by homologous recombination involving the viral LTRs. Cell 33:379–387, 1983.

Engels WR: The P family of transposable elements in Drosophila. Annu Rev Genet 17:315–344, 1983.

Engels WR: A *trans*-acting product needed for P factor transposition in Drosophila. Science 226:1194–1196, 1984.

Flavell AJ: Role of reverse transcription in the generation of extrachromosomal *copia* mobile genetic elements. Nature 310:514–516, 1984.

Jackson JA, Fink GR: Meiotic recombination between duplicated genetic elements in *Saccharomyces cerevisiae.* Genetics 109:303–332, 1985.

Jaenisch R, Harbers K, Schnieke A, Lohler J, Chumakov I, Jahner D, Grotkopp D, Hoffmann E: Germline integration of Moloney murine leukemia virus at the *Mov-13* locus leads to recessive lethal mutation and early embryonic death. Cell 32:209–216, 1983.

Jenkins NA, Copeland NG: High frequency germ line acquisition of ecotropic MuLV proviruses in SWR/J-RF/J hybrid mice. Cell 43:811–819, 1985.

Jenkins NA, Copeland NG, Taylor BA, Lee BK: Dilute (*d*) coat-color mutation of DBA/2J mice is associated with the site of integration of an ecotropic MuLV genome. Nature 293:370–374, 1981.

Johns MA, Mottinger J, Freeling M: A low copy number, *copia*-like transposon in maize. EMBO J 4:1093–1102, 1985.

Kidwell MG: Hybrid dysgenesis in *Drosophila melanogaster:* The relationship between the P-M and the I-R interaction systems. Genet Res 33:205–217, 1979.

Kidwell MG, Kidwell JF, Sved JA: Hybrid dysgenesis in *Drosophila melanogaster.* A syndrome of aberrant traits including mutation, sterility and male recombination. Genetics 86:813–833, 1977.

Martin MA, Bryan T, Rasheed S, Kahn AS: Identification and cloning of endogenous retroviral sequences present in human DNA. Proc Natl Acad Sci USA 78:4982–4986, 1981.

McClintock B: Cytogenetic studies of maize and neurospora. Carnegie Inst Washington Yearbook 44:108–112, 1945.

McClintock B: Maize genetics. Carnegie Inst Washington Yearbook 45:176–186, 1946.

Nadeau JH, Taylor BA: Lengths of chromosomal segments conserved since divergence of man and mouse. Proc Natl Acad Sci USA 81:814–818, 1984.

O'Brien SJ: A compilation of linkage maps of genetically studied organisms. In: "Genetic Maps." Cold Spring Harbor, NY: Cold Spring Harbor Laboratory, 1984.

O'Connell C, O'Brien S, Nash WG, Cohen M: ERV3, a full-length human endogenous provirus: Chromosomal localization and evolutionary relationships. Virology 138:225–235, 1984.

O'Hare K, Rubin GM: Structures of P transposable elements and their sites of insertion and excision in the *Drosophila melanogaster* genome. Cell 34:25–35, 1983.

Potter SS: DNA sequence of a foldback transposable element in *Drosophila.* Nature 297:201–204, 1982.

Quint W, Vander Putten H, Janssen F, Berns A: Mobility of endogenous murine leukemia viral genomes within mouse chromosomal DNA and integration of a mink cell focus-forming virus-type recombinant provirus in the germline. J Virol 41:901–908, 1982.

Rinchik EM, Russell LB, Copeland NG, Jenkins NA: Molecular genetic analysis of the *dilute-short ear (d-se)* region of the mouse. Genetics 112:321–342, 1986.

Rowe WP, Kozak CA: Germline reinsertions of AKR murine leukemia virus genomes in *AKv-1* congenic mice. Proc Natl Acad Sci USA 77:4871–4874, 1980.

Saigo K, Kugimiya W, Matsuo Y, Inouye S, Yoshioka K, Yuki S: Identification of the coding sequence for a reverse transcriptase-like enzyme in a transposable genetic element in *Drosophila melanogaster.* Nature 312:659–661, 1984.

Schnieke A, Harbers K, Jaenisch R: Embryonic lethal mutation in mice induced by retrovirus insertion into the αl (1) collagen gene. Nature 304:315–320, 1983.

Shapiro JA: "Mobile Genetic Elements." New York: Academic Press, 1983.

Silvers WK: "The Coat Colors of Mice." New York: Springer-Verlag, 1979.

Steffen DL, Taylor BA, Weinberg RA: Continuing germline integration of AKV proviruses during the breeding of AKR mice and derivative recombinant inbred strains. J Virol 42:165–175, 1982.

Swaroop A, Paco-Larson ML, Garen A: Molecular genetics of a transposon-induced dominant mutation in the *Drosophila* locus Glued. Proc Natl Acad Sci USA 82:1751–1755, 1985.

Maternal Genes: Mitochondrial Diseases

Douglas C. Wallace, PhD

Departments of Biochemistry, Pediatrics, and Anthropology, Emory University, Atlanta, GA 30322

The mitochondrial DNA (mtDNA) is the only human genomic element that does not follow the laws of Gregor Mendel. Unlike mendelian genes, each cell contains thousands of mtDNAs, and these are located within the mitochondria of the cytoplasm. The mtDNA constitute genes that code for elements essential to energy metabolism and thus are required for the survival and health of all cells. Yet these small, closed circular molecules have a number of extraordinary genetic properties that set them apart from other known genetic systems. As compared with nuclear DNA, mtDNAs have a highly compact structure and an altered genetic code. They are inherited exclusively from the mother and can segregate during mitotic growth. Mt-DNAs have a high mutation rate, yet encode highly conserved proteins. Because of these traits, analysis of mtDNA variation has provided a powerful new tool in the search for human origins and yielded new insights into the inheritance of a number of devastating neuromuscular degenerative diseases.

MITOCHONDRIAL GENES AND BIOGENESIS

Structure and Function

There are hundreds of pleomorphic mitochondria in each human cell [Johnson et al, 1980] and thousands of mtDNAs [Shmookler Reis and Gold-

This work was supported by NIH grants GM33022 and NS21328, NSF grant BNS 8508497, ACS grant NP 460, a Clinical Research Grant from the Muscular Dystrophy Foundation, and a grant from the Huntington Foundation.

Birth Defects: Original Article Series, Volume 23, Number 3, pages 137–190

stein, 1983; Bogenhagen and Clayton, 1974]. The most important function of the double-membraned mitochondrion is the synthesis of ATP by the process of oxidative phosphorylation (OXPHOS).

OXPHOS involves five multiple polypeptide enzyme complexes embedded in the mitochondrial inner membrane (Fig. 1). Complexes I to IV (I, NADH-CoQ oxidoreductase; II, succinate-CoQ oxidoreductase; III, CoQ-cytochrome c oxidoreductase; IV, cytochrome c oxidase) are concerned with the oxidation of electrons from NADH or succinate by O_2 (electron transport). This process pumps protons out of the mitochondrion, generating a mitochondrial inner membrane electrochemical gradient. Complex V (ATP synthetase) exploits this proton gradient to synthesize ATP [Tzagoloff, 1982].

The biogenesis of OXPHOS complexes I, III, IV, and V is under the combined control of genes in both the nuclear DNA and the mtDNA (Fig. 1). At least 69 separate polypeptides are known to be required for OXPHOS, only 13 of which are coded by the mtDNA.

Chromosomal OXPHOS genes are replicated within the nucleus and transcribed by the α-aminitin sensitive RNA polymerase II (Fig. 2). The resulting heterogeneous nuclear RNA is processed and the mature mRNA transported to the cytosol. This cytosol mRNA is translated on the cycloheximide and emetine-sensitive (S) but chloramphenicol (CAP)-resistant (R) cytosol ribosomes [Tzagoloff, 1982]. The resulting precursor polypeptides are vectorially transferred into the mitochondrion where they are cleaved into their mature forms [Doonan et al, 1984]. By contrast, the small, closed circular mtDNA is replicated by a separated ethidium bromide and berenil-S replication apparatus [Clayton, 1982; Fairfield et al, 1985] (Fig. 2). The mtDNA genes are transcribed by a mitochondria-specific RNA polymerase [Shuey and Attardi, 1985] into rRNAs, tRNAs, and mRNAs [Clayton, 1984]. The rRNAs are assembled into CAP-S, but cycloheximide and emetine-R, mitochondrial ribosomes (mitoribosomes) using nuclear encoded ribosomal proteins [Schieber and O'Brien, 1985]. Mitoribosomes then translate the mitochondrial mRNAs into OXPHOS polypeptides.

The OXPHOS polypeptides are assembled into the complexes in precise stoichiometric ratios yet the ploidy of the responsible nuclear and mitochondrial DNA genes differs by a thousand-fold. Consequently, to understand the role of the energy genes in human biology and health, it is essential that we understand the structure, expression, inheritance, and variation of the genes of the human mtDNA.

The Mitochondrial DNA

The 16,569 nucleotide pair (np) human mtDNA [Anderson et al, 1981] (Fig. 3) codes for a small (12S) and a large (16S) rRNA, 22 tRNAs, and 13

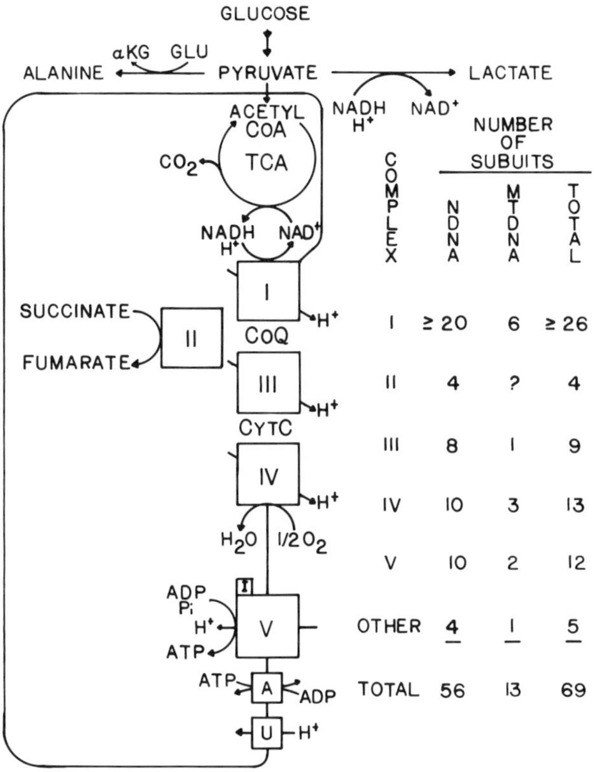

Fig. 1. OXPHOS: Function and biogenesis [Tzagoloff, 1982]. I, Complex I: total subunits [Heron et al, 1979], mtDNA subunits [Chomyn et al, 1985]. II, Complex II: total subunits range 4 to 5 [Ackrell et al, 1980; Capaldi, 1982; Kenney et al, 1976]. No evidence for mtDNA encoded subunits. III, Complex III: total subunits range 8 to 10 [Bell and Capaldi, 1976; Capaldi, 1982; Marres and Slater, 1977; Nalecz et al, 1985; Rieske, 1976], mtDNA subunits [Anderson et al, 1981; Gellerfors and Nelson, 1981; Wilson et al, 1981]. IV, Complex IV: total subunits range 11 to 13 [Capaldi, 1982; Kadenbach and Merle, 1981; Kadenbach et al, 1983; Kuhn-Nentwig and Kadenbach, 1984], mtDNA subunits [Anderson et al, 1981; Ching and Attardi, 1982; Hare et al, 1980]. V, Complex V: total subunits range 12 to 18 [Amzel and Pedersen, 1983; Capaldi, 1982; Glaser et al, 1980; Hatefi et al, 1979], mtDNA subunits [Anderson et al, 1981; Macreadie et al, 1983]. CYTC, cytochrome c, nDNA [Scarpulla et al, 1981, 1982]. I, ATP synthetase (Complex V) inhibitor protein; nDNA [Frangione et al, 1981; Pedersen et al, 1981]. A, adenine nucleotide translocator, nDNA [Aquila et al, 1982]. U, uncoupler protein, nDNA [Nicholls and Locke, 1984].

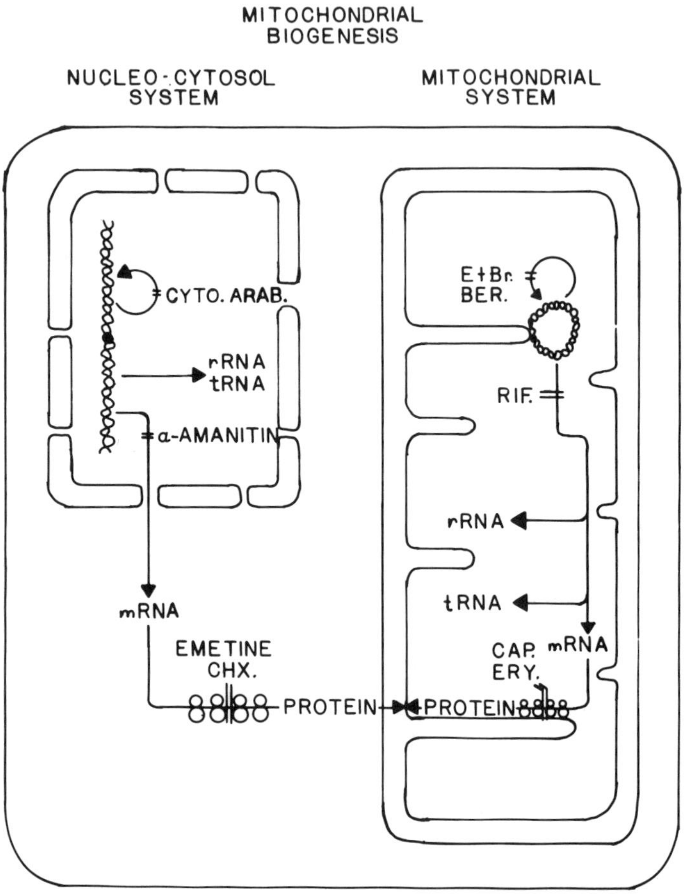

Fig. 2. Abbreviations and references for differential inhibitors. α-AMANITIN [Roeder et al, 1970]; BER, berenil [Fairfield et al, 1985]; CAP, chloramphenicol [Lederman and Attardi, 1970; Loeb and Hubby, 1968; Oliver and Wallace, 1982; Wallace, 1982], CHX, cycloheximide [Loeb and Hubby, 1968; Wallace, 1982b]; CYTO ARAB, cytosine arabinoside (1-β-D-arabinofuranosyl cytosine) [Mattoccia and Roberti, 1974] and adenine arabinoside (9 β-D-arabinofuranosyl adenine) [Shipman et al, 1972]; EMETINE [Perlman and Penman, 1970; Wallace, 1982b]; ERY, erythromycin [Doersen and Stanbridge, 1979; Towers et al, 1972]; EtBr, ethidium bromide [replication inhibition, Mattoccia and Roberti, 1974; Meyer and Simpson, 1969; Radsak et al, 1971]; [transcription inhibition, Lederman and Attardi, 1970; Zylber et al, 1969]; mRNA, messenger RNA; RIF, rifampicin or derivatives [Gadaleta and Saccone, 1974; Gallerani and Saccone, 1974; Kroon and DeVries, 1970; Wu and Dawid, 1972]; rRNA, ribosomal RNA; tRNA, transfer RNA.

polypeptides (via open reading frames [ORFs]). Virtually all rRNAs and ORFs are separated by one or more tRNA genes with few if any nucleotides separating the coding regions of adjacent genes. Consequently, almost every nucleotide in the mtDNA has either a coding function or participates in the control of replication or transcription.

The base compositon of the two mtDNA strands is highly asymmetric, yielding a G-rich heavy (H) and a C-rich light (L) strand [Berk and Clayton, 1974; Clayton et al, 1970]. The replication of these two strands proceeds from two different unidirectional origins separated by 66% of the genome. H-strand synthesis initiates at the displacement (D)-loop region, a triple-stranded region involving a short nascent H-strand molecule (7S DNA) base-paired with the L-strand template. The 7S strand of H-strand synthesis is primed by RNA molecules from the L-strand promoter that terminate at 3 downstream conserved sequence blocks (CSBs) within the D-loop. The 7S DNA synthesis initiates after these terminated primers and continues through a termination associated sequence (TAS) about 655 nps from the start site (Fig. 3) [Chang and Clayton, 1985; Doda et al, 1981; Crews et al, 1979; Tapper and Clayton, 1981]. Complete H-strand synthesis continues from the 7S DNA two-thirds of the way around the L-strand template. The L-strand origin (O_L) is then exposed, and L-strand synthesis initiates and proceeds back along the free H-strand. Eventually, both strands are completed and the molecules separate [Clayton, 1982].

Transcriptional promoters for both the H- and L-strand are located between the D-loop and the 12S rRNA. Transcription proceeds around the molecule symmetrically [Murphy et al, 1975], with the L-strand transcription starting at np 407 and H-strand transcription starting primarily between np 558 and 561 (Fig. 3). The tRNAs then fold within the nascent transcript and are excised by an RNAase P-like nuclease [Clayton, 1984; Doerson et al, 1985; Ojala et al, 1981b] liberating the rRNAs and mRNAs.

Mitochondrial mRNA molecules have little or no 5′ noncoding sequence, the RNA generally beginning at or within a few nucleotides of the methionine initiation codons (ATG or ATA) [Anderson et al, 1981; Montoya et al, 1981]. Mitochondrial mRNAs also terminate at or within the termination codon UAA. The termination codon is completed by the posttranscriptional addition of a poly-A tail [Anderson et al, 1981; Ojala et al, 1981b].

The most extraordinary feature of the human mtDNA is its modified genetic code. In contrast to virtually all other genetic systems known, human, animal, and fungal mtDNAs use the universal UGA stop codon to code for tryptophan. Further, in human and animal mtDNAs AUA is used for methionine instead of isoleucine and AGA and AGG for stop codons instead of for

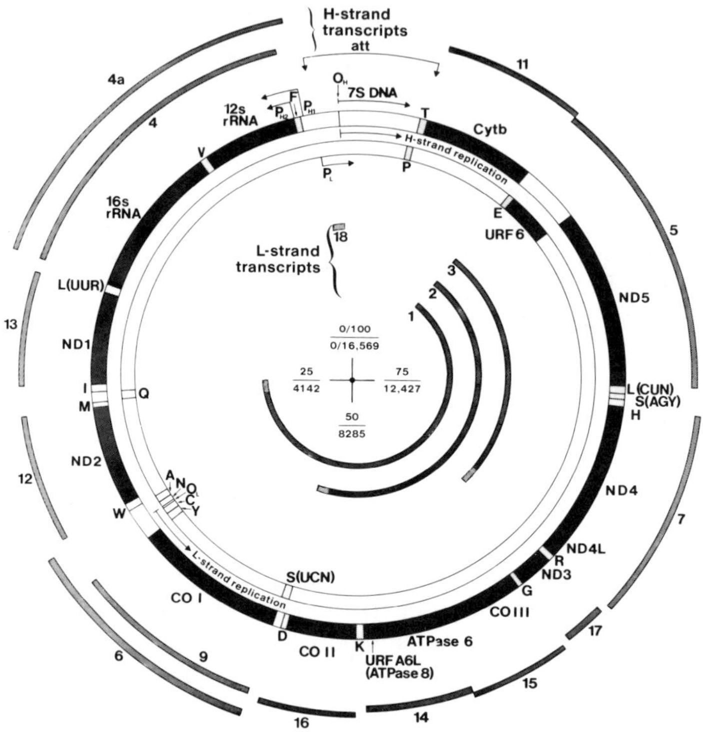

Fig. 3. The mitochondrial DNA (mtDNA) map is based on the mtDNA sequence (data are from Anderson et al [1981] and updated from Wallace [1984]). Transcripts and their locations are from Attardi et al [1982b], Attardi and Montoya [1983], Montoya et al [1981, 1982] and Ojala et al [1981b]. The O_H, heavy strand origin, is incorporated within the D-loop. Four major and three minor 7S DNAs, base paired to the L-strand, have been observed. The most prevalent starts at np 191 [Crews et al, 1979] and the others at approximately 110, 174, 169, 219, 310, and 441 [Chang and Clayton, 1985; Tapper and Clayton, 1981]. Three of these correspond to L-strand transcription stop sequences at conserved sequence blocks I to III (CSB-I = 213–235 np, CSB-II = 299–315, and CSB-III = 346–363) [Chang and Clayton, 1985]. All 7S DNA molecules end at nucleotides 16,104 to 16,107 just past the termination associated sequence (TAS) at 16,157 to 16,172 nps [Doda et al, 1981]. O_L, light strand origin, including sequences 5,729 to 5,805 np. Replication starts at 5,761 np and at 5,798 np [Clayton, 1982; Tapper and Clayton, 1981]. P_L, light strand promoter, including nucleotides from about 392 to 435 nps with transcription starting at 407 np [Chang and Clayton, 1984; Clayton, 1984; Hixson and Clayton, 1985; Montoya et al, 1982; Ojala et al, 1981a; Walberg and Clayton, 1983]. A minor L-strand start site occurs at 561 np [Chang and Clayton, 1985]. P_{H1}, major heavy strand promoter [Montoya et al, 1983], includes nucleotides from about 545 to 567 np [Bogenhagen et al, 1984; Chang and Clayton, 1984; Hixson and Clayton, 1985] with transcription starting at 558 $\pm$ 2 [Yoza and Bogenhagen, 1984] or 561 $\pm$ 2 [Chang and Clayton, 1984; Hixson and Clayton, 1985]. P_{H2}, minor heavy strand start site at about 645 bp [Montoya et al, 1982, 1983; Yoza and Bogenhagen, 1984]. F, tRNA Phen (577–647 np); 12S rRNA (648–1,601 np); V, tRNA Val (1,602–1,670 np); 16S rRNA (1,671–3,229 np); L

arginine [Anderson et al, 1981, 1982; Bibb et al, 1981; Barrell et al, 1979; Pepe et al, 1983; Wallace, 1982a]. The entire human mtDNA genetic code can be read using only 22 tRNAs, rather than the traditional 32 [Anderson et al, 1981]. This is accomplished by inclusion of an unmodified U in the tRNA wobble position that can recognize all four member of a codon family [Heckman et al, 1980; Wallace, 1982a].

The small size and altered genetic code of the mtDNA has been proposed to be a consequence of its intracellular environment [Wallace, 1982a]. Since smaller mtDNAs replicate faster than larger ones, mtDNAs that lost unnecessary genes (ie, redundant tRNAs) or transferred genes to the nucleus would have a selective advantage. Since the nuclear-cytosol and mitochondrial genetic systems are in such intimate association, the altered codon rules would serve to reduce accidental information transfer between the two systems. The presence of UGA codons in mitochondrial mRNAs would make them untranslatable in the cytosol. The absence of a mitochondrial AGA/AGG tRNA would make cytosol mRNAs untranslatable in the mitochondrion. Once the altered mtDNA genetic code became established, gene transfer to the nucleus would be blocked. This could account for the extraordinary stability of animal mtDNAs today [Wallace, 1982a].

The function of most of the mtDNA genes is now known. Every gene identified has been found to function in OXPHOS. Genes for COI, COII, COIII, cytb, ATPase 6 [Anderson et al, 1981], and ATPase 8 (URFA6L)

(UUA/G), tRNA Leu (3,230–3,304 np); ND1, NADH dehydrogenase subunit (3,307–4,262 np); I, tRNA Ile (4,263–4,331 np); Q, tRNA G1N (4,329–4,400 np); M, tRNA Met (4,402–4,469 np); ND2, NADH dehydrogenase subunit (4,470–5,511 np); W, tRNA Trp (5,512–5,576 np); A, tRNA Ala (5,587–5,655 np); N, tRNA Asn (5,657–5,729 np); C, tRNA Cys (5,761–5,826 np); Y; tRNA Tyr (5,826-5,891 np); COI, cytochrome c oxidase subunit I (5,904–7,444 np); S (UCN), tRNA Ser (7,445–7,516 np); D, tRNA Asp (7,518–7,585 np); COII, cytochrome c oxidase subunit II (7,586–8,262 np); K, tRNA Lys (8,295–8,364 np); URFA6L, ATPase subunit 8 (8,366–8,572 np) [Macreadie et al, 1983]; ATPase 6, ATPase subunit 6 (8,527–9,207 np); COIII, cytochrome c oxidase subunit III (9,207–9,990 np); G, tRNA Gly (9,991–10,058 np); ND3, NADH dehydrogenase subunit (10,059–10,404 np); R, tRNA Arg (10,405–10,469 np); ND4L, NADH dehydrogenase subunit (10,470–10,766 np); ND4, NADH dehydrogenase subunit (10,760–12,137 np); H, tRNA His (12,138–12,206 np); S (AGU/C), tRNA Ser (12,207–12,265 np); L (CUN), tRNA Leu (12,266–12,336 np); ND5, NADH dehydrogenase subunit (12,337–14,148 np); URF6, unidentified reading frame 6 (14,149–14,673 np); E, tRNA Glu (14,674–14,742 np); Cytb, cytochrome b (14,747–15,887 np); T, tRNA Thr (15,888–15,953 np); P, tRNA Pro (15,955–16,023 np); att, membrane attachment site (about 15,925–499) [Albring et al, 1977]. ND1 to ND5 have been identified as NADH dehydrogenase subunits by Chomyn et al [1985]. A possible 5S rRNA-like sequence (3,206–3,229 np) has been identified [Nierlich, 1982].

[Macreadie et al, 1983] were identified through sequence homology with known proteins or yeast mtDNA genes. The functions of the remaining genes were determined by assigning mitochondrial translation products to each gene and then identifying the enzyme complexes incorporating those mitochondrial proteins.

Proteins translated from mitochondrial mRNAs can be selectively labeled by growth of the cells in ^{35}S-methionine and the cytosol ribosome inhibitors emetine or cycloheximide (Fig. 2). The labeled proteins can then be resolved by SDS-polyacrylamide gel electrophoresis (SDS-PAGE), followed by fluorography [Oliver and Wallace, 1982; Ching and Attardi, 1982]. Two different methods have been used to assign the roughly 13 mitochondrial translation products [Oliver and Wallace, 1982] to genes. In the first, peptide maps of the mitochondrial proteins were determined using a new two-dimensional proteolytic digest (2D-PD) gel system [Oliver et al, 1984]. These peptide maps were then matched with computer predictions of the peptide maps of the mtDNA ORFs. By this procedure, a protein was matched with each ORF [Oliver et al, 1983; Wallace et al, 1986]. In the second procedure, the mitochondrial translation products were precipitated using antibodies prepared either to the respiratory complex or to an oligopeptide prepared from the ORF sequences predicted from the mtDNA [Chomyn et al, 1983, 1985; Hare et al, 1980; Mariottini et al, 1983]. The results of these two approaches are in good agreement (Table 1). Those differences that do exist may well reflect differences in the SDS-PAGE systems used. For instance, the cytochrome b assignments differ, yet this protein is known to migrate differently in different gel systems [Beattie et al, 1979].

The protein assignments confirm that all of the mtDNA genes are expressed within the mitochondrion and correlate with the observed mtDNA mRNAs (Fig. 3). The only transcript that has not been assigned a protein function is the 7S RNA of the L-strand. This transcript is unique in having a 5′ noncoding sequence, a complete termination codon, and two overlapping ORFs. One of these codes for 23 or 24 amino acids and corresponds to the mtDNA genetic code. The other is 45 amino acids long and corresponds to the cytosol genetic code [Clayton, 1984; Greenberg et al, 1983; Ojala et al, 1981a; Wallace, 1982a]. It is not known if this transcript is translated; however, analysis of mitochondrial translation products by SDS-PAGE reveals 14 bands, the smallest of which has not been assigned to a gene [Wallace et al, 1986].

The mitochondrial translation products for NADH dehydrogenase subunits 1, 2, 3, 4L, 4, and 5 have been shown to copurify and coimmunoprecipitate with respiratory Complex I [Chomyn et al, 1985]. Similar results have been

TABLE 1. Gene Assignments of Mitochondrial Proteins

		SDS-PAGE Mobility			
	Predicted	Tris-Glycine‡		PO4 = Urea¶	
Gene	Mol Wt†	No.	MWt	No.	MWt
ND5	66,600	1	43.5		
COI	57,000	2	39.5	2,3	45
ND4	51,400	3	36.5	4,5	39,36
Cytb	42,700	6	27.5	7–10	29
ND2	38,900	4	31.5	11	25
ND1	35,600	5	29.5	12	24
COIII	30,000	8	22.5	15,16	18
COII	25,500	7	23.6	13,14	20
ATPase 6	24,800	9	21.6	17,18	16
URF6	18,600	10	16.7		
ND3	13,200	12	13.5	23,24	6
ND4L	10,700	11?	14.8	26	3.5
ATPase 8	7,900	13 + 13a	9.8	25	4.5

†Data are from Anderson et al [1981].

‡Data on labeling of mitochondrial proteins and gel systems are from Oliver and Wallace [1982] and Oliver et al [1984]. Mitochondrial proteins are numbered from slowest (1) to fastest (13a) mobility. Apparent molecular weights, in kilodaltons, are from Oliver and Wallace [1982] as revised by Wallace et al [1986]. Assignments of URF6 and ND3 are from Oliver et al [1983]. Remaining assignments are from Wallace et al [1986].

¶Data on labeling of mitochondrial proteins, gel systems, protein band numbering, and apparent molecular weights, in kilodaltons, are from Ching and Attardi [1982]. Assignments for COI, COII, and COIII are from Ching and Attardi [1982] and Hare et al [1980]. The additional COII assignment and the ATPase 8 assignment are from Mariottini et al [1983]. Assignments for ND1, ATPase 6 and ND3 are from Chomyn et al [1983]. Assignments for ND4, Cytb, ND2, and ND4L are from Chomyn et al [1985].

found in *Neurospora* [Ise et al, 1985]. Hence, six mitochondrial translation products appear to be subunits of Complex I. Therefore, all 12 of the mtDNA H-strand genes are components of OXPHOS. The remaining gene (URF6) is the only polypeptide encoded by the L-strand.

Analysis of the relative levels of expression of the mitochondrial polypeptides has led to several interesting speculations on OXPHOS regulation [Wallace et al, 1986]. The expression of the URF6 polypeptide is more variable than that of the other proteins, possibly reflecting its transcription from a separate promoter. The proteins assigned to ATPase 6 and COIII have been found to vary reciprocally in their labeling intensity. Similar changes have been observed in oligomycin-resistant (OLG-R) Chinese hamster cells [Shew and Breen, 1985]. OLG resistance is caused by an ATPase 6 mutation [Slott et al, 1983]. These observations might reflect the fact that ATPase 6 and COIII are the only two adjacent mtDNA genes not separated by a tRNA, yet translated from separate mRNAs. In the DNA, the termination codon of the ATPase 6 gene overlaps the start codon of COIII (UA*AUG*). Cleavage of the ATPase 6-COIII transcript (14+15, Fig. 3) between the two As followed by polyadenylation would yield active mRNAs for both proteins. However, if the precursor transcript were not cleaved only ATPase 6 would be translated. Since COIII is essential for electron transport and ATPase 6 for the ATP synthetase, the regulation of the processing of the ATPase 6-COIII transcript could provide an effective mechanism for coordinating OXPHOS.

Unique structural arrangements are also seen for two other pairs of genes, ATPase 8 with ATPase 6 (transcript 14) and ND4L with ND4 (transcript 7). Both pairs of genes are adjacent, not separated by a tRNA, and have overlapping coding sequences. In each pair, the two genes function in the same enzyme complex. Translation of the 5′ protein precludes the translation of the 3′ protein. Consequently, the ratio of synthesis of the two proteins could be regulated by the relative efficiency of ribosome initiation at the 5′ and internal AUGs of these linked cistrons.

PRINCIPLES OF HUMAN MtDNA GENETICS

The behavior of human mtDNA genes has now been investigated at the cellular, family, and population levels. These studies have revealed a set of genetic phenomena totally unique to the mtDNA.

Cellular Genetics

Genetics is the study of variation. Three types of mitochondrial gene variants have been analyzed in human cells: drug-R mutants, mtDNA restriction fragment length polymorphisms (RFLPs), and protein electrophoretic polymorphisms [Wallace, 1982b]. Studies on the behavior of these various mutants have provided considerable insight into the factors controlling the behavior and expression of mtDNA genes in human cells.

Cytoplasmic transmission. Mutants affecting mitochondrial function can occur in either nuclear or cytoplasmic genes. The cybrid fusion technique was developed to differentiate between these two types of mutants in drug-R cells (Fig. 4).

Five types of drug-R mutants have been isolated in human cells: CAP-R [Mitchell and Attardi, 1978; Munro et al, 1978; Spolsky and Eisenstadt, 1973; Wallace et al, 1975; Wallace, 1981], CAP-R with respiratory deficiency [Wiseman and Attardi, 1978], erythromycin-R [Doersen and Stanbridge, 1979], oligomycin-R [Webster et al, 1982], and antimycin-R [Munro

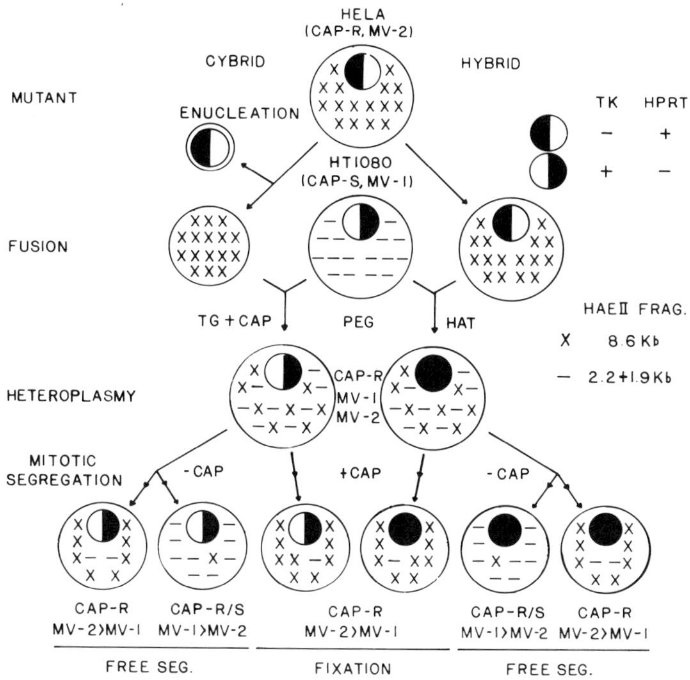

Fig. 4. Hypothetic series of somatic cell genetic experiments involving HeLa and HT1080 mtDNA genes. Experiments are patterned after Oliver and Wallace [1982] and Wallace [1981, 1982b, 1986] and are described in the text. X and - represent HeLa and HT1080 mtDNAs as identified by their *Hae*II restriction fragments 8.6 kb and 2.2 + 1.9 kb, respectively. CAP-R and CAP-S (chloramphenicol resistant and sensitive) are cellular phenotypes associated with mutations at the CAP-R locus on the 16S rRNA gene [Blanc et al, 1981a,b]. CAP-R/S indicates that this phenotype is variable in cells with small percentages of CAP-R mtDNAs. MV-1 and MV-2 are the HT1080 and HeLa polymorphic proteins associated with the HeLa ND3 mutations [Oliver et al, 1983]. MV-1 > MV-2 indicates the relative expression of their proteins in heteroplasmic cells labeled with ^{35}S-methionine in emetine [Oliver and Wallace, 1982]. The double arrows for "Mitotic Segregation" indicate prolonged mitotic propagation with or without CAP. Fixation, fixing the CAP-R gene by selection; Free Seg, permitting the mtDNAs to segregate randomly without selection; TK, thymidine kinase; HPRT, hypoxanthine phosphoribosyl transferase; TG, 6-thioguanine; HAT, hypoxanthine + aminopterin + thymidine; PEG, polyethylene glycol. Half-filled nuclei are defined in the upper right-hand corner.

et al, 1983]. In the cybrid test, the nuclei of these drug-R cells were physically removed by treatment with cytochalasin B and suspension in a centrifugal field (enucleation). The cytochalasin B broke down the cellular cytoskeleton, and the centrifugal field pulled the nuclei from the cells. The resulting membrane-bound cytoplasmic fragments (cytoplasts) were collected and fused to drug-S recipient cells. The fusion mixture was then cultured in the drug to which the mitochondrial donor was resistant and in agents that selected for the recipient cell nucleus. If viable colonies (cybrids) appeared in such fusions, then the drug-R marker was assumed to be cytoplasmic. If colonies were not found, the mutant was assumed to be nuclear [Bunn et al, 1974; Wallace et al, 1975; Wallace, 1982b].

Cybrid analysis showed that the CAP-R, CAP-R with respiratory deficiency, erythromycin-R, and antimycin-R mutants were cytoplasmic, while the oligomycin-R mutant was nuclear. The nuclear inheritance of oligomycin resistance was confirmed by assignment of this phenotype to human chromosome 10 using human-mouse hybrids [Webster et al, 1982]. The cybrid test has been extended to nonselectable markers by demonstrating their cotransfer with selectable markers such as CAP resistance [Oliver and Wallace, 1982].

RFLP linkage. Cytoplasmic genetic markers can be directly assigned to the mtDNA by demonstrating their linkage to mtDNA RFLPs [Ho and Coon, 1979]. The mtDNAs of human HeLa and HT1080 cells can be distinguished with a number of restriction endonucleases. When HeLa and HT1080 hybrids and cybrids were selected in CAP, the mtDNA of the CAP-R parent was enriched (Fig. 4, FIXATION). Thus, CAP resistance was linked to the mtDNA [Wallace, 1981]. This assignment was confirmed by identifying the specific nucleotide changes in the mtDNA 16S rRNA gene responsible for CAP resistance [Blanc et al, 1981a,b; Kearsey and Craig, 1981; Wallace, 1982b]. Surprisingly, even after prolonged selection, some CAP-S mtDNAs were still retained. Hence, the CAP resistance phenotype appears to reflect the average of the cellular mtDNA genotype rather than the total presence or absence of alleles as seen for nuclear genes [Wallace, 1981].

RFLP linkage analysis has also been used to assign nonselectable markers to the mtDNA. HeLa cell mitochondria synthesize a polymorphic protein not found in other cells [Yatscoff et al, 1978]. The common human protein (MV-1) has a molecular weight of about 15,000 daltons (d), while the HeLa variant (MV-2) is 14,000 d [Oliver and Wallace, 1982]. In HeLa-HT1080 hybrids and cybrids the expression of MV-2 was found to be directly proportional to the intracellular percentage of the HeLa mtDNAs as determined by RFLPs. Hence MV-1 and MV-2 were linked to the mtDNA [Oliver and Wallace,

1982]. This assignment was confirmed by assigning MV-1 and MV-2 to the ND3 gene using the 2D-PD peptide mapping strategy and by identifying the HeLa ND3 gene missense mutation responsible for the polymorphism [Oliver et al, 1983].

Mitotic segregation. Because of the cytoplasmic location and high copy number of mtDNAs, mtDNA genes behave differently from nuclear genes at mitosis. For cells containing a mixture of mtDNAs (heteroplasmic), the proportion of the mtDNAs and their associated mitochondrial phenotypes can change during mitotic growth (Fig. 4, FREE SEG.) [Howell et al, 1984; Kearsey et al, 1985; Wallace, 1981, 1986; White and Bunn, 1984].

This process of "mitotic segregation" has been demonstrated best in hybrids between CAP-R and CAP-S HeLa and HT1080 cells isolated by HAT selection (Fig. 4). When such hybrids were grown mitotically and monitored for parental mtDNAs using RFLPs, the proportion of HeLa and HT1080 mtDNAs was found to drift randomly. The mtDNAs of some hybrids drifted toward predominantly HeLa mtDNA, others toward predominantly HT1080, while still others retained both mtDNAs. Concurrent analysis of the ability of these hybrids to clone with and without CAP revealed that the gradual changes in mtDNA genotype were often correlated with marked changes in the CAP resistance phenotype. In one hybrid, CAP resistance was lost and later abruptly reacquired as the proportion of CAP-R mtDNAs dropped to 12% and then increased to 22% (CAP-R) [Wallace, 1986]. Such large changes in cellular phenotype associated with only small changes in mtDNA genotype have been designated the "threshold effect" [Howell, 1983; Wallace, 1986]. Though the exact relationship between the CAP-R mtDNAs and the CAP resistance phenotype is still unclear, these results clearly demonstrated that the segregation of the mtDNA genotype can have major effects on the cellular phenotype.

Mitochondrial complementation. The HeLa-HT1080 genetic studies suggested that the mitochondrial phenotype reflects some form of intracellular averaging of the mtDNA genotype. Hybrids and cybrids selected in CAP became enriched for the CAP-R mtDNAs but never lost all CAP-S mtDNAs. Hybrids that segregated CAP-R mtDNAs remained resistant until about 90% of the mtDNAs were CAP-S; then they switched to CAP-S. This phenotypic averaging is now known to result from intracellular mtDNA complementation.

MtDNA complementation was demonstrated by examining the expression of MV-1 and MV-2 in HeLa-HT1080 hybrids and cybrids. In these hybrids and cybrids, one mtDNA was CAP-R and carried one protein allele, while the other was CAP-S and carried the second allele. The proportion of the parental mtDNAs was quantitated using RFLPs. The relative synthesis of MV-1 and MV-2 was then examined in cells labeled in the presence of emetine with or without CAP. Two alternative hypotheses were envisioned: Either the HeLa and HT1080 mtDNAs would remain separated in different mitochondria or the two mtDNAs would interact, presumably within the same mitochondrion (Fig. 5). If the mtDNAs remained separated, labeling in CAP would suppress expression of the protein allele linked to the CAP-S mtDNA. If they interacted, then the CAP-S mtDNA protein would be expressed using the CAP-R ribosomes of the CAP-R mtDNA (Fig. 5). The latter alternative proved to be the case. Labeling the hybrids and cybrids in emetine or emetine + CAP yielded the same protein patterns regardless of which mtDNA was CAP-R. MtDNAs within cells must then interact and complement each other [Oliver and Wallace, 1982].

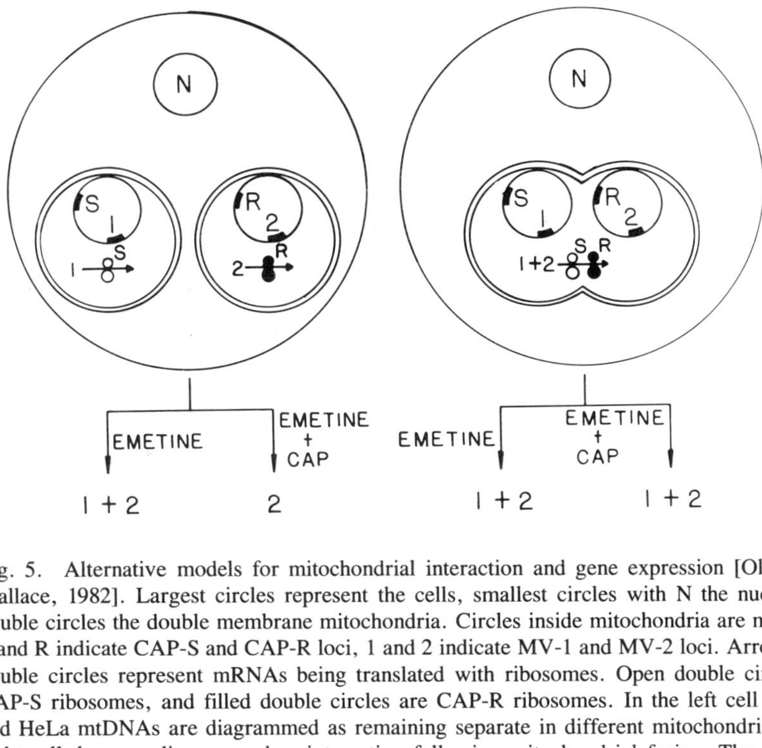

Fig. 5. Alternative models for mitochondrial interaction and gene expression [Oliver and Wallace, 1982]. Largest circles represent the cells, smallest circles with N the nuclei, and double circles the double membrane mitochondria. Circles inside mitochondria are mtDNAs. S and R indicate CAP-S and CAP-R loci, 1 and 2 indicate MV-1 and MV-2 loci. Arrows with double circles represent mRNAs being translated with ribosomes. Open double circles are CAP-S ribosomes, and filled double circles are CAP-R ribosomes. In the left cell HT1080 and HeLa mtDNAs are diagrammed as remaining separate in different mitochondria. In the right cell they are diagrammed as interacting following mitochondrial fusion. The expected expression of MV-1 and MV-2 in cells labeled in emetine and emetine plus CAP is shown below each cell.

Each mitochondrion contains multiple mtDNAs (there are thousands of mtDNAs but only hundreds of mitochondria). Assuming that mtDNA interaction occurs by mitochondrial fusion and fission, binomial modeling of the expression of MV-1 and MV-2 suggested that the average mitochondrion contained two to ten mtDNAs [Oliver and Wallace, 1982].

The retention of CAP-S mtDNAs in CAP-R cells can now be explained. Mitochondria maintain a 50-fold excess of ribosomes to mRNAs [Attardi et al, 1982a]. Thus, above a certain percentage of CAP-R mtDNAs there would be sufficient CAP-R ribosomes to translate all mRNAs fully. At this point there would be no further selective advantage for loss of the CAP-S mtDNAs. The threshold effect can also be explained. Below a certain percentage of CAP-R mtDNAs, most of the mRNA is translated by CAP-S ribosomes. Hence, mitochondrial translation in CAP would be insufficient to support growth.

Mitochondrial recombination. Though mtDNAs interact within a mitochondrion, they do not appear to recombine. Though two early reports suggested that recombination occurred [Horak et al, 1974; Hayashi et al, 1981], subsequent studies have not substantiated these observations [De Francesco et al, 1980; Giles et al, 1980b; Hayashi et al, 1982; Ho and Coon, 1979; Wallace, 1981, 1986; Wallace and Eisenstadt, 1979; Zuckerman et al, 1984]. Thus, even though the cellular phenotype is an integrated reflection of the mtDNA genotype, each individual mtDNA apparently remains distinct.

Familial Genetics

The unique intracellular behavior of mtDNAs is paralleled by an equally distinctive familial inheritance pattern. All nuclear genes are inherited equally from the two parents, but the mtDNA is inherited exclusively from the mother. Apparently, the father makes no mtDNA contribution. This has been demonstrated by tracing the inheritance of the mtDNAs in a number of families in whom the parental mtDNAs differed in their restriction fragment patterns. In every case, all of the children inherited only their mother's mtDNA. No evidence of paternal transmission was detected (Fig. 6). The daughters passed the mtDNA on to the next generation but the sons did not [Case and Wallace, 1981; Giles et al, 1980a] (Fig. 7).

Identical results have been reported in a variety of animals [Avise et al, 1979; Dawid and Blackler, 1972; Francisco et al, 1979; Hayashi et al, 1978; Hutchison et al, 1974; Kroon et al, 1978; Reilly and Thomas, 1980]. In fact, in one invertebrate study, 91 generations of backcrosses failed to detect any paternal input of mtDNAs into the maternal germ line [Lansman et al, 1983].

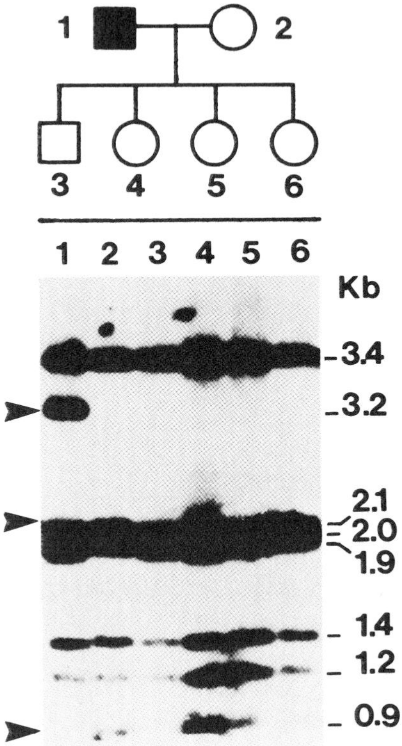

Fig. 6. Absence of paternal transmission of human mtDNA. Southern blot of *Hinc*II-digested blood cell mtDNAs [Wallace et al, 1982]. The father's mtDNA (1) is *Hinc*II morph 1, the mother's (2) is morph 3. *Hinc*II morph 3 differs from morph 1 in having two additional restriction sites that split the 3.2 kb fragment into 2.1, 0.9, and 0.3 kb fragments. No evidence of paternal transmission of the 3.2 kb fragment is seen in the progeny (3 to 6). (Reproduced from Wallace et al, 1985b, with permission of the publisher.)

To a great extent, the exclusive maternal transmission of mtDNA is a consequence of its cytoplasmic location. Each mammalian oocyte has a large cytoplasm, numerous mitochondria, and about 2.6×10^5 mtDNAs [Michaels et al, 1982]. By contrast, the sperm has few mitochondria and these are confined to the midpiece. Though the sperm midpiece has been observed to penetrate the egg [Anderson and Perotti, 1975; Gresson, 1940, 1941], the frequency of this event is unclear. Sperm mtDNAs that do enter the egg might be eliminated by any one of a number of possible mechanisms. Their inclusion in the embryo could be minimized by their dilution in the egg, their being sequestered in regions of the egg destined to be extraembryonic tissues,

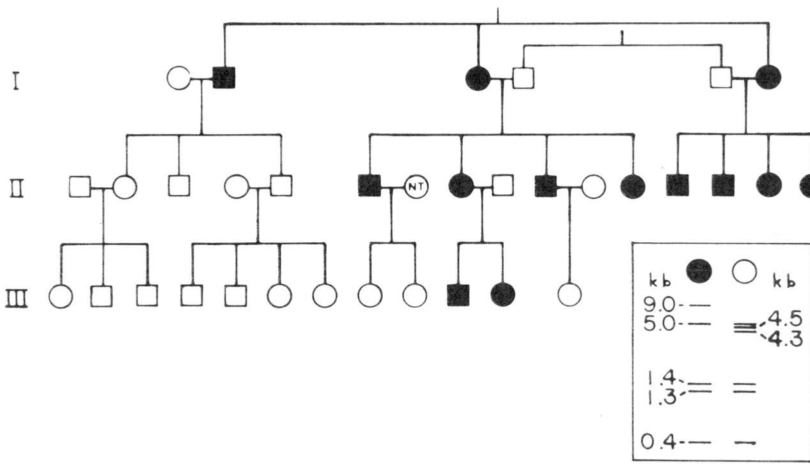

Fig. 7. Maternal inheritance of human mtDNA in a three-generation pedigree. The *Hae*II restriction patterns (lower right insert) for each individual were identified by Southern blots of blood platelet mtDNAs [Giles et al, 1980]. The open symbols indicate individuals whose mtDNA is *Hae*II morph 1, while the filled symbols indicate those with mtDNA *Hae*II morph 2. *Hae*II restriction fragment sizes have subsequently been corrected slightly [Johnson et al, 1983]. NT, not tested. (Reproduced from Giles et al, 1980, with permission of the publisher.)

or their mitotic segregation during development. Alternatively, sperm mtDNAs might be eliminated by mitochondrial debilitation, replication block [Vaughn et al, 1980], or active enzymatic destruction [Sager and Kitchin, 1975].

Whatever the mechanism, current evidence indicates that animal mtDNAs are inherited exclusively from the mother. Consequently, the mtDNA has proven a valuable marker for the maternal cytoplasm. In one study, the mtDNA was used to investigate the origin of complete hydatidiform moles. Complete moles are aberrant pregnancies that contain exclusively paternal chromosomes. Analysis of molar mtDNAs, however, revealed that they were exclusively maternal. Hence, moles must be the product of the fusion of anucleate eggs with one or more sperm [Edwards et al, 1984; Wallace et al, 1982].

Intraindividual mtDNA variation. Restriction digestion of mtDNAs from human blood samples yields unambiguous patterns as if each individual has only one type of mtDNA. Yet each cell contains thousands of mtDNAs and thus has the potential for harboring an enormous number of mtDNA variants.

Sequence analysis of numerous mtDNAs from single individuals has revealed extraordinary homogeneity, though rare nucleotide substitutions have been found [Greenberg et al, 1983; Monnat and Loeb, 1985]. Intraindividual small insertion-deletion mutations have also been detected in mouse [Brown et al, 1983], cow [Hauswirth et al, 1984], frogs [Monnerot et al, 1984], and *Drosophila* [Solignac et al, 1983]. Consequently, animal mtDNAs are maintained in a highly homogeneous state, though some intraindividual variation does occur.

The mechanism by which mtDNA homogeneity is maintained is unknown, but the sorting process acts quickly. In cattle, mtDNA types have been observed to switch in only a few generations [Hauswirth and Laipis, 1982; Olivo et al, 1983]. Mammalian eggs (bovine and mouse) contain between 100,000 and 260,000 mtDNA molecules, roughly 100 times the mtDNA molecules of somatic cells [Michaels et al, 1982; Piko and Matsumoto, 1976]. Since animal mtDNAs do not replicate until the blastocyst stage [Piko, 1970; Rinaldi et al, 1979; Pollak and Sutton, 1980] and only three of the 64 blastocyst cells participate in assembling the embryo [Markert and Petters, 1978], only a small proportion of the oocyte mtDNAs are passed on to the next generation [Hauswirth and Laipis, 1982]. This periodic expansion of the mtDNA population in the oocyte, followed by a random sampling of a small number of mtDNAs for inclusion in the embryo, would result in rapid shifts in the ratios of mtDNA mixtures in successive generations. Such intracellular "genetic drift" might account for the apparent homogeneity of cellular mtDNAs.

Population Studies

Though the mtDNAs within individuals are relatively uniform, mtDNAs between individuals are extraordinarily variable. Numerous RFLPs [Blanc et al, 1983; Bonne-Tamir et al, 1986; Brega et al, 1987a,b; Brown, 1980; Brown and Goodman, 1979; Cann et al, 1982, 1984; Denaro et al, 1981; Johnson et al, 1983; Horai et al, 1984; Wallace, 1983; Wallace et al, 1985a] and small insertion-deletion mutations [Cann and Wilson, 1983; Greenberg et al, 1983] have been identified in human mtDNA. In fact, comparison of variation between mtDNAs and nuclear DNAs has led to the estimate that the mammalian mtDNA fixes mutations seven to ten times faster than the nuclear DNA [Brown et al, 1979, 1982; Miyata et al, 1982]. This is consistent with an observation in *Drosophila* that infertility between isolated populations appears first in a maternally inherited genetic element [Kilias and Alahiotis, 1982].

The high mutation rate and exclusively matrilineal inheritance of the human mtDNA make the mtDNA a potentially powerful tool for tracing human lineages. Because of the high mutation rate, even closely related individuals and groups can be distinguished. Because of the maternal inheritance, mtDNAs never mix and thus cannot recombine. This means that mtDNA markers always remain linked (total linkage disequilibrium). Each mtDNA must thus diverge from all others by the sequential fixation of mutations along radiating maternal lineages. Consequently, it should be possible to reconstruct the human female lineage by analyzing mtDNA variation.

This concept has been supported by a worldwide survey of the RFLPs of several hundred human blood cell mtDNAs. Digestion with the restriction endonuclease *Hpa*I (GTTAAC) identified six different restriction fragment patterns (mtDNA morphs) of between two to four fragments each (Fig. 8). All six patterns could be interrelated by sequential restriction site changes,

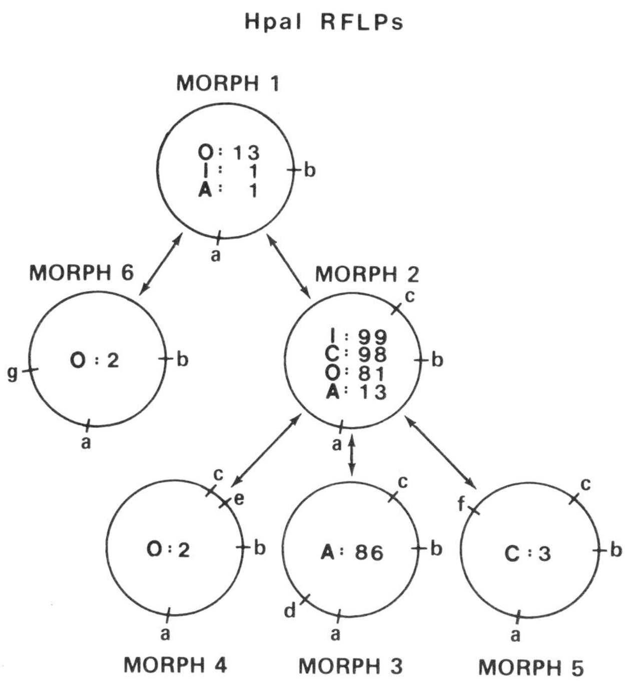

Fig. 8. *Hpa*I restriction maps of mtDNA morphs 1 to 6 from Africans (A), Caucasians (C), Amerindians (I), and Orientals (O). Each circle is an mtDNA morph. Small letters equal restriction sites. Numbers inside each circle indicate the percentage of individuals from A, C, I, or O who have mtDNAs with that morph. Interconnecting arrows indicate single restriction site changes. [Modified from Denaro et al, 1981.]

and each mtDNA morph was found to correlate with the geographic and ethnic origin of the blood sample. MtDNA *Hpa*I-morph 3 was found in 86% of the Africans examined but not in mtDNAs from any other continent. *Hpa*I-morph 1 was found primarily in Asians, while *Hpa*I-morph 2 was found in virtually all Europeans and Amerindians and the majority of Asians [Denaro et al, 1981; Wallace et al, 1985a].

Comparable results were obtained in an analysis of European and Asian mtDNAs with the related enzyme *Hinc*II (GTPyrPurAC). *Hinc*II, like *Hpa*I, distinguished Asian mtDNA-morph 1 from Eurasian morph 2. All of the *Hinc*II morphs were also interrelated by single site changes and correlated with the ethnic origin of the sample [Blanc et al, 1983].

Comparison of the human *Hinc*II and *Hpa*I morphs with those of other primates suggested that mtDNAs with *Hinc*II-*Hpa*I morph 1 might be closest to the root of the *Homo sapiens* mtDNA tree [Denaro et al, 1981; Wallace, 1983]. This conclusion was supported by demonstrating that the human morph 1 sequence was equivalent to that of other primates and differed from morph 2 in a mutation that altered the amino acid sequence of the ND5 protein [Blanc et al, 1983].

These studies were extended to encompass a total of six informative restriction enzymes and applied to blood samples from Europe, Asia, Africa [Johnson et al, 1983; Brega et al, 1987a,b], and America [Wallace et al, 1985a] (Fig. 9). Again, the restriction patterns for each enzyme were found to be interrelated by single nucleotide changes and generally to correlate with the continental origin of the sample. Further, each individual's mtDNA type, as defined by the combined restriction data of all six enzymes, was found to be related to all other mtDNAs by single site changes and to correlate with the individual's ethnic and geographic origin.

Over 50 mtDNA types have been identified (Fig. 9). Two mtDNA types (large grey circles) were found in most or all of the continents. The most common mtDNA (mtDNA type 1-2, largest grey circle) was *Hpa*I-*Hinc*II morph 2. This mtDNA was found to be most prevalent in Europe, followed in decreasing order by Asia, America, and Africa. All other mtDNAs appear to radiate from this central mtDNA by sequential restriction site changes (larger white circles). The smaller grey circle (mtDNA type 8-1) differed from mtDNA type 1 in being *Hpa*I-*Hinc*II morph 1. This mtDNA together with other *Hpa*I-*Hinc*I morph 1 mtDNAs were found predominantly in Asia, though also in Africa and America.

One possible interpretation of these data is that the sequential mutations summarized in Figure 9 recapitulate the radiation of *Homo sapiens* female lineages. If so, the worldwide distribution of mtDNA types 1-2 and 8-1

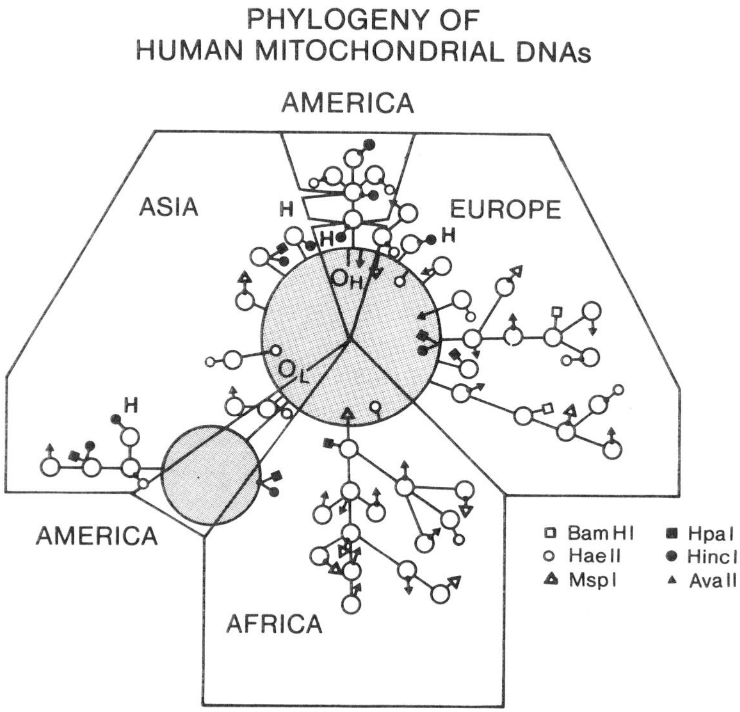

Fig. 9. Phylogeny of human mtDNAs based on RFLPs [results derived from Blanc et al, 1983; Denaro et al, 1981; Johnson et al, 1983; Wallace, 1983; Wallace et al, 1986]. Each larger white or grey circle represents a different mtDNA type as defined by six informative restriction enzymes (*Hpa*I, *Bam*HI, *Hae*II, *Hpa*II, *Msp*I, *Ava*II, and *Hinc*II). The locations of the diagnostic restriction sites for the central mtDNA 1-2 (2-1-1-1-2, largest grey circle) are indicated by the symbols on the inside of the circle positioned relative to the heavy and light strand origins of replication (O_H and O_L, respectively). All other mtDNAs radiate from the central mtDNA by single sequential restriction site changes, indicated by connecting lines. Moving out from the center, the restriction site symbol attached to the circle indicates the site change that differentiates it from the previous mtDNA. Restriction site symbols whose stem connects the mtDNA circle without a dot indicate site gains, while those that are connected with a dot indicate site losses. The position of the connecting symbols indicates the position of the altered site. For example, the second largest grey circle is mtDNA 8-1 (1-1-1-1-1-1), which differs from type 1-2 (2-1-1-1-1-2) by the loss of a combined *Hpa*I (GTAAC) and *Hinc*II (GTPyrPurAC) site. This is indicated by filled circle and square intersecting at a dot on the circle at about 3 o'clock [Blanc et al, 1983]. The sectors represent the continents on which each mtDNA type was found. Circles that overlap lines were found on the continents that the circle contains. See Wallace et al [1986] for additional discussion. (Reproduced from Wallace et al, 1986, with permission of the publisher.)

suggest that they existed before this radiation and represent the roots of the tree. This is consistent with mtDNA type 8-1 being *Hpa*I-*Hinc*II morph 1. Since all of the mtDNA types carrying *Hpa*I-*Hinc*II morph 1 are found in Asia it is possible that the human mtDNA lineage may have originated there, though alternative hypotheses such as an African origin are possible [Johnson et al, 1983]. The highly successful mtDNA type 1-2 might then have evolved from 8-1 in Asia, accounting for its high frequency in Asia, Europe, and America. Females carrying mtDNA type 1-2 may then have migrated from Asia onto the different continents. Once geographically isolated, these mtDNAs continued to radiate by sequential mutation fixation, giving rise to the present day lineage of continent-specific mtDNA types.

Assuming that these conjectures are true, the human mtDNA seems to represent a unique biologic history book of women.

MITOCHONDRIAL GENES AND DISEASE

The high mutation rate and compact organization of the human mtDNA suggest that variants in mtDNA proteins should be common. Yet the amino acid sequences of mammalian mtDNA polypeptides are highly conserved between species [Anderson et al, 1981, 1982; Bibb et al, 1981; Pepe et al, 1983], and the vast majority of nucleotide sequence changes identified in mtDNAs are due to silent mutations [Brown et al, 1982; Brown and Simpson, 1982; Cann et al, 1984; Johnson et al, 1983]. Since mutations are random, amino acid substitution mutations must also occur at a high frequency. Therefore, their absence in the population indicates that they are deleterious and eliminated by selection. Individuals with these mtDNA mutations must then have lethal congenital diseases of oxidative metabolism. Such individuals might make up a significant proportion of patients with undiagnosed metabolic disorders.

Genetic diseases of OXPHOS could result from either nuclear or mitochondrial DNA mutations (Fig. 1). The relative frequency of mutations in these two genomes should be related to the number of genes in each genome and the relative rates of mutation fixation (Table 2). The nucleus encodes far more OXPHOS subunits than the mtDNA, and a number of these genes may have tissue-specific isozymes (estimated for this calculation at three isozymes per gene) [Jarausch and Kadenbach, 1982; Kadenbach et al, 1982, 1983; Merle and Kadenbach, 1982; Scarpulla et al, 1982; Schultheiss and Klingenberg, 1984, 1985]. However, the high mutation rate of the mtDNA offsets the numerical superiority of the nuclear OXPHOS genes. Hence, the frequency of OXPHOS defects resulting from mtDNA mutations may be at least comparable to that of nuclear mutations.

TABLE 2. Relative Mutation Frequencies

Protein coding site	No. of proteins†	Isozymes‡	Mutation fixation rate¶	Relative mutation rate
Nuclear DNA	56	~3	~1	~168
MtDNA	13	1	~10	~130

†From Figure 1.
‡The number of isozymes is unknown. This is thought to be a high estimate.
¶This estimate of mtDNA mutation rate is from Brown et al [1979].

Putative hereditary diseases resulting from mtDNA mutations can be identified by the distinctive genetic characteristics of the mtDNA. Maternal inheritance provides the best criterion for detecting such diseases [Case and Wallace, 1981; Fine 1978; Giles et al, 1980a; Wallace et al, 1985b]. Recently a list of diseases showing possible maternal transmission has been published [Merril and Harrington, 1985]. However, large pedigrees are required to establish maternal inheritance, while only a single case of paternal transmission is sufficient to eliminate the hypothesis.

Mitotic segregation of heteroplasmic mtDNA populations and "threshold" expression provide a second criterion for identifying mtDNA mutations. The phenotypic effects of mitotic segregation might be manifest at both the tissue and pedigree levels, that is, different tissues in one individual may show varying degrees of respiratory deficiency and different individuals in a maternal lineage may show varying severities of symptoms. Thus, in pedigrees apparently normal individuals might appear in otherwise affected maternal lineages or affected individuals might appear in otherwise apparently normal maternal lineages.

This latter class of "spontaneous" or "recurrent" lethal disease is potentially of considerable theoretic and clinical importance. Deleterious mtDNA mutations present at frequencies below their "threshold level" presumably have no adverse effects on the cellular phenotype. Consequently, such mutations could be maintained in the female germ line for long periods. Further, because of the masking effect of mtDNA complementation, a maternal lineage could sustain a number of different deleterious mtDNA mutations. Each would complement the other such that the number of mutant copies of each gene would be low, but the proportion of mutant mtDNAs would be high. The survival of such a maternal lineage would ultimately be limited by the increasing probability that one or another of the mutant mtDNAs would become enriched and kill the lineage. Indeed, the control of mtDNA genetic load may have been one of the primary factors in the evolution of the cyclic expansion and contraction of mtDNAs along the female germ line.

Though the current changes in mtDNA numbers in the female germ line keep the mtDNA reasonably homogeneous, the process is by no means complete (see above). Consequently, latent mtDNA mutations may still be an important source of "spontaneous" cases of metabolic diseases of oxidative metabolism.

Twelve of the 13 mtDNA genes are known to function in OXPHOS. Hence, patients with mtDNA mutations might be expected to display chronic symptoms similar to those seen when OXPHOS is inhibited by environmental factors such as oxygen limitation (hypoxia) or metabolic poisons.

Hypoxia can result from asphyxiation or ischemia (restricted blood flow). The brain is the organ most sensitive to hypoxia, showing neuron death in as little as 2 minutes. The kidney and heart are the next most sensitive organs, undergoing necrosis after about 30 minutes of anoxia [Purtilo, 1978]. Hypoxia also adversely affects the liver. Hypoxic lesions of the brain have been observed in the globus pallidus, striatum, thalamus, and hypothalamus, while those of ischemia have been found in the cerebral cortex, Purkinje cells, and striatum [Janssen, 1984]. Subcellular changes in ischemic brain include marked neuronal mitochondrial swelling and condensation of the cytoplasm and nucleoplasm [Garcia et al, 1977; Kalimo et al, 1977]. Mitochondrial swelling is also characteristic of ischemic heart [Ganote et al, 1975], and kidney [Glaumann et al, 1975].

Though skeletal muscle, as an organ, is not as sensitive to ischemia as brain, heart, kidney, and liver, ischemia results in a very characteristic muscle pathology. Skeletal muscle is composed of two major cell types, the mitochondria-rich and oxidative slow-twitch-fibers (type I) and the mitochondria-poor and glycolytic fast-twitch-fibers (type II). Ischemia specifically and rapidly damages the type I fibers leading to characteristic ragged-red fibers (RRFs). These RRFs are detected in muscle sections by the Gomori-modified trichrome stain. The type I fibers appear "moth-eaten" and contain red material clustered around the sarcolemmal membrane and scattered throughout the cytoplasm. The type II fibers stain a uniform blue-green. The red clumps in the type I fibers are aggregates of highly abnormal mitochondria that can be swollen or contain granules or crystals [Hanzlikova and Schiaffino, 1977; Heffner and Barron, 1978; Karpati et al, 1974; Sarnat, 1983].

Inhibitors of oxidative metabolism also preferentially affect nervous and muscle tissue. Injection of monkeys with cyanide, an inhibitor of Complex IV, rapidly leads to unconsciousness. Chronic exposure to sublethal doses results in demyelination of the cerebral white matter or necrosis of the cerebral and cerebellar cortex [Hurst, 1942]. In rats, sublethal doses of cyanide produce epileptic spike and wave activity in EEG, grand mal seizures, and myoclonus (periodic jerking) [Brierley et al, 1976], while in man cyanide intoxication results in unconsciousness and lactic acidosis (elevated serum lactate). Chronic cyanide poisoning in humans has been associated with optic atrophy, deafness, ataxia, and altered EEG. Such poisoning can result from ingestion of the cyanogenic glycoside of cassava (Nigerian nutritional ataxic neuropathy) or from heavy tobacco smoking (tobacco amblyopia) [Vogel and Sultan, 1981; Smith, 1964].

Chronic injection of monkeys with sodium azide (a Complex IV inhibitor) has a most singular effect. The optic nerve undergoes demyelination and necrosis from the center outward, motor abnormalities approximating dystonic rigidity are observed, and there is marked selective cell loss from the caudate nucleus and putamen (neostriatum) [Hurst, 1942]. Bilateral striatal necrosis is also the most prominent pathologic finding in rats chronically poisoned with sodium azide [Miyoshi, 1967] and in mice poisoned with the 3-nitropropionate [Gould and Gustine, 1982], a Complex II inhibitor [Coles et al, 1979].

Perfusion of rats with respiratory inhibitors also affects striated muscle. The OXPHOS uncouplers 2,4-dinitrophenol and carbonylcyanide-m-chlorophenylhydrazone have the most marked effect. These poisons induce RRFs, mitochondrial abnormalities, hypermetabolism, muscle rigidity, and lactic acidosis within 90 minutes. Interestingly, chloramphenicol blocks the formation of RRFs but not mitochondrial ultrastructural changes. This implies that mitochondrial gene expression is required for the accumulation of mitochondria, presumably during mitochondrial proliferation [Melmed et al, 1975]. Cyanide also induces RRFs, though not as effectively [Melmed et al, 1974].

These toxicologic studies suggest that mutations in mitochondrial OXPHOS genes should affect the nervous system, muscle, heart, liver, and kidney. RRFs, abnormal mitochondria, and lactic acidosis should also be commonly observed.

Mitochondrial Genetic Diseases: Maternal Versus Mendelian

Two heterogeneous classes of hereditary diseases display a number of the characteristics expected for mtDNA mutations. These are 1) neurodegenerative diseases associated with Leber hereditary optic neuropathy (LHON) and 2) neuromuscular degenerative diseases (mitochondrial myopathies [MM]) associated with RRF disease.

Leber Hereditary Optic Neuropathy

LHON pedigrees provide the most convincing example of a maternally inherited disease. LHON patients have a late onset, acute or subacute, and bilateral, central vision loss (scotoma), usually without pain. Peripheral vision is retained. The median age of onset is 20 to 24 years with a range from adolescence to late adulthood. During the acute phase the retinal disk is swollen, the retinal vessels tortuous, and the nerve fiber layer thick and opaque. Later, as the central scotoma develops, the retinal disk becomes pale and the nerve fibers disappear [Nikoskelainen, 1984]. Pathologic analysis has revealed central optic nerve demyelination and degeneration as well as cell loss in the lateral geniculate body [Went, 1972]. In some cases, the optic atrophy is associated with other neurologic, psychiatric, cardiac, or skeletal signs [Adams et al, 1966; Nikoskelainen et al, 1985; Wallace, 1970; Went, 1972; Wilson, 1963, 1965b]. Elevated serum lactate is not generally seen in LHON patients, though enlargement of subsarcolemmal mitochondria has been reported [Nikoskelainen et al, 1984].

LHON pedigrees are frequently very large. Invariably, the symptoms are transmitted through the maternal lineage. No instance of paternal transmission of blindness has been documented, though paternal transmission of subclinical vision perturbations has been reported in one family [Carrol and Mastaglia, 1979]. In one large pedigree, virtually every male and female maternal descendant was affected [Wallace, 1970], though in many pedigrees there is a strong bias toward males expressing the disease [Egger and Wilson, 1983; Erickson, 1972; Fine, 1980b; Nikoskelainen, 1984; Went, 1972].

LHON can be associated with the much more severe disease, infantile bilateral striatal necrosis (IBSN). IBSN is an early onset (2 to 4 years) neurodegenerative disease associated with variable muscle weakness, generalized dystonic rigidity, dysarthria (impaired speech), short stature, and reduced intelligence [Novotny et al, 1986; Went, 1964a,b]. Muscle biopsies have shown no overt muscle degenerative disease, and elevated serum lactate levels have not been found. Computer tomography and magnetic resonance imaging, by contrast, showed marked cell loss in the neostriatum presumably accounting for the neurologic symptoms [Novotny et al, 1986; Wallace et al, 1985b].

In LHON + IBSN pedigrees, both LHON and IBSN are transmitted along the maternal lineage [Novotny et al, 1986; Went, 1964a,b]. One five-generation pedigree (Fig. 10) of 79 members had 21 affected individuals, all maternal progeny. Eight of these had LHON (six males and two females), while 14 had IBSN (five males and nine females). The proportion of LHON patients peaked in this pedigree in generation III at 33%, but the proportion of IBSN patients increased progressively (0, 13%, and 38%) from generations II to IV [Novotny et al, 1986]. Since LHON and IBSN occur in the same pedigree and some individuals can express both, it seems probable

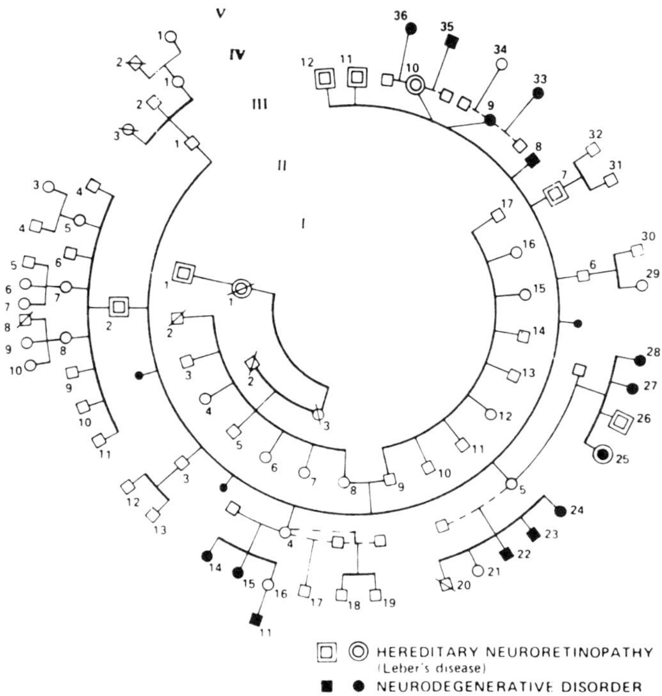

Fig. 10. Pedigree showing the maternal inheritance of LHON (double line symbols) and IBSN (filled symbols). The three small circles in generation III indicate death from unknown cause. (Reproduced from Novotny et al, 1986, with permission of the publishers.)

within any one pedigree that both phenotypes are the product of the same mutation, LHON being the mild form and IBSN the severe form. Consequently, IBSN + LHON pedigrees show maternal inheritance but with variable expression as expected for an mtDNA genetic disease [Novotny et al, 1986; Wallace et al, 1985b].

The inheritance of LHON and LHON + IBSN together with the associated changes in the muscle mitochondria observed in LHON patients is consistent with the hypothesis that these diseases are the result of mtDNA mutations. Several reports have suggested that LHON may result from reduced mitochondrial rhodanese (thiosulfate-sulphur transferase) [Cagianut et al, 1981; Egger and Wilson, 1983; Wilson, 1965a, 1971], though this has not been consistently observed [Nikoskelainen et al, 1984]. Since all mtDNA proteins are hydrophobic and rhodanese is water soluble, rhodanese does not appear to be a good candidate for the primary defect in LHON. Rather, an mtDNA mutation would imply that the primary defect is in one of the OXPHOS subunits. This supposition is supported by several lines of evidence. Optic atrophy has been correlated with heavy smoking and long-term cassava consumption [Vogel and Sultan, 1981] and has been accidentally induced by irrigation of the lacrimal ducts of one patient with chloramphenicol [Rothkoff et al, 1979]. More importantly, intoxication of rodents and primates with sodium azide and 3-nitropropionate closely mimic the symptoms and pathology of LHON + IBSN [Hurst, 1942]. Thus, LHON and IBSN are most probably the result of OXPHOS defects.

To determine if LHON-associated diseases are the product of gross mtDNA abnormalities, mtDNA was purified from the cultured lymphoblasts of two IBSN patients of the pedigree shown in Figure 10. These mtDNAs were digested with 12 restriction endonucleases and the fragments analyzed for insertions, deletions, or rearrangements using agarose and polyacrylamide gels. Though a number of RFLPs were identified, no structural alterations of greater than 50 base pairs were detected [Novotny et al, 1986; Wallace et al, 1985b]. Thus, if LHON + IBSN is the product of an mtDNA mutation, the mutation is too subtle to be detected with restriction analyses alone.

Mitochondrial Myopathies

In contrast to LHON, pedigrees with MM tend to be small. Still, maternal inheritance has been suggested in some cases.

Mitochondrial myopathies associated with OXPHOS defects. MMs are muscle degenerative diseases associated with mitochondrial defects. RRFs and abnormal mitochondria are characteristic of one subclass of these diseases, and biochemical analysis of muscle mitochondria from these patients has frequently revealed defects in OXPHOS [DiMauro et al, 1985; Rowland, 1983; Sengers et al, 1984; Stumpf, 1979]. Thus, congenital diseases of OXPHOS seem to yield the same muscle pathology as hypoxia and intoxication.

RRFs and abnormal mitochondria have generally become accepted as diagnostic of the OXPHOS MMs. Often such patients also have elevated serum lactate (lactic acidosis). A broad range of other symptoms affecting the central nervous system, heart, liver, and kidney can be associated with these diseases.

The first RRF disease reported (Luft hypermetabolic syndrome) was associated with hypermetabolism and abnormal skeletal muscle mitochondria. Two isolated patients have been reported. These patients had loosely coupled mitochondria consistent with their high caloric needs, hyperrespiration, and high heat generation [Afifi et al, 1972; DiMauro et al, 1972;

Edelman et al, 1975; Ernster and Luft, 1963; Haydar et al, 1971; Luft et al, 1962]. The hypermetabolism has been correlated with an inability of muscle mitochondria to retain calcium [DiMauro et al, 1976] and with impaired succinate and α-glycerophosphate utilization [Bonilla et al, 1977].

A variety of neurodegenerative symptoms involving the spongy deterioration of the brain have also been associated with RRF disease and lactic acidosis. A portion of these have recently been grouped into three phenotypically distinct syndromes. These are myoclonic epilepsy and RRFs (MERRF); Kearns-Sayre syndrome (KSS); and mitochondrial encephalopathy, myopathy, lactic acidosis, and stroke-like episodes (MELAS) [DiMauro et al, 1985].

The neurologic disturbances in MERRF include myoclonus, epilepsy, ataxia, bilateral hearing loss, and dementia [Fukuhara et al, 1980; Rosing et al, 1985; Tsairis et al, 1973]. Patients with MERRF are distinguished from those with other RRF syndromes in not having ophthalmoplegia, cortical blindness, or hemiparesis. The only MERRF patient examined biochemically lacked muscle mitochondrial cytochrome b [Morgan-Hughes et al, 1982].

The symptoms associated with KSS patients include external ophthalmoplegia (EO; limited eye movement), ptosis, ataxia, heart block, high cerebrospinal fluid protein, and cataract [Berenberg et al, 1977; DiMauro et al, 1973; Drachman, 1968; Egger et al, 1981; Karpati et al, 1973; Kearns and Sayre, 1958]. Myoclonus, hemiparesis, and cortical blindness are not characteristic of this group of patients. Biochemical analysis of muscle mitochondria from a number of EO and KSS patients have shown defects in Complex IV [DiMauro et al, 1985] as well as Complexes I and III [Morgan-Hughes et al, 1979, 1982; Sengers et al, 1984].

Common neurologic findings in MELAS patients include variable hemiparesis (partial paralysis) associated with fluctuating low density areas in CT scan [Yamamoto et al, 1984], seizures, stroke-like episodes, cortical blindness, and hemianopsia (blindness in part of the visual field). Myoclonus and ophthalmoplegia are not found [DiMauro et al, 1985]. Biochemical analysis of four MELAS patients failed to reveal a clear-cut biochemical defect. One patient had a 50% reduction in Complex IV and reduced Complex II and III activities, a second had reduced Complex IV activity, and another had reduced Complex II and III activity [Pavlakis et al, 1984].

The term *juvenile mitochondrial myopathy* has been used to define a diffuse group of infants with RRFs, hypotonia, mitochondrial myopathy, lactic acidosis, and growth failure. A number of these cases with both lethal and benign course have been associated with Complex IV deficiency [Bresolin et al, 1985; DiMauro et al, 1983, 1985; Zeviani et al, 1985]. However, one well-characterized case of lethal infantile neuromuscular disease associated with lactic acidosis was found to have a generalized deficiency in the iron-sulfur clusters of Complex I [Moreadith et al, 1984].

The childhood disease subacute necrotizing encephalomyelopathy (Leigh syndrome) has also been associated with RRF [Crosby and Chou, 1974; Egger et al, 1982]. This disease has been correlated with Complex IV deficiency [Bresolin et al, 1985; DiMauro et al, 1985; Willems et al, 1977] as well as other enzymatic defects [Sengers et al, 1984].

Facioscapulohumeral muscular dystrophy (FSH) associated with MM (FSH-MM) has been reported [Hudgson et al, 1972; Kamieniecka, 1976]. Symptoms included FSH, diabetes mellitus, cerebellar ataxia, electromyographic perturbations, and lactic acidosis under stress. EO and ptosis were not observed. Mitochondria from the patients of one large family were found to be highly abnormal. These mitochondria had reduced respiratory control ratios, implying that they were uncoupled [Bradley et al, 1978; Hudgson et al, 1972; Mechler et al, 1981; Worsfold et al, 1973].

Four adult cases of mitochondrial myopathy with RRF but no apparent neurologic symptoms have been reported. Two cases had muscle weakness associated with ATP synthetase alterations [Muller-Hocker et al, 1985; Schotland et al, 1976]. Two others were reported to show reduced muscle Complex III and diminished cytochrome b [Darley-Usmar et al, 1983; Darley-Usmar and Watanabe, 1985; Kennaway et al, 1984; Morgan-Hughes et al, 1977]. The muscle of one of the Complex III-deficient patients was analyzed immunochemically and found to lack several Complex III subunits including cytochrome b [Darley-Usmar et al, 1983; Eleff et al, 1984].

Cardiomyopathy has also been associated with RRF disease [Bogousslavsky et al, 1982; Neustein et al, 1979] as well as with familial OXPHOS defects [Barth et al, 1983]. Mitochondria from the latter patients showed multiple cytochrome deficiency, reduced P:O ratio, and reduced ATPase activity [Barth et al, 1983].

Finally, alterations in OXPHOS have been reported for X-linked Duchenne muscular dystrophy. The most marked deficiencies were observed in Complexes I and III [Scholte et al, 1985].

Two approaches for treatment of OXPHOS MMs have met with some success. First, high doses of cofactors for mitochondrial enzymes have been given in an attempt to increase the efficiency of OXPHOS. Riboflavin treatment yielded improvement in a patient with muscle Complex I deficiency [Arts et al, 1983], while coenzyme Q was beneficial for several KSS patients [Ogasahara et al, 1986]. Second, exogenous alternative electron donors have been given in an attempt to bypass the metabolic block. Menadione and ascorbate ingestion markedly improved muscle performance in a case of Complex III deficiency [Eleff et al, 1984].

Though RRFs are associated with defects in OXPHOS, there seems to be a conspicuous lack of correlation between the clinical symptoms and the OXPHOS complex affected [Sengers et al, 1984]. One possible explanation for this disturbing situation could be that the symptoms are not associated with a specific enzyme defect but rather with the net decrease in cellular ATP production caused by the mutation.

Genetics of mitochondrial myopathies. The analysis of the inheritance of the RRF-related diseases has proven to be complicated. Some pedigrees appear to show maternal inheritance while others are clearly mendelian.

The most convincing evidence for maternal inheritance of RRF disease comes from a large MERRF pedigree [Rosing et al, 1985] (Fig. 11). In this pedigree, all affected individuals examined had RRF disease, but their neurologic symptoms varied from mild bilateral hearing loss to severe progressive neurodegenerative disease and dementia. In every case the RRF disease was transmitted from the mother to both male and female progeny. The daughters, in turn, passed the disease on, but the sons did not. This pattern of transmission is inconsistent with any known mendelian inheritance pattern, but it is fully consistent with maternal inheritance. The only possible excep-

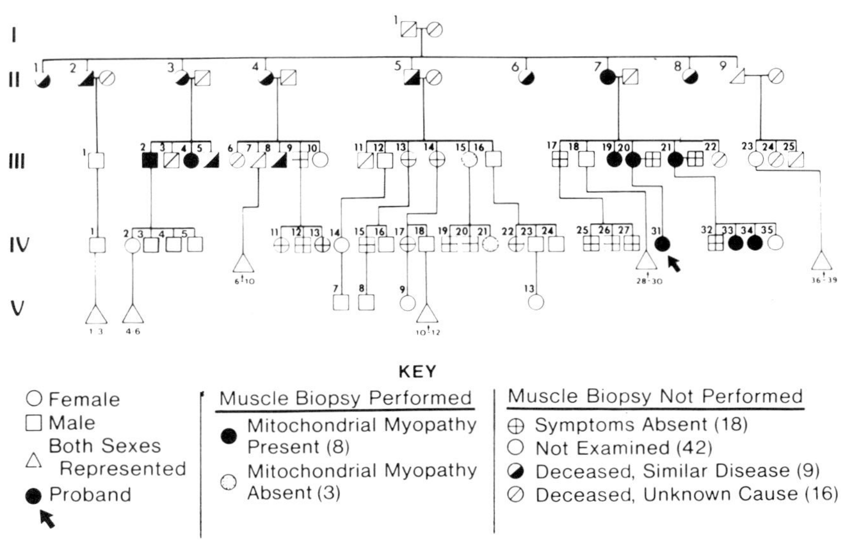

Fig. 11. Pedigree showing maternal inheritance of MERRF. For those individuals for whom muscle biopsy was not performed, "symptoms" include bilateral deafness; alterations of EEG, VER, and SSER; myoclonus; and epilepsy. (Reproduced from Rosing et al, 1985, with permission of the publisher.)

tions to maternal inheritance in this pedigree were two descendants of an affected male who had mild neurologic symptoms but no evidence of RRF disease. This pedigree could be explained best by hypothesizing that it is the product of an mtDNA mutation maintained in a heteroplasmic state with the germ line. MtDNA inheritance would cause the maternal transmission while mitotic segregation could yield the variable expression [Rosing et al, 1985].

Three other published MERRF pedigrees also show maternal transmission. However, these pedigrees are too small (two generation) to make a convincing case for maternal inheritance [Fitzsimmons et al, 1981; Fukuhara et al, 1980; Tsairis et al, 1973].

The mtDNAs from three affected individuals in the large MERRF family (Fig. 11) were examined for gross DNA changes. The mtDNAs were purified from cultured lymphoblasts, digested with 12 restriction endonucleases, and the fragments analyzed by gel electrophoresis. Though RFLPs were observed, no insertion-deletion or rearrangement mutations of greater than 50 base pairs were detected [Wallace et al, 1985b]. Thus, if the symptoms in this MERRF family are due to an mtDNA mutation, the mtDNA change must be small.

The first pedigree proposed to show maternal inheritance of RRF disease was a large family with FSH-MM and uncoupled mitochondria [Hudgson et al, 1972]. All severely affected individuals were related through the maternal lineage, and expression was variable. However, this pedigree was strongly biased toward reproductive females (15 females with descendants versus five males), so there were few opportunities to detect paternal transmission. Recent studies have revealed that three individuals with subclinical symptoms (abnormal electromyograms) are related to the severely affected patients through their mildly affected father [Mechler et al, 1981]. Consequently, this pedigree may be maternal or may involve mendelian genes. Maternal transmission has also been observed for a smaller two-generation FSH-MM pedigree [Kamieniecka, 1976], but this pedigree is too small to establish maternal inheritance.

A number of small EO and KSS pedigrees have been reported that show maternal transmission. Five three-generation pedigrees and one two-generation pedigree have been examined in which at least one individual had documented RRF disease. Associated symptoms ranged from complete KSS through chronic nephropathy to mild muscle weakness. In these pedigrees, all affected individuals were related through the maternal lineage [Egger and Wilson, 1983]. Similarly, in another study a great grandmother, grandmother, mother, and daughter all had ptosis. The mother and daughter also had documented EO + RRFs. However, 14 other maternal relatives were not overtly affected [Iannaccone et al, 1974]. Maternal transmission has also been observed in a small three-generation EO pedigree [Pepin et al, 1980] and four small two-generation EO pedigrees [Hart et al, 1977; Leshner et al, 1978; Levic et al, 1975; Tassin et al, 1980]. Further, maternal transmission has been observed in three small pedigrees with mitochondrial myopathy associated with neurologic and/or cardiac signs but without EO [Monnens et al, 1975; Shapira et al, 1975; Walter et al, 1981]. Though these scattered pedigrees indicate that maternal transmission is common in neuromuscular degenerative diseases associated with RRFs, none is large enough to prove maternal inheritance. Hence, each of these pedigrees could reflect either maternal or autosomal dominant inheritance.

Maternal inheritance has been proposed for a 23-member pedigree with hereditary cardiac septal defects. All seven affected members occurred on the maternal lineage, and two of these were female. The severity of the disease was variable. Hence, this pedigree appears to show maternal inheritance and has the characteristics expected for an mtDNA mutation [Sherman et al, 1985].

Finally, in one pedigree of cardiac and skeletal muscle myopathy associated with reduced intelligence and abnormal mitochondria, both parents and all six sons were affected. Inheritance in this pedigree might have involved maternal factors or possibly multiple nuclear genes [Lochner et al, 1981; Przybojewski et al, 1981].

At least two of the above pedigrees provide strong support for the hypothesis that RRF diseases can be maternally inherited. Further, many of the smaller pedigrees are consistent with this hypothesis. However, this in no way implies that all RRF diseases are maternally inherited. In fact, the inheritance pattern of a number of other RRF pedigrees can only be explained by mendelian genetics. Several EO and KSS pedigrees have been reported in which ptosis is transmitted through the males. Paternal transmission of ptosis was observed in ten of 12 offspring in one three-generation pedigree; five of seven offspring in a second five-generation pedigree [Bastiaensen et al, 1979]; and in three independent small four-generation pedigrees [Barron et al, 1979; Berenberg et al, 1977; McAuley, 1956]. Further, paternal transmission of mitochondrial myopathy has been documented in cases of cytochrome b deficiency [Spiro et al, 1970]; general muscle weakness and lactic acidosis [Buscaino et al, 1970]; and muscle weakness, paralysis, ataxia, and salt craving [Shy et al, 1966]. Paternal transmission has also been reported for two female infants who died with multisystem failure. These patients were related through a maternal grandfather and showed a deficiency in cytochrome c oxidase and in cytochromes $a+a_3$ and b. In one patient only muscle was involved, but in the other only liver was affected [Boustany et al, 1983].

Finally, clear mendelian inheritance has been observed in two large pedigrees of cardiomyopathy associated with RRFs and abnormal mitochondrial structure and function. In these pedigrees, only males were affected. All affected males were related through asymptomatic females, and approximately 50% of all males were affected. These pedigrees thus show classic X-linked inheritance [Barth et al, 1983; Neustein et al, 1979].

In most cases of RRF disease, there is inadequate family history to establish the inheritance pattern. In some cases, other sibs are affected, suggesting a familial disease [Buscaino et al, 1970; Okamura et al, 1976; Shy et al, 1966] while in others the affected individual represents a "sporadic" case. This latter situation is often seen in lethal RRF diseases such as "fatal infantile mitochondrial myopathy" [Bresolin et al, 1985; DiMauro et al, 1983, 1985; Moreadith et al, 1984; Zeviani et al, 1985], KSS and EO patients [DiMauro et al, 1973; Drachman, 1968; Karpati et al, 1973; Kearns and Sayre, 1958], other severe RRF diseases [Askanas et al, 1978; McLeod et al, 1975; Schotland et al, 1976], and Luft syndrome [Haydar et al, 1971; Luft et al, 1962]. Such "spontaneous" and "recurrent" cases could be due either to the mitotic enrichment of latent mtDNA mutations or to inherited recessive or new dominant nuclear mutations.

Since mtDNA genes are common to all tissues of the body, but nuclear OXPHOS genes can have tissue-specific isozymes, tissue specificity offers one possible criterion for differentiating between nuclear and mtDNA mutations in spontaneous cases. For instance, in one adult case of Complex III deficiency resulting in muscle weakness, the defect was expressed in muscle but not in skin fibroblasts, suggesting a nuclear mutation [Darley-Usmar et al, 1983]. Similarly, in the Luft hypermetabolic syndrome the enzymatic defect was expressed in muscle but not skin, liver, or intestine [Afifi et al, 1972; Luft et al, 1962]. However, caution must be exercised in concluding that all tissue-specific diseases are due to nuclear mutations, for mitotic segregation of mutant mtDNAs might also result in differential tissue effects. Muscle from patients with Luft disease for example, has been observed to contain two types of mitochondria: large and defective, and small and possibly normal [Bonilla et al, 1977; Ernster and Luft, 1963]. This could reflect either differential nuclear gene expression within different muscle cells or the segregation of a population of mutant and normal mtDNAs.

Developmentally controlled OXPHOS isozymes could offer an explanation for the time of onset of infantile mitochondrial myopathy. It is possible that certain nuclear encoded OXPHOS genes may have fetal and adult isozymes similar to those of the globin genes. The transition from fetal to adult genes shortly after birth could turn on a defective adult gene, leading to "fatal infantile mitochondrial myopathy" [Bresolin et al, 1985; Heiman-Patterson et al, 1982; Moreadith et al, 1984; Sengers et al, 1984], or turn off a defective fetal gene, leading to "benign infantile mitochondrial myopathy" [DiMauro et al, 1983].

The accumulated pedigree and biochemical data strongly indicate that RRF diseases associated with abnormal mitochondria can result from both mtDNA and nuclear DNA mutations [Harding et al, 1985]. This is fully consistent with the genomic distribution of OXPHOS genes. Therefore, the recent proposal that all RRF diseases be grouped together under the term *mitochondrial cytopathy* and assumed to be maternally inherited [Egger and Wilson, 1983] is not supported by the available data. Even if particular disease phenotypes could be correlated with specific respiratory complex defects, it would be ill advised to attempt to propose a general inheritance pattern for those symptoms. After all, each respiratory complex is composed of both nuclear and cytoplasmic subunits, loss of any one of which would destroy the complex. Thus, to avoid providing damaging misinformation to patients, it currently seems advisable for each case of RRF disease to be viewed as independent and the inheritance of each RRF disease pedigree be evaluated on its own merits.

Nuclear-Cytoplasmic Interactions

Though maternal and mendelian models of inheritance explain many features of the transmission of mitochondria-related diseases, other anomalies cannot readily be understood by single gene hypotheses. These anomalies include the striking tissue-specific expression of LHON + IBSN; the strong male bias in LHON; and the mild electrophysiologic aberrations in some descendants of males with LHON [Carroll and Mastaglia, 1979], MERFF [Rosing et al, 1985], and FSH-RRF [Mechler et al, 1981] pedigrees. Anomalies such as these could be explained as aberrations in the interaction of the nuclear and cytoplasmic OXPHOS genes.

The predilection of maternally inherited diseases toward involvement of nervous tissue, heart, liver, kidney, and skeletal muscle clearly reflects the high reliance of these tissues on energy production by oxidative metabolism. However, since the mtDNAs of all tissues appear to be the same, the tissue-specific differences in oxidative metabolism must be the direct consequence of differential expression of tissue-specific nuclear isozyme genes. Thus, the marked sensitivity of optic and neostriatal nerve cells to electron transport inhibition must reflect the expression of nuclear genes in those tissues that make them particularly reliant on electron transport. If so, then LHON and IBSN could be the result of a mutant mtDNA protein that interacts adequately with the OXPHOS ioszymes of most tissues but not those expressed in the optic nerve and neostriatum. If this optic nerve isozyme were X-linked, this hypothesis could explain the bias toward male specificity of LHON. If this X-linked gene had multiple alleles, then the disease might affect only males who received both the deleterious mtDNA mutation and deleterious X-chromosome allele. Alternatively, if this gene escaped X-chromosome inactivation then the differential expression in males might simply reflect the difference in X-chromosome number in males and females.

Similar arguments could be used to explain the occasional paternal transmission of subclinical symptoms in matrilineal pedigrees. The primary defect in these pedigrees must be in the mtDNA. However, the severity of expression of this mutation might be modulated by nuclear gene alleles. When paired with normal mtDNAs, the nuclear alleles would have little adverse effect, the most deleterious allele yielding only subclinical neurologic symptoms. However, when paired with the mutant mtDNA, the deleterious nuclear allele would yield the most severe clinical symptoms. Since severely affected males would transmit their mildly deleterious nuclear allele but not their mtDNA, half of their progeny should show mild subclinical symptoms.

Incompatible nuclear-cytoplasmic interactions might also explain some of the unusual characteristics of other MM pedigrees such as KSS and EO families. If the expression of the MM requires the simultaneous presence of both a nuclear gene allele and a particular mtDNA, then pedigrees that combine both would express the disease along the maternal lineage but with variable expression. Further, if the MM nuclear gene were rare but the MM mtDNA were common, then a small proportion of MM males would marry women with the MM mtDNA, resulting in the apparent paternal transmission of MM. Finally, in some instances a male carrying the nuclear MM gene masked by a normal mtDNA might marry a female with the MM mtDNA masked by normal nuclear genes. Half of their children would show "spontaneous" MM.

In such maternally inherited diseases, the determining factor would be the deleterious mtDNA mutation. Expression of this mutation would then be modulated by otherwise "normal" nuclear genes. However, the converse situation might also occur. The determining factor in a disease might be a deleterious nuclear gene mutation whose expression was modulated by otherwise "normal" mtDNAs. Huntington disease may be an example of this phenomenon.

Huntington disease is an autosomal dominant neurodegenerative disease caused by a mutation in a chromosome 4 gene [Gusella et al, 1983]. The disease is associated with shrinkage of the caudate and putamen, motor abnormalities, rigidity and/or chorea, severe psychiatric symptoms, and death (traits similar to those found in IBSN and chronic sodium azide poisoning). Ultrastructural analysis of Huntington patient brain tissue has revealed a variety of mitochondrial structural changes [Bruyn et al, 1979; Barr et al, 1978; Tellez-Nagel et al, 1973]. These changes have been correlated with a significant decrease in caudate cell respiration rate, cytochrome c oxidase, and cytochromes $a+a_3$ [Brennan et al, 1985].

Huntington disease has two general forms, a late onset (35 to 70 years) form that has little effect on reproductive fitness and an early onset (5 to 35 years) form that is a partial lethal. Surprisingly, the age of onset of Huntington disease correlates with the sex of the affected parent. Virtually all of the juvenile-onset cases (ages 5 to 19 years) come from affected fathers, while 70% to 80% of the late-onset cases come from affected mothers [Myers et al, 1983].

To explain this phenomenon, it has been hypothesized that the expression of the Huntington gene is modulated by an mtDNA factor [Myers et al, 1983], though alternative hypotheses have been proposed [Boehnke et al, 1983; Erickson, 1985; Myers et al, 1985]. According to the mtDNA hypothesis, when the Huntington nuclear gene is combined with one mtDNA gene allele, the Huntington nuclear gene functions close to normal, resulting in the reproductively successful late-onset form. However, when it is combined with a different mtDNA gene allele, the nuclear gene is fully expressed, yielding lethal early-onset disease. Since late-onset mothers will transmit both their Huntington nuclear gene and their mtDNA to their affected offspring, they will breed true and have late-onset children. Huntington males, by contrast, will transmit only their nuclear gene to their children. The severity of their children's disease will depend on their mother's mtDNA.

Myotonic dystrophy shows a somewhat similar phenomenon. This autosomal dominant (chromosome 19), type I fiber disease [Galdi, 1984; Harper et al, 1985] also has an early and late onset form. The early-onset form is almost invariably inherited from affected mothers, but, contrary to expectations for the mtDNA, it does not breed true along maternal lineages [Harper, 1975; Harper and Dyken, 1972; Tanaka et al, 1981]. Consequently, if the variability in the expression of myotonic dystrophy is related to the mtDNA, additional factors must also modulate expression of this nuclear gene.

Genetic Counseling and Genetic Therapy

Current evidence strongly suggests that genetic diseases of mitochondrial OXPHOS may result from mtDNA mutations or be influenced by mtDNA genes. However, nuclear gene mutations and nuclear-cytoplasmic interactions undoubtedly also play a major role in these diseases. Consequently, broad generalizations on the inheritance of these diseases are premature, and accurate genetic counseling is totally dependent on the available pedigree data and, to a certain extent, on the tissue specificity of the enzyme defect.

It thus seems that accurate genetic counseling of most OXPHOS diseases must wait until new in vitro genetic and molecular genetic procedures are developed for differentiating between nuclear and cytoplasmic genes. Fortunately, new hope has been provided by the isolation of the first nuclear OXPHOS genes [Jacobsson et al, 1985; Lomax et al, 1984; Parimoo et al, 1984; Ohta and Kagawa, 1986] and by the rapid development of new procedures for identifying and characterizing mtDNA mutants.

The identification and isolation of the genes involved in OXPHOS disease will raise the possibility of gene therapy. Gene therapy for nuclear OXPHOS mutants seems unlikely in the foreseeable future because of the complexities of tissue-specific expression, accurate gene delivery, and proper OXPHOS gene expression [Anderson, 1984]. However, gene therapy of mtDNA mutants may soon be feasible at the germ line level. In families with maternally inherited diseases, the mitochondria of oocytes from affected women could be supplemented or even replaced by microinjection of normal mitochondria. Alternatively, the nuclei of affected eggs could be removed with micropipettes and injected into enucleated oocytes from families with normal mitochondria. The resulting oocytes could then be cultured and the early embryos reimplanted in the uterus for normal development and birth. Such mitochondrial transfer therapy would have the notable advantage of not interfering with the basic structure or regulation of the patient's genome.

Thus, recent, rapid advances in our knowledge of the genetics and biochemistry of mitochondrial genes and neuromuscular diseases provides us with exciting new opportunities for the accurate diagnosis and treatment of OXPHOS diseases. We can only hope that we will soon be in a position to help alleviate some of the awful suffering associated with these dread maladies.

REFERENCES

Ackrell BAC, Ball MB, Kearney EB: Peptides from complex II active in reconstitution of succinate-ubiquinone reductase. J Biol Chem 255:2761–2769, 1980.

Adams JH, Blackwood W, Wilson J: Further clinical and pathological observations on Leber's optic atrophy. Brain 89:15–26, 1966.

Afifi AK, Ibrahim MZM, Bergman RA, Abu Haydar N, Mire J, Bahuth N, Kaylani F: Morphologic features of hypermetabolic mitochondrial disease. A light microscopic, histochemical and electron microscope study. J Neurol Sci 15:271–290, 1972.

Albring M, Griffith J, Attardi G: Association of a protein structure of a probable membrane derivation with HeLa cell mitochondrial DNA near its origin of replication. Proc Natl Acad Sci USA 74:1348, 1977.

Amzel LM, Pedersen PL: Protein ATPases: Structure and mechanism. Annu Rev Biochem 52:801–824, 1983.

Anderson S, Bankier AT, Barrell BG, de Bruijn MHL, Coulson AR, Drouin J, Eperon IC, Nierlich DP, Roe BA, Sanger F, Schreier PH, Smith AJH, Staden R, Young IG: Sequence and organization of the human mitochondrial genome. Nature 290:457–465, 1981.

Anderson S, de Bruijn MHL, Coulson AR, Eperon IC, Sanger F, Young IG: Complete sequence of bovine mitochondrial DNA: Conserved features of the mammalian mitochondrial genome. J Mol Biol 156:683–717, 1982.

Anderson WA, Perotti ME: An ultracytochemical study of the respiratory potency, integrity, and fate of the sea urchin sperm mitochondria during early embryogenesis. J Cell Biol 66:367–376, 1975.

Anderson WF: Prospects for human gene therapy. Science 226:401–409, 1984.

Aquila H, Misra D, Eulitz M, Klingenberg M: Complete amino acid sequence of the ADP/ATP carrier from beef heart mitochondria. Hoppe-Seyler's Z Physiol Chem 363:345–349, 1982.

Arts WFM, Scholte HR, Bogaard JM, Kerrebijn KF, Luyt-Houwen IEM: NADH-CoQ reductase deficient myopathy: Successful treatment with riboflavin. Lancet 2:581–582, 1983.

Askanas V, Engel WK, Britton DE, Adornato BT, Eiben RM: Reincarnation in cultured muscle of mitochondrial abnormalities. Two patients with epilepsy and lactic acidosis. Arch Neurol 35:801–809, 1978.

Attardi G, Cantatore P, Chomyn A, Crew S, Gelfand R, Merkel C, Montoya J, Ojala D: A comprehensive view of mitochondrial gene expression in human cells. In Slonimski P, et al (eds): "Mitochondrial Genes." Cold Spring Harbor, NY: Cold Spring Harbor Laboratory, 1982a, pp 51–71.

Attardi G, Chomyn A, Montoya J, Ojala D: Identification and mapping of human mitochondrial genes. Cytogenet Cell Genet 32:85–98, 1982b.

Attardi G, Montoya J: Analysis of human mitochondrial RNA. Methods Enzymol 97:435–468, 1983.

Avise JC, Lansman RA, Shade RO: The use of restriction endonucleases to measure mitochondrial DNA sequence relatedness in natural populations. I. Population structure and evolution in the genus peromyscus. Genetics 92:279–295, 1979.

Barr AN, Heinze WJ, Dobben GD, Valvassori GE, Sugar O: Bicaudate index in computerized tomography of Huntington's disease and cerebral atrophy. Neurology 28:1197–1200, 1978.

Barrell BG, Bankier AT, Drouin J: A different genetic code in human mitochondria. Nature 282:189–194, 1979.

Barron SA, Heffner RR, Zwirecki R: A familial mitochondrial myopathy with central defect in neural transmission. Arch Neurol 36:553–556, 1979.

Barth PG, Scholte HR, Berden JA, Van Der Klei-Van Moorsel JM, Luyt-Houwen IEM, Van 'T Veer-Korthof E Th, Van Der Harten JJ, Sobotka-Plojhar MA: An X-linked mitochondrial disease affecting cardiac muscle, skeletal muscle and neutrophil leucocytes. J Neurol Sci 62:327–355, 1983.

Bastiaensen LAK, Jaspar HHJ, Stadhouders AM: Ophthalmoplegia-plus. Doc Ophthalmol 46(2):365–380, 1979.

Beattie DS, Chen Y-C, Clejan L, Hou Lin L-F: Electrophoretic behavior of cytochrome b in a partially purified preparation and evidence for high molecular weight associated mitochondrial translation products. Biochemistry 18:2400–2406, 1979.

Bell RL, Capaldi RA: The polypeptide composition of ubiquinone-cytochrome c reductase (complex III) from beef heart mitochondria. Biochemistry 15:996–1001, 1976.

Berenberg RA, Pellock JM, DiMauro S, Schotland DL, Bonilla E, Eastwood A, Hays A, Vicale CT, Behrens M, Chutorian A, Rowland LP: Lumping or splitting? "Ophthalmoplegia-plus" or Kearns-Sayre syndrome? Ann Neurol 1:37–54, 1977.

Berk AJ, Clayton DA: Mechanism of mitochondrial DNA replication in mouse L-cells: Asynchronous replication of strands, segregation of circular daughter molecules, aspects of topology and turnover of an initiation sequence. J Mol Biol 86:801–824, 1974.

Bibb MJ, Van Etten RA, Wright CT, Walberg MW, Clayton DA: Sequence and gene organization of mouse mitochondrial DNA. Cell 26:167–180, 1981.

Blanc H, Adams CA, Wallace DC: Differential nucleotide changes on the large rRNA gene of the mitochondrial DNA confer chloramphenicol resistance to two human cell lines. Nucleic Acids Res 9:5785–5795, 1981a.

Blanc H, Chen K-H, D'Amore MA, Wallace DC: Amino acid change associated with the major polymorphic *Hinc* II site of Oriental and Caucasian mitochondrial DNAs. Am J Hum Genet 35:167–176, 1983.

Blanc H, Wright CT, Bibb MJ, Wallace DC, Clayton DA: Mitochondrial DNA of chloramphenicol-resistant mouse cells contains a single nucleotide change in the region encoding the 3′ end of the large ribosomal RNA. Proc Natl Acad Sci USA 78:3789–3793, 1981b.

Boehnke M, Conneally PM, Lange K: Two models for a maternal factor in the inheritance of Huntington disease. Am J Hum Genet 35:845–860, 1983.

Bogenhagen DF, Applegate EF, Yoza BK: Identification of a promoter for transcription of the heavy strand of human mtDNA: In vitro transcription and deletion mutagenesis. Cell 36:1105–1113, 1984.

Bogenhagen DF, Clayton DA: The number of mitochondrial deoxyribonucleic acid genomes in mouse L and human HeLa cells. J Biol Chem 249:7991–7995, 1974.

Bogousslavsky J, Perentes E, Deruaz JP, Regli F: Mitochondrial myopathy and cardiomyopathy with neurodegenerative features and multiple brain infarcts. J Neurol Sci 55:351–357, 1982.

Bonilla E, Schotland DL, Di Mauro S, Lee C-P: Luft's disease: An electron cytochemical study. J Ultrastruct Res 58:1–9, 1977.

Bonne-Tamir B, Johnson MJ, Natali A, Wallace DC, Cavalli-Sforza L-L: Human mitochondrial DNA types in two Israeli populations. A comparative study of the DNA level. Am J Hum Genet 38:341–351, 1986.

Boustany RN, Aprille JR, Halperin J, Levy H, DeLong GR: Mitochondrial cytochrome deficiency presenting as a myopathy with hypotonia, external ophthalmoplegia, and lactic acidosis in an infant and as fatal hepatopathy in a second cousin. Ann Neurol 14:462–470, 1983.

Bradely WG, Tomlinson BE, Hardy M: Further studies on mitochondrial and lipid storage myopathies. J Neurol Sci 35:201–210, 1978.

Brega A, Gardella R, Semino O, Morpurgo G, Astaldi-Ricotti GB, Wallace DC, Santachiara-Benerecetti AS: Genetic studies on the Tharu population of Nepal: Restriction endonuclease polymorphisms of mitochondrial DNA. Am J Hum Genet 39:502–512, 1987a.

Brega A, Scozzari R, Maccioni L, Jodice C, Wallace DC, Bianco I, Cao A, Santachiara-Benerecetti AS: Mitochondrial DNA polymorphisms in Italy. I. Population data from Sardinia and Rome. Ann Hum Genet 50:327–338, 1987b.

Brennan WA Jr, Bird ED, Aprille JR: Regional mitochondrial respiratory activity in Huntington's disease brain. J Neurochem 44:1948–1950, 1985.

Bresolin N, Zeviani M, Bonilla E, Miller RH, Leech RW, Shanske S, Nakagawa M, DiMauro S: Fatal infantile cytochrome c oxidase deficiency: Decrease of immunologically detectable enzyme in muscle. Neurology 35:801–812, 1985.

Brierley JB, Brown AW, Calverley J: Cyanide intoxication in the rat: Physiological and neuropathological aspects. J Neurol Neuosurg Psychiatry 39:129–140, 1976.

Brown GG, Bell G, Desrosiers L, Prussick R: Variation in animal mitochondrial genes and gene products. In Schenk HEA, et al (eds): "Endocytobiology II." New York: Walter de Gruyter, 1983, pp 247–261.

Brown GG, Simpson MV: Novel features of animal mtDNA evolution as shown by sequence of two rat cytochrome oxidase subunit II genes. Proc Natl Acad Sci USA 79:3246–3250, 1982.

Brown WM: Polymorphism in mitochondrial DNA of humans as revealed by restriction endonuclease analysis. Proc Natl Acad Sci USA 77:3605–3609, 1980.

Brown WM, George M Jr, Wilson AC: Rapid evolution of animal mitochondrial DNA. Proc Natl Acad Sci USA 76:1967–1971, 1979.

Brown WM, Goodman MM: Quantitation of intrapopulation variation by restriction endonuclease analysis of human mitochondrial DNA. In Cummings DJ, et al (eds): "Extrachromosomal DNA." New York: Academic Press, 1979, pp 485–499.

Brown WM, Prager EM, Wang A, Wilson AC: Mitochondrial DNA sequences in primates: Tempo and mode of evolution. J Mol Evol 18:225–239, 1982.

Bruyn GW, Bots G Th AM, Dom R: Huntington's chorea: Current neuropathological status. In Chase TN, et al (eds): "Advances in Neurology, Vol 23: Huntington's Disease." New York: Raven Press, 1979, pp 83–93.

Bunn CL, Wallace DC, Eisenstadt JM: Cytoplasmic inheritance of chlormaphenicol resistance in mouse tissue culture cells. Proc Natl Acad Sci USA 71:1681–1685, 1974.

Buscaino GA, De Giacomo P, Mazzarella L: Two cases of so-called "mitochondrial myopathy." Preliminary report. In Wallon JN, et al (eds): "Muscle Diseases." Proceedings of an International Congress, Milan, May 19–21, 1969. International Congress Series No. 199. Amsterdam: Excerpta Medica, 1970, pp 112–115.

Cagianut B, Rhyner K, Furrer W, Schnebli HP: Thiosulphate-sulphur transferase (Rhodanese) deficiency in Leber's hereditary optic atrophy. Lancet 2:981–982, 1981.

Cann RL, Brown WM, Wilson AC: Evolution of human mitochondrial DNA: A preliminary report. In Bonné-Tamir B, et al (eds): "Human Genetics, Part A: The Unfolding Genome." New York: Alan R. Liss, Inc., 1982, pp 157–165.

Cann RL, Brown WM, Wilson AC: Polymorphic sites and the mechanism of evolution in human mitochondrial DNA. Genetics 106:479–499, 1984.

Cann RL, Wilson AC: Length mutations in human mitochondrial DNA. Genetics 104:699–711, 1983.

Capaldi RA: Arrangement of proteins in the mitochondrial inner membrane. Biochim Biophys Acta 694:291–306, 1982.

Carroll WM, Mastaglia FL: Leber's optic neuropathy. A chemical and visual evoked potential study of affected and asymptomatic members of a six generation family. Brain 102:559–580, 1979.

Case JT, Wallace DC: Maternal inheritance of mitochondrial DNA polymerase in cultured human fibroblasts. Somatic Cell Genet 7:103–108, 1981.

Chang DD, Clayton DA: Precise identification of individual promoters for transcription of each strand of human mitochondrial DNA. Cell 36:635–643, 1984.

Chang DD, Clayton DA: Priming of human mitochondrial DNA replication occurs at the light-strand promoter. Proc Natl Acad Sci USA 82:351–355, 1985.

Ching E, Attardi G: High-resolution electrophoretic fractionation and partial characterization of the mitochondrial translation products from HeLa cells. Biochemistry 21:3188–3195, 1982.

Chomyn A, Mariottini P, Cleeter MWJ, Ragan CI, Matsuno-Yagi A, Hatefi Y, Doolittle RF, Attardi G: Six unidentified reading frames of human mitochondrial DNA encode components of the respiratory-chain NADH dehydrogenase. Nature 314:592–597, 1985.

Chomyn A, Mariottini P, Gonzalez-Cadavid N, Attardi G, Strong DD, Trovato D, Riley M, Doolittle RF: Identification of the polypeptides encoded in the ATPase 6 gene and in the unassigned reading frames 1 and 3 of human mtDNA. Proc Natl Acad Sci USA 80:5535–5539, 1983.

Clayton DA: Replication of animal mitochondrial DNA. Cell 28:693–705, 1982.

Clayton DA: Transcription of the mammalian mitochondrial genome. Annu Rev Biochem 53:573–594, 1984.

Clayton DA, Davis RW, Vinograd J: Homology and structural relationships between the dimeric and monomeric circular forms of mitochondrial DNA from human leukemic leukocytes. J Mol Biol 47:137–153, 1970.

Coles CJ, Edmondson DE, Singer TP: Inactivation of succinate dehydrogenase by 3-nitropropionate. J Biol Chem 254:5161–5167, 1979.

Crews S, Ojala D, Posakony J, Nishiguchi J, Attardi G: Nucleotide sequence of a region of human mitochondrial DNA containing the precisely identified origin of replication. Nature 277:192–198, 1979.

Crosby TW, Chou SM: "Ragged-red" fibers in Leigh's disease. Neurology 24:49–54, 1974.

Darley-Usmar VM, Kennaway NG, Buist NRM, Capaldi RA: Deficiency in ubiquinone cytochrome c reductase in a patient with mitochondrial myopathy and lactic acidosis. Proc Natl Acad Sci USA 80:5103–5106, 1983.

Darley-Usmar VM, Watanabe M: Analyses of muscle proteins in a patient with a mitochondrial myopathy. J Biochem 97:1767–1775, 1985.

Dawid IB, Blackler AW: Maternal and cytoplasmic inheritance of mitochondrial DNA in Xenopus. Dev Biol 29:152–161, 1972.

De Francesco L, Attardi G, Croce CM: Uniparental propagation of mitochondrial DNA in mouse-human cell hybrids. Proc Natl Acad Sci USA 77:4079–4083, 1980.

Denaro M, Blanc H, Johnson MJ, Chen KH, Wilmson E, Cavalli-Sforza LL, Wallace DC: Ethnic variation in *Hpa*I endonuclease cleavage patterns of human mitochondrial DNA. Proc Natl Acad Sci USA 78:5768–5772, 1981.

DiMauro S, Bonilla E, Lee C-P, Schotland DL, Scarpa A, Conn H Jr, Chance B: Luft's disease. Further biochemical and ultrastructural studies of skeletal muscle in the second case. J Neurol Sci 27:217–232, 1976.

DiMauro S, Bonilla E, Zeviani M, Nakagawa M, DeVivo DC: Mitochondrial myopathies. Ann Neurol 17:521–538, 1985.

DiMauro S, Nicholson JF, Hays AP, Eastwood AB, Papadimitriou A, Koenigsberger R, DeVivo DC: Benign infantile mitochondrial myopathy due to reversible cytochrome c oxidase deficiency. Ann Neurol 14:226–234, 1983.

DiMauro S, Schotland DL, Bonilla E, Lee C-P, Gambetti P, Rowland LP: Progressive ophthalmoplegia, glycogen storage, and abnormal mitochondria. Arch Neurol 29:170–179, 1973.

DiMauro S, Schotland DL, Lee C-P, Bonilla E, Conn H Jr: Biochemical and ultrastructural studies of mitochondria in Luft's disease: Implications for "mitochondrial myopathies." Trans Am Neurol Assoc 97:265–267, 1972.

Doda JN, Wright CT, Clayton DA: Elongation of displacement-loop strands in human and mouse mitochondrial DNA is arrested near specific template sequences. Proc Natl Acad Sci USA 78:6116–6120, 1981.

Doersen C-J, Guerrier-Takada C, Altman S, Attardi G: Characterization of an RNase P activity from HeLa cell mitochondria: Comparison with the cytosol RNase P activity. J Biol Chem 260:5942–5949, 1985.

Doersen C-J, Stanbridge EJ: Cytoplasmic inheritance of erythromycin resistance in human cells. Proc Natl Acad Sci USA 76:4549–4553, 1979.

Doonan S, Marra E, Passarella S, Saccone C, Quagliariello E: Transport of proteins into mitochondria. Int Rev Cytol 91:141–186, 1984.

Drachman DA: Ophthalmoplegia plus. The neurodegenerative disorders associated with progressive external ophthalmoplegia. Arch Neurol 18:654–674, 1968.

Edelman NH, Santiago TV, Conn HL Jr: Luft's syndrome: O_2 cost of exercise and chemical control of breathing. J Appl Physiol 39:857–859, 1975.

Edwards Y-H, Jeremiah SJ, McMillan SL, Povey S, Fisher RA, Lawler SD: Complete hydatiform moles combine maternal mitochondria with a paternal nuclear genome. Ann Hum Genet 48:119–127, 1984.

Egger J, Lake BD, Wilson J: Mitochondrial cytopathy. A multisystem disorder with ragged red fibres on muscle biopsy. Arch Dis Child 56:741–752, 1981.

Egger J, Wilson J: Mitochondrial inheritance in a mitochondrially mediated disease. N Engl J Med 309:142–146, 1983.

Egger J, Wynne-Williams CJE, Erdohazi M: Mitochondrial cytopathy or Leigh's syndrome? Mitochondrial abnormalities in spongiform encephalopathies. Neuropediatrics 15:219–224, 1982.

Eleff S, Kennaway NG, Buist NRM, Darley-Usmar VM, Capaldi RA, Bank WJ, Change B: 31P NMR study of improvement in oxidative phosphorylation by vitamin K3 and C in a patient with a defect in electron transport at complex III in skeletal muscle. Proc Natl Acad Sci USA 81:3529–3533, 1984.

Erickson RP: Leber's optic atrophy, a possible example of maternal inheritance. Am J Hum Genet 24:348–349, 1972.

Erickson RP: Chromosomal imprinting and the parent transmission of specific variation in expressivity of Huntington disease. Am J Hum Genet 37:827–828, 1985.

Ernster L, Luft R: Further study of a population of human skeletal muscle mitochondria lacking respiratory control. Exp Cell Res 32:26–35, 1963.

Fairfield FR, Bauer WR, Simpson MV: Studies on mitochondrial type I topoisomerase and on its function. Biochim Biophys Acta 824:45–57, 1985.

Fine PEM: Mitochondrial inheritance and disease. Lancet 1:659–661, 1980a.

Fine PEM: Leber's optic neuropathy. Br Med J 280:1097–1098, 1980b.

Fitzsimmons RB, Clifton-Bligh P, Wolfenden WH: Mitochondrial myopathy and lactic acidaemia with myoclonic epilepsy, ataxia and hypothalamic infertility: A variant of Ramsay-Hunt syndrome? J Neurol Neurosurg Psychiatry 44:79–82, 1981.

Francisco JF, Brown GG, Simpson MV: Further studies on types A and B rat MtDNAs: Cleavage maps and evidence for cytoplasmic inheritance in mammals. Plasmid 2:426–436, 1979.

Frangione B, Rosenwasser E, Penefsky HS, Pullman ME: Amino acid sequence of the protein inhibitor of mitochondrial adenosine triphosphatase. Proc Natl Acad Sci USA 78:7403–7407, 1981.

Fukuhara N, Tokiguchi S, Shirakawa K, Tsubaki T: Myoclonus epilepsy associated with ragged-red fibres (mitochondrial abnormalities): Disease entity or a syndrome? J Neurol Sci 47:117–133, 1980.

Gadaleta MN, Saccone C: Effect of rifampicin on macromolecular synthesis in rat liver mitochondria. In Kroon AM, et al (eds): "The Biogenesis of Mitochondria." New York: Academic Press, 1974, pp 85–88.

Galdi AP: "Diagnosis and Management of Muscle Disease." New York: Spectrum Publications, 1984.

Gallerani R, Saccone C: The DNA-dependent RNA polymerase from rat liver mitochondria. In Kroon AM, et al (eds): "The Biogenesis of Mitochondria." New York: Academic Press, 1974, pp 59–69.

Ganote CE, Seabra-Gomes R, Nayler WG, Jennings RB: Irreversible myocardial injury in anoxic perfused rat hearts. Am J Pathol 80:419–450, 1975.

Garcia JH, Kalimo H, Kamijyo Y, Trump BF: Cellular events during partial cerebral ischemia. I. Electron microscopy of feline cerebral cortex after middle-cerebral-artery occlusion. Virchows Arch B [Cell Pathol] 25:191–206, 1977.

Gellerfors P, Nelson BD: Biogenesis of the cytochrome bc complex in isolated rat hepatocytes. Biochem Biophys Res Commun 99:170–175, 1981.

Giles RE, Blanc H, Cann HM, Wallace DC: Maternal inheritance of human mitochondrial DNA. Proc Natl Acad Sci USA 77:6715–6719, 1980a.

Giles RE, Stroynowski I, Wallace DC: Characterization of mitochondrial DNA in chloramphenicol-resistant interspecific hybrids and a cybrid. Somatic Cell Genet 6:543–554, 1980b.

Glaser E, Norling B, Ernster L: Reconstitution of mitochondrial oligomycin and dicyclohexylcarbodiimide-sensitive ATPase. Eur J Biochem 110:225–235, 1980.

Glaumann B, Glaumann H, Berezesky IK, Trump BF: Studies on the pathogenesis of ischemic cell injury. II. Morphological changes of the pars convoluta (P1 and P2) of the proximal tubule of the rat kidney made ischemic in vivo. Virchows Arch B [Cell Pathol] 19:281–302, 1975.

Gould DH, Gustine DL: Basal ganglia degeneration, myelin alterations, and enzyme inhibition induced in mice by the plant toxin 3-nitropropanoic acid. Neuropathol Appl Neurobiol 8:377–392, 1982.

Greenberg BD, Newbold JE, Sugino A: Intraspecific nucleotide sequence variability surrounding the origin of replication in human mitochondrial DNA. Gene 21:33–49, 1983.

Gresson RAR: Presence of the sperm middle-piece in the fertilized egg of the mouse (*Mus musculus*). Nature 3672:425, 1940.

Gresson RAR: A study of the cytoplasmic inclusions during maturation, fertilization and the first cleavage division of the egg of the mouse. Q J Microsc Sci 83:35–60, 1941.

Gusella JF, Wexler NS, Conneally PM, Naylor SL, Anderson MA, Tanzi RE, Watkins PC, Ottina K, Wallace MR, Sakaguchi AY, Young AB, Shoulson I, Bonilla E, Martin JB: A polymorphic DNA marker genetically linked to Huntington's disease. Nature 306:234–238, 1983.

Hanzlikova V, Schiaffino S: Mitochondrial changes in ischemic skeletal muscle. J Ultrastruct Res 60:121–133, 1977.

Harding AE, Petty RKH, Morgan-Hughes J-A: Mitochondrial myopathy: A genetic study. J Med Genet 22:138, 1985.

Hare JF, Ching E, Attardi G: Isolation, subunit composition, and site of synthesis of human cytochrome oxidase. Biochemistry 19:2023–2030, 1980.

Harper PS: Congenital myotonic dystrophy in Britain. I. Clinical aspects. Arch Dis Child 50:505–521, 1975.

Harper PS, Dyken PR: Early-onset dystrophia myotonica. Evidence supporting a maternal environmental factor. Lancet 2:53–55, 1972.

Harper PS, Shaw D, Meredith L, Sarfarazi M, Borrk D, Huson S: Gene mapping and myotonic dystrophy. Prog Clin Biol Res 177:61–76, 1985.

Hart ZH, Chang C-H, Perrin EVD, Neerunjun JS, Ayyar R: Familial poliodystrophy, mitochondrial myopathy, and lactate acidemia. Arch Neurol 34:180–185, 1977.

Hatefi Y, Galante YM, Stiggall DL, Ragan CI: Proteins, polypeptides, prosthetic groups, and enzymic properties of complexes I, II, III, IV, and V of the mitochondrial oxidative phosphorylation system. Methods Enzymol 56:577–602, 1979.

Hauswirth WW, Laipis PJ: Mitochondrial DNA polymorphism in a maternal lineage of Holstein cows. Proc Natl Acad Sci USA 79:4686–4690, 1982.

Hauswirth WW, Van de Walle MJ, Laipis PJ, Olivo PO: Heterogeneous mitochondrial DNA. D-loop sequences in bovine tissue. Cell 37:1001–1007, 1984.

Hayashi J-I, Gotoh O, Tagashira Y-K, Tosu M, Sekiguchi T-Z: Analysis of mitochondrial DNA species in interspecific hybrid somatic cells using restriction endonucleases. Identification of recombinant mtDNA molecules. Exp Cell Res 131:458–462, 1981.

Hayashi J-I, Gotoh O, Tagashira Y, Tosu M, Sekiguchi T, Toshida MC: Mitochondrial DNA analysis of mouse-rat hybrid cells: Effect of chloramphenicol selection on the relative amounts of parental mitochondrial DNAs. Somatic Cell Genet 8:67–81, 1982.

Hayashi J-I, Yonekawa Y, Gotoh O, Watanabe J, Tagashira Y: Strictly maternal inheritance of rat mitochondrial DNA. Biochem Biophys Res Commun 83:1032–1038, 1978.

Haydar NA, Conn HL Jr, Afifi A, Wakid N, Ballas S, Fawaz K: Severe hypermetabolism with primary abnormality of skeletal muscle mitochondria. Functional and therapeutic effects of chloramphenicol treatment. Ann Intern Med 74:548–558, 1971.

Heckman JE, Sarnoff J, Alzner-DeWeerd B, Yin S, RajBhandary UL: Novel features in the genetic code and codon of reading patterns in Neurospora crassa mitochondria based on sequences of six mitochondrial tRNAs. Proc Natl Acad Sci USA 77:3159–3163, 1980.

Heffner RR, Barron SA: The early effects of ischemia upon skeletal muscle mitochondria. J Neurol Sci 38:295–315, 1978.

Heiman-Patterson TD, Bonilla E, DiMauro S, Foreman J, Schotland DL: Cytochrome-c-oxidase deficiency in a floppy infant. Neurology 32:898–900, 1982.

Heron C, Smith S, Ragan CI: An analysis of the polypeptide composition of bovine heart mitochondrial NADH-ubiquinone oxidoreductase by two-dimensional polyacrylamide-gel electrophoresis. Biochem J 181:435–443, 1979.

Hixson JE, Clayton DA: Initiation of transcription from each of the two human mitochondrial promoters requires unique nucleotides at the transcriptional start sites. Proc Natl Acad Sci USA 82:2660–2664, 1985.

Ho C, Coon HG: Restricted mitochondrial DNA fragments as genetic markers in cytoplasmic hybrids. In Cummings DJ, et al (eds): "Extrachromosomal DNA. ICN-UCLA Symp Mol Cell Biol XV." New York: Academic Press, 1979, pp 501–514.

Horai S, Gojobori T, Matsunaga E: Mitochondrial DNA polymorphism in Japanese. I. Analysis with restriction enzymes of six base pair recognition. Hum Genet 68:324–332, 1984.

Horak I, Coon HG, Dawid IB: Interspecific recombination of mitochondrial DNA molecules in hybrid somatic cells. Proc Natl Acad Sci USA 71:1828–1832, 1974.

Howell N: Origin, cellular expression, and cybrid transmission of mitochondrial CAP-R, PYR-IND, and OLI-R mutant phenotypes. Somatic Cell Genet 9:1–24, 1983.

Howell N, Huang P, Kolodner RD: Origin, transmission, and segregation of mitochondrial DNA dimers in mouse hybrid and cybrid cell lines. Somatic Cell Mol Genet 10:259–274, 1984.

Hudgson P, Bradley WG, Jenkison M: Familial "mitochondrial" myopathy: A myopathy associated with disordered oxidative metabolism in muscle fibres. Part 1. Clinical, electrophysiological and pathological findings. J Neurol Sci 16:343–370, 1972.

Hurst EW: Experimental demyelination of the central nervous system. 3. Poisoning with potassium cyanide, sodium azide, hydroxylamine, narcotics, carbon monoxide, etc., with some consideration of bilateral necrosis occurring in the basal nuclei. Aust J Exp Biol Med Sci 20:297–305, 1942.

Hutchison CA III, Newbold JA, Potter SS, Edgell MH: Maternal inheritance of mammalian mitochondrial DNA. Nature 251:536–538, 1974.

Iannaccone ST, Griggs RC, Markesbery WR, Joynt RJ: Familial progressive external ophthalmoplegia and ragged-red fibers. Neurology 24:1033–1038, 1974.

Ise W, Haiker H, Weiss H: Mitochondrial translation of subunits of the rotenone-sensitive NADH: Ubiquinone reductase in Neurospora crassa. EMBO J 4:2075–2080, 1985.

Jacobsson A, Stadler U, Glotzer MA, Kozak LP: Mitochondrial uncoupling protein from mouse brown fat: Molecular cloning, genetic mapping, and mRNA expression. J Biol Chem 260:16250–16254, 1985.

Janssen W: "Forensic Histopathology." [Forster S, translator.] New York: Springer-Verlag, 1984.

Jarausch J, Kadenbach B: Tissue-specificity overrides species-specificity in cytoplasmic cytochrome c oxidase polypeptides. Hoppe-Seyler's Z Physiol Chem 363:1133–1140, 1982.

Johnson LV, Walsh ML, Chen LB: Localization of mitochondria in living cells with rhodamine 123. Proc Natl Acad Sci USA 77:990–994, 1980.

Johnson MJ, Wallace DC, Ferris SD, Rattazzi MC, Cavalli-Sforza LL: Radiation of human mitochondria DNA types analyzed by restriction endonuclease cleavage patterns. J Mol Evol 19:255–271, 1983.

Kadenbach BR, Hartmann R, Glanville R, Buse G: Tissue-specific genes code for polypeptide VIa of bovine liver and heart cytochrome c oxidase. FEBS Lett 138:236–238, 1982.

Kadenbach B, Merle P: On the function of multiple subunits of cytochrome c oxidase from higher eukaryotes. FEBS Lett 135:1–11, 1981.

Kadenbach B, Ungibauer M, Jarausch J, Buge U, Kuhn-Nentwig L: The complexity of respiratory complexes. TIBS 8:398–400, 1983.

Kalimo H, Garcia JH, Kamijyo Y, Tanaka J, Trump BF: The ultrastructure of "brain death." II. Electron microscopy of feline cortex after complete ischemia. Virchows Arch B [Cell Pathol] 25:207–220, 1977.

Kamieniecka Z: Myopathies with abnormal mitochondria. A clinical, histological, and electrophysiological study. Acta Neurol Scand 55:57–75, 1976.

Karpati G, Carpenter S, Larbrisseau A, Lafontaine R: The Kearns-Shy syndrome. A multisystem disease with mitochondrial abnormality, demonstrated in skeletal muscle and skin. J Neurol Sci 19:133–151, 1973.

Karpati G, Carpenter S, Melmed C, Eisen AA: Experimental ischemic myopathy. J Neurol Sci 23:129–161, 1974.

Kearns TP, Sayre GP: Retinitis pigmentosa, external ophthalmoplegia, and complete heart block. Arch Ophthalmol 60:280–289, 1958.

Kearsey SE, Craig I: Altered ribosomal RNA genes in mitochondria from mammalian cells with chloramphenicol resistance. Nature 290:607–608, 1981.

Kearsey SE, Munro E, Craig IW: Studies of heterogeneous mitochondrial populations in a mouse cell line: The effects of selection for or against mitochondrial genomes that confer chloramphenicol resistance. Proc R Soc Lond [Biol] 224:315–323, 1985.

Kennaway NG, Buist NRM, Darley-Usmar VM, Papadimitriou A, DiMauro S, Kelley RI, Capaldi RA, Blank NK, D'Agostino A: Lactic acidosis and mitochondrial myopathy associated with deficiency of several components of complex III of the respiratory chain. Pediatr Res 18:991–999, 1984.

Kenney WMC, Mowers PC, Seng RL, Singer TP: Localization of the substrate and oxalacetate binding site of succinate dehydrogenase. J Biol Chem 251:2369–2373, 1976.

Kilias G, Alahiotis SN: Genetic studies on sexual isolation and hybrid selection in long term cage populations of *Drosophila melanogaster*. Evolution 36:121–131, 1982.

Kroon AM, de Vos WM, Bakker H: The heterogeneity of rat-liver mitochondrial DNA. Biochim Biophys Actra 519:269–273, 1978.

Kroon AM, De Vries H: Antibiotics: A tool in the search for the degree of autonomy of mitochondria in higher animals. In Miller PL (ed): "Control of Organelle Development." Symp Soc Exp Biol, Vol 24. New York: Academic Press, 1970, p 181–199.

Kuhn-Nentwig L, Kadenbach B: Immmunological identification of four different polypeptides in "subunit VII" of mammalian cytochrome c oxidase. FEBS Lett 172:189–192, 1984.

Lansman RA, Avise JC, Huettel MD: Critical experimental test of the possibility of "paternal leakage" of mitochondrial DNA. Proc Natl Acad Sci USA 80:1969–1971, 1983.

Lederman M, Attardi GA: In vitro protein synthesis in mitochondrial fraction from HeLa cells: Sensitivity to antibiotics and ethidium bromide. Biochem Biophys Res Commun 40:1492–1500, 1970.

Leshner RT, Spector RH, Seybold M, Romine J, Sipe J, Kelts A: Progressive external ophthalmoplegia (PEO) with ragged red fibers: An intrafamilial study. Neurology 28:364–365, 1978.

Levic ZM, Stefanovic BS, Nikolic MZ, Pistejic DT: Progressive nuclear ophthalmoplegia associated with mental deficiency, lingua scrotalis, and other neurologic and ophthalmologic signs in a family. Neurology 25:68–71, 1975.

Lochner A, Hewlett RH, O'Kennedy A, Van Der Walt JJ, Tiedt FAC, Hoffman H, De Graaf AS, Przybojewski JZ, Torrington M: A study of a family with inherited disease of cardiac and skeletal muscle. Part II. Skeletal morphology and mitochondrial oxidative phosphorylation. S Afr Med J 59:453–461, 1981.

Loeb JN, Hubby BG: Amino acid incorporation by isolated mitochondria in the presence of cycloheximide. Biochim Biophys Acta 166:745–748, 1968.

Lomax MI, Bachman NJ, Nasoff MS, Caruthers MH: Isolation and characterization of a cDNA clone for bovine cytochrome c oxidase substrate IV. Proc Natl Acad Sci USA 81:6295–6299, 1984.

Luft R, Ikkos D, Palmieri G, Ernster L, Afzelius B: A case of severe hypermetabolism of nonthyroid origin with a defect in the maintenance of mitochondrial respiratory control: A correlated clinical, biochemical, and morphological study. J Clin Invest 41:1776–1804, 1962.

Macreadie IG, Novitski CE, Maxwell RJ, John U, Ooi B-G, McMullen GL, Lukins HB, Linnane AW, Nagley P: Biogenesis of mitochondria: The mitochondrial gene (aapl) coding for mitochondrial ATPase subunit 8 in Saccharomyces cerevisiae. Nucleic Acids Res 11:4435–4451, 1983.

Mariottini P, Chomyn A, Attardi G, Trovato D, Strong DD, Doolittle RF: Antibodies against synthetic peptides reveal that the unidentified reading frame A6L, overlapping the ATPase 6 gene, is expressed in human mitochondria. Cell 32:1269–1277, 1983.

Markert CL, Petters RM: Manufactured hexaparental mice show that adults are derived from three embryonic cells. Science 202:56–58, 1978.

Marres CAM, Slater EC: Polypeptide composition of purified QH2: Cytochrome c oxidoreductase from beef-heart mitochondria. Biochim Biophys Acta 462:531–548, 1977.

Mattoccia LP, Roberti S: Effect of specific inhibitors on mitochondrial DNA replication in HeLa cells. Biochem Biophys Res Commun 60:882–893, 1974.

McAuley FD: Progressive external ophthalmoplegia. Br J Ophthalmol 40:686–690, 1956.

McLeod JG, Baker W De C, Shorey CD, Kerr CB: Mitochondrial myopathy with multisystem abnormalities and normal ocular movements. J Neurol Sci 24:39–52, 1975.

Mechler F, Fawcett PRW, Mastaglia FL, Hudgson P: Mitochondrial myopathy: A study of clinically affected and asymptomatic members of a six-generation family. J Neurol Sci 50:191–200, 1981.

Melmed C, Karpati G, Carpenter S: Experimental mitochondrial myopathy with "ragged red" fibers. Neurology 24:353, 1974.

Melmed C, Karpati G, Carpenter S: Experimental mitochondrial myopathy produced by in vivo uncoupling of oxidative phosphorylation. J Neurol Sci 26:305–318, 1975.

Merle P, Kadenbach B: Kinetic and structural differences between cytochrome c oxidases from beef liver and heart. Eur J Biochem 125:239–244, 1982.

Merril CR, Harrington MG: The search for mitochondrial inheritance of human diseases. Trends Genet 1:140–144, 1985.

Meyer RR, Simpson MV: DNA biosynthesis in mitochondria: Partial purification of a distinct DNA polymerase from isolated rat liver mitochondria. Proc Natl Acad Sci USA 61:130–137, 1969.

Michaels GS, Hauswirth WW, Laipis PJ: Mitochondrial DNA copy number in bovine oocytes and somatic cells. Dev Biol 94:246–251, 1982.

Mitchell CH, Attardi G: Cytoplasmic transfer of chloramphenicol resistance in a human cell line. Somatic Cell Genet 4:737–744, 1978.

Miyata T, Hayashida H, Kikuno R, Hasegawa M, Kobayashi M, Koike K: Molecular clock of silent substitution: At least six-fold preponderance of silent changes in mitochondrial genes over those in nuclear genes. J Mol Evol 19:28–35, 1982.

Miyoshi K: Experimental striatal necrosis induced by sodium azide. A contribution to the problem of selective vulnerability and histochemical studies of enzymatic activity. Acta Neuropathol (Berl) 9:199–216, 1967.

Monnat RJ Jr, Loeb LA: Nucleotide sequence preservation of human mitochondrial DNA. Proc Natl Acad Sci USA 82:2895–2899, 1985.

Monnens L, Gabreels F, Willems J: A metabolic myopathy associated with chronic lactic acidemia, growth failure, and nerve deafness. J Pediatr 86:983, 1975.

Monnerot M, Monolou J-C, Solignac M: Intra-individual length heterogeneity of Rana escarlinta mitochondrial DNA. Biol Cell 52:213–218, 1984.

Montoya J, Christianson T, Levens D, Rabinowitz M, Attardi G: Identification of initiation sites for heavy-strand and light-strand transcription in human mitochondrial DNA. Proc Natl Acad Sci USA 79:7195–7199, 1982.

Montoya J, Gaines GL, Attardi G: The pattern of transcription of the human mitochondrial rRNA genes reveals two overlapping transcription units. Cell 34:151–159, 1983.

Montoya J, Ojala D, Attardi G: Distinctive features of the 5′-terminal sequences of the human mitochondrial mRNAs. Nature 290:465–470, 1981.

Moreadith RW, Batshaw ML, Ohnishi T, Kerr D, Knox B, Jackson D, Hruban R, Olson J, Reynafarje B, Lehninger AL: Deficiency of the iron-sulfur clusters of mitochondrial reduced nicotinamide-adenine dinucleotide-ubiquinone oxidoreductase (complex I) in an infant with congenital lactic acidosis. J Clin Invest 74:685–697, 1984.

Morgan-Hughes JA, Darveniza P, Kahn SN, Landon DN, Sherratt RM, Land JM, Clark JB: A mitochondrial myopathy characterized by a deficiency in reducible cytochrome b. Brain 100:617–640, 1977.

Morgan-Hughes JA, Darveniza P, Landon DN, Clark JB: A mitochondrial myopathy with a deficiency of respiratory chain NADH-CoQ reductase activity. J Neurol Sci 43:27–46, 1979.

Morgan-Hughes JA, Hayes DJ, Clark JB, Landon DN, Swash M, Stark RJ, Rudge P: Mitochondrial encephalomyopathies: Biochemical studies in two cases revealing defects in the respiratory chain. Brain 105:553–582, 1982.

Muller-Hocker J, Paetzke I, Pongratz D, Hubner G: Mitochondrial myopathy with diffuse activation and focal deficiency of mitochondrial ATPase and carnitine deficiency. Virchows Arch B [Cell Pathol] 48:185–196, 1985.

Munro E, Siegal RL, Craig IW, Sly WS: Cytoplasmic transfer of a determinant for chloramphenicol resistance between mammalian cell lines. Proc R Soc London [Biol] 201:73–85, 1978.

Munro E, Webb M, Kearsey SE, Craig IW: The isolation of an antimycin A-resistant human cell line. Exp Cell Res 147:329–339, 1983.

Murphy WI, Attardi B, Tu C, Attardi G: Evidence for complete symmetrical transcription in vivo of mitochondrial DNA in HeLa cells. J Mol Biol 99:809–814, 1975.

Myers RH, Cupples LA, Schoenfeld M, D'Agostino RB, Terrin NC, Goldmakher N, Wolf PA: Maternal factors in onset of Huntington disease. Am J Hum Genet 37:511–523, 1985.

Myers RL, Goldman D, Bird ED, Sax DS, Merril CR, Schoenfeld M, Wolf PA: Maternal transmission in Huntington's disease. Lancet 1:208–210, 1983.

Nalecz MJ, Bolli R, Azzi A: Molecular conversion between monomeric and dimeric states of the mitochondrial cytochrome b-cl complex: Isolation of active monomers. Arch Biochem Biophys 236:619–628, 1985.

Neustein HB, Lurie PR, Dahms B, Takahashi M: An X-linked recessive cardiomyopathy with abnormal mitochondria. Pediatrics 64:24–29,1979.

Nicholls DG, Locke RM: Thermogenic mechanisms in brown fat. Physiol Rev 64:1–64, 1984.

Nierlich DP: Fragmentary 5S rRNA gene in the human mitochondrial genome. Mol Cell Biol 2:207–209, 1982.

Nikoskelainen E: New aspects of the genetic, etiologic, and clinical puzzle of Leber's disease. Neurology 34:1482–4, 1984.

Nikoskelainen E, Hassinen IE, Paljkarvi L, Lang H, Kalimo H: Leber's hereditary optic neuroretinopathy, a mitochondrial disease? Lancet 2:1474, 1984.

Nikoskelainen E, Wanne O, Dahl M: Pre-excitation syndrome and Leber's hereditary optic neuroretinopathy. Lancet 1:696, 1985.

Novotny EJ Jr, Singh G, Wallace DC, Dorfman LJ, Louis A, Sogg RL, Steinman L: Leber's disease and dystonia: A mitochondrial disease. Neurology 36:1053–1060, 1986.

Ogasahara S, Nishikawa Y, Yorifuji S, Soga F, Nakamura Y, Takahashi M, Hashimoto S, Kono N, Tarui S: Treatment of Kearns-Sayre syndrome with coenzyme Q10. Neurology 36:45–53, 1986.

Ohta S, Kagawa Y: Human F1-ATPase: Molecular cloning at cDNA for the beta subunit. J Biochem 99:135–141, 1986.

Ojala D, Crews S, Montoya J, Gelfand R, Attardi G: A small polyadenylated RNA (7SRNA), containing a putative ribosome attachment site, maps near the origin of human mitochondrial DNA replication. J Mol Biol 150:303–314, 1981a.

Ojala D, Montoya J, Attardi G: tRNA punctuation model of RNA processing in human mitochondria. Nature 290:470–474, 1981b.

Okamura K, Santa T, Nagae K, Omae T: Congenital oculoskeletal myopathy with abnormal muscle and liver mitochondria. J Neurol Sci 27:79–91, 1976.

Oliver N, McCarthy J, Wallace DC: Comparison of mitochondrially synthesized polypeptides of human, mouse, and monkey cell lines by a two-dimensional protease gel system. Somatic Cell Mol Genet 10:639–643, 1984.

Oliver NA, Greenberg BD, Wallace DC: Assignment of a polymorphic polypeptide to the human mitochondrial DNA unidentified reading frame 3 gene by a new peptide mapping strategy. J Biol Chem 258:5834–5839, 1983.

Oliver NA, Wallace DC: Assignment of two mitochondrially synthesized polypeptides to human mitochondrial DNA and their use in the study of intracellular mitochondrial interaction. Mol Cell Biol 2:30–41, 1982.

Olivo PD, Van de Walle MJ, Laipis PJ, Hauswirth WW: Nucleotide sequence evidence for rapid genotypic shifts in the bovine mitochondrial DNA D-loop. Nature 306:400–402, 1983.

Parimoo S, Seelan RS, Desai S, Buse G, Padmanaban G: Construction of a cDNA clone for a nuclear-coded subunit of cytochrome c oxidase from rat liver. Biochem Biophys Res Commun 118:902–909, 1984.

Pavlakis SG, Phillips PC, DiMauro S, De Vivo DC, Rowland LP: Mitochondrial myopathy, encephalopathy, lactic acidosis, and strokelike episodes: A distinctive clinical syndrome. Ann Neurol 16:481–488, 1984.

Pedersen PL, Schwerzmann K, Cintron N: Regulation of the synthesis and hydrolysis of ATP in biological systems: Role of peptide inhibitors of H^+-ATPase. In Sanadi DR (ed): "Current Topics in Bioenergetics." New York: Academic Press, 1981, Vol 11, pp 149–197.

Pepe G, Holtrop H, Gadaleta G, Kroon AM, Cantatore P, Gallerani R, de Benedetto C, Quagliariello C, Sbisa E, Saccone C: Nonrandom patterns of nucleotide substitutions and codon strategy in the mammalian mitochondrial genes coding for identified and unidentified reading frames. Biochem Int 6:553–563, 1983.

Pepin B, Mikol J, Golstein B, Aron JJ, Lebuisson DA: Familial mitochondrial myopathy with cataract. J Neurol Sci 45:191–203, 1980.

Perlman S, Penman S: Mitochondrial protein synthesis: Resistance to emetine and response to RNA synthesis inhibitors. Biochem Biophys Res Commun 40:941–948, 1970.

Piko L: Synthesis of macromolecules in early mouse embryos cultured in vitro: RNA, DNA, and a polysaccharide component. Dev Biol 21:257–279, 1970.

Piko L, Matsumoto L: Number of mitochondria and some properties of mitochondrial DNA in the mouse egg. Dev Biol 49:1–10, 1976.

Pollak JK, Sutton R: The differentiation of animal mitochondria during development. TIBS 5:23–27, 1980.

Przybojewski JZ, Hoffman H, De Graaf AS, Van Der Walt JJ, Tiedt FAC, O'Kennedy A, Torrington M, Lochner A, Hewlett R: A study of a family with inherited disease of cardiac and skeletal muscle. Part I. Clinical, electrocardiographic, echocardiographic, haemodynamic, electrophysiological and electron microscopic studies. S Afr Med J 59:363–373, 1981.

Purtilo DT: "A Survey of Human Diseases." Reading, MA: Addison-Wesley, 1978.

Radsak K, Kato K, Sato N, Koprowski H: Effect of ethidium bromide on mitochondrial DNA and cytochrome synthesis in HeLa cells. Exp Cell Res 66:410–416, 1971.

Reilly JG, Thomas CA Jr: Length polymorphisms, restriction site variation, and maternal inheritance of mitochondrial DNA of *Drosophila melanogaster*. Plasmid 3:109–115, 1980.

Rieske JS: Composition, structure, and function of complex III of the respiratory chain. Biochim Biophys Acta 456:195–247, 1976.

Rinaldi AM, De Leo G, Arzone A, Salcher I, Storage A, Mutolo V: Biochemical and electron microscopic evidence that cell nucleus negatively controls mitochondrial genomic activity in early sea urchin development. Proc Natl Acad Sci USA 76:1916–1920, 1979.

Roeder RG, Reeder RH, Brown DD: Multiple forms of RNA polymerase in Xenopus laevis: Their relationship to RNA synthesis in vivo and their fidelity of transcription in vitro. Cold Spring Harbor Symp Quant Biol 35:727–735, 1970.

Rosing HS, Hopkins LC, Wallace DC, Epstein CM, Weidenheim K: Maternally inherited mitochondrial myopathy and myoclonic epilepsy. Ann Neurol 17:228–237, 1985.

Rothkoff L, Biedner B, Shoham K, Blumenthal M: Optic atrophy after irrigation of the lacrimal ducts with chloramphenicol. Ann Ophthalmol 11:105–106, 1979.

Rowland LP: Molecular genetics, pseudogenetics, and clinical neurology. Neurology 33:1179–1195, 1983.

Sager R, Kitchin R: Selective silencing of eukaryotic DNA. Science 189:426–433, 1975.

Sarnat HB: "Muscle Pathology and Histochemistry." Chicago: American Society of Clinical Pathology Press, 1983.

Scarpulla RC, Agne KM, Wu R: Isolation and structure of a rat cytochrome c gene. J Biol Chem 256:6480–6486, 1981.

Scarpulla RC, Agne KM, Wu R: Cytochrome c gene-related sequences in mammalian genomes. Proc Natl Acad Sci USA 79:739–743, 1982.

Schieber GL, O'Brien TW: Site of synthesis of the proteins of mammalian mitochondrial ribosomes: Evidence from cultured bovine cells. J Biol Chem 260:6367–6372, 1985.

Scholte HR, Luyt-Houwen IEM, Busch HFM, Jennekens FGI: Muscle mitochondria from patients with Duchenne muscular dystrophy have a normal beta oxidation, but an impaired oxidative phosphorylation. Neurology 35:1396–1397, 1985.

Schotland DL, DiMauro S, Bonilla E, Scarpa A, Lee C-P: Neuromuscular disorder associated with a defect in mitochondrial energy supply. Arch Neurol 33:475–479, 1976.

Schultheiss H-P, Klingenberg M: Immunochemical characterization of the adenine nucleotide translocator. Organ specificity and confirmation specificity. Eur J Biochem 143:599–605, 1984.

Schultheiss H-P, Klingenberg M: Immunoelectrophoretic characterization of the ADP/ATP carrier from heart, kidney, and liver. Arch Biochem Biophys 239:273–279, 1985.

Sengers RCA, Stadhouders AM, Trijbels JMF: Mitochondrial myopathies: Clinical, morphological and biochemical aspects. Eur J Pediatr 141:192–207, 1984a.

Sengers RCA, Trijbels JMF, Bakkeren JAJM, Ruitenbeek W, Fischer JC, Janssen ATM, Stadhouders AM, terLaak HJ: Deficiency of cytochrome b and aa3 in muscle from a floppy infant with cytochrome oxidase deficiency. Eur J Pediatr 141:178–180, 1984b.

Shapira Y, Cederbaum SD, Cancilla PA, Nielsen D, Lippe BM: Familial poliodystrophy, mitochondrial myopathy, and lactate acidemia. Neurology 25:614–621, 1975.

Sherman J, Angulo M, Boxer R, Gluck RA: Possible mitochondrial inheritance of congenital cardiac septal defects. N Engl J Med 313:186–187, 1985.

Shew J-Y, Breen GAM: Two cytoplasmically inherited oligomycin-resistant Chinese hamster cell lines exhibit an altered mitochondrial translation product. Somatic Cell Mol Genet 11:103–108, 1985.

Shipman C Jr, Smith SH, Drach JC: Selective inhibition of nuclear DNA by 9-β-D-arabinofuranosyl adenine in rat cells transformed by Rous sarcoma virus. Proc Natl Acad Sci USA 69:1753–1757, 1972.

Shmookler Reiss RJ, Goldstein S: Mitochondrial DNA in mortal and immortal human cells: Genome number, integrity, and methylation. J Biol Chem 258:9078–9085, 1983.

Shuey DA, Attardi G: Characterization of an RNA polymerase activity from HeLa cell mitochondria, which initiates transcription at the heavy strand rRNA promoter and the light strand promoter in human mitochondrial DNA. J Biol Chem 260:1952–1958, 1985.

Shy GM, Gonatas NK, Perez M: Two childhood myopathies with abnormal mitochondria. I. Megaconial myopathy. II. Pleoconial myopathy. Brain 89:133–158, 1966.

Slott EF Jr, Shade RO, Lansman RA: Sequence analysis of mitochondrial DNA in a mouse cell line resistant to chloramphenicol and oligomycin. Mol Cell Biol 3:1694–1702, 1983.

Smith ADM: Cyanide encephalopathy in man? Lancet 2:668–670, 1964.

Solignac M, Monnerot M, Mounolou J-C: Mitochondrial DNA heteroplasmy in *Drosophila mauritiana*. Proc Natl Acad Sci USA 80:6942–6946, 1983.

Spiro AJ, Moore CL, Prineas JW, Strasberg PM, Rapin I: A cytochrome-related inherited disorder of the nervous system and muscle. Arch Neurol 23:103–112, 1970.

Spolsky CM, Eisenstadt JM: Chloramphenicol-resistant mutants of human HeLa cells. FEBS Lett 25:319–324, 1973.

Stumpf DA: Mitochondrial multisystem disorders: Clinical, biochemical, and morphologic features. In Tyler HR, et al (eds): "Current Neurology." Boston: Houghton Mifflin Professional Publications Medical Division, Vol 11, 1979, p 117–149.

Tanaka K, Takeshita K, Takita, M: Deoxycholic acid, a candidate for the maternal intrauterine factor in early-onset myotonic dystrophy. Lancet 1:1046–1047, 1981.

Tapper DP, Clayton DA: Mechanism of replication of human mitochondrial DNA: Localization of the 5′ ends of nascent daughter strands. J Biol Chem 256:5109–5115, 1981.

Tassin S, Walter GF, Brucher JM, Rousseau JJ: Histochemical and ultrastructural analysis of the mitochondrial changes in a familial mitochondrial myopathy. Neuropathol Appl Neurobiol 6:337–347, 1980.

Tellez-Nagel I, Johnson AB, Terry RD: Ultrastructural and histochemical study of cerebral biopsies in Huntington's chorea. In Barbeau A, et al (eds): "Advances in Neurology, Vol 1: Huntington's Chorea. 1872–1972." New York: Raven Press, 1973, pp 387–398.

Towers NR, Dixon H, Kellerman GM, Linnane AW: Biogenesis of mitochondria. 22. The sensitivity of rat liver mitochondria to antibiotics; a phylogenetic difference between a mammalian system and yeast. Arch Biochem Biophys 151:361–369, 1972.

Tsairis P, Engel WK, Kark P: Familial myoclonic epilepsy syndrome associated with skeletal muscle mitochondrial abnormalities. Neurology 23:408, 1973.

Tzagoloff A: "Mitochondria." New York: Plenum Press, 1982.

Vaughn KC, DeBonte LR, Wilson KG, Schaffer GW: Organelle alteration as a mechanism for maternal inheritance. Science 208:196–198, 1980.

Vogel SN, Sultan TR: Cyanide poisoning. Clin Toxicol 18 (3):367–383, 1981.
Walberg MW, Clayton DA: In vitro transcription of human mitochondrial DNA. J Biol Chem 258:1268, 1983.
Wallace DC: A new manifestation of Leber's disease and a new explanation for the agency responsible for its unusual pattern of inheritance. Brain 93:121–132, 1970.
Wallace DC: Assignment of the chloramphenicol resistance gene to mitochondrial deoxyribonucleic acid and analysis of its expression in cultured human cells. Mol Cell Biol 1:697–710, 1981.
Wallace DC: Structure and evolution of organelle genomes. Microbiol Rev 46:208–240, 1982a.
Wallace DC: Cytoplasmic inheritance of chloramphenicol resistance in mammalian cells. In Shay JW (ed): "Techniques in Somatic Cell Genetics." New York: Plenum Press, 1982b, pp 159–187.
Wallace DC: Structure and evolution of organelle DNAs. In Schenk HEA, et al (eds): "Endocytobiology, Vol II: Intracellular Space as Oligogenetic Ecosystem." New York: Walter de Gruyter, 1983, pp 87–100.
Wallace DC: Human mitochondrial DNA (Homo sapiens). In O'Brien SJ (ed): "Genetic Maps/1984." Cold Spring Harbor, NY: Cold Spring Harbor Laboratory, Vol 3, 1984.
Wallace DC: Mitotic segregation of mitochondrial DNAs in human cell hybrids and expression of chloramphenicol resistance. Somatic Cell Mol Genet 12:41–49, 1986.
Wallace DC, Bunn CL, Eisenstadt JM: Cytoplasmic transfer of chloramphenicol resistance in human tissue culture cells. J Cell Biol 67:174–188, 1975.
Wallace DC, Eisenstadt JM: Expression of cytoplasmically inherited genes for chloramphenicol resistance in interspecific somatic cell hybrids and cybrids. Somatic Cell Genet 5:373–396, 1979.
Wallace DC, Garrison K, Knowler WC: Dramatic founder effects in Amerindian mitochondrial DNAs. Am J Phys Anthropol 68:149–155, 1985a.
Wallace DC, Singh G, Hopkins LC, Novotny EJ: Maternally inherited diseases of man. In Quagliariello E, et al (eds): "Achievements and Perspectives of Mitochondrial Research, Vol II: Biogenesis." Amsterdam, the Netherlands: Elsevier Science Publishers, 1985b, pp 427–436.
Wallace DC, Surti U, Adams CW, Szulman AE: Complete moles have paternal chromosomes but maternal mitochondrial DNA. Hum Genet 61:145–147, 1982.
Wallace DC, Yang J, Ye J, Lott MT, Oliver NA, McCarthy J: Computer prediction of peptide maps: Assignment of polypeptides to human and mouse mitochondrial DNA genes by analysis of two dimensional-proteolytic digest gels. Am J Hum Genet 38:320–350, 1986.
Walter GF, Tassin S, Brucher JM: Familial mitochondrial myopathies. Acta Neuropathol (Suppl) VII:283–286, 1981.
Webster KA, Oliver NA, Wallace DC: Assignment of an oligomycin-resistance locus to human chromosome 10. Somatic Cell Genet 8:223–244, 1982.
Went LN: A sex-linked heredo-degenerative neurological disorder associated with Leber's optic atrophy: Genetical aspects. Acta Genet 14:220–239, 1964a.
Went LN: A sex-linked heredo-degenerative neurological disorder associated with Leber's optic atrophy. Part 2. Laboratory investigations. J Neurol Sci 1:81–87, 1964b.
Went LN: Leber disease and variants. In Vinken PJ, et al (eds): "Handbook of Clinical Neurology." Amsterdam: North-Holland Publishing Co, 1972, Vol 13, pp 94–110.
White FA, Bunn CL: Segregation of mitochondrial DNA in human somatic cell hybrids. Mol Gen Genet 197:453–460, 1984.

Willems JL, Monnens LAH, Trijbels JMF, Veerkamp JH, Meyer AEFH, van Dam K, van Haelst U: Leigh's encephalomyelopathy in a patient with cytochrome *c* oxidase deficiency in muscle tissue. Pediatrics 60:850–857, 1977.

Wilson G, Hodges R, Hare JF: Site of synthesis of the mitochondrial cytochrome in hepatocyte cultures. J Biol Chem 256:5197–5203, 1981.

Wilson J: Leber's hereditary optic atrophy: Some clinical and aetiological considerations. Brain 86:347–362, 1963.

Wilson J: Leber's hereditary optic atrophy: A possible defect of cyanide metabolism. Clin Sci 29:505–515, 1965a.

Wilson J: Skeletal manifestations in Leber's hereditary optic atrophy: A possible disorder of cyanide metaboism. Ann Physiol Med 8:91–95, 1965b.

Wilson J: Plasma-cobalamins in neuro-ophthalmological diseases. Lancet 1:259–261, 1971.

Wiseman A, Attardi G: Cytoplasmically inherited mutations of a human cell-line resulting in difficult mitochondrial protein synthesis. Somatic Cell Genet 5:241–262, 1978.

Worsfold M, Park DC, Pennington RJ: Familial "mitochondrial" myopathy: A myopathy associated with disordered oxidative metabolism in muscle fibres. Part 2. Biochemical findings. J Neurol Sci 19:261–274, 1973.

Wu G-J, Dawid IB: Purification and properties of mitochondrial deoxyribonucleic acid dependent ribonucleic acid polymerase from ovaries of Xenopus laevis. Biochemistry 11:3589–3595, 1972.

Yamamoto T, Beppu H, Tsubaki T: Mitochondrial encephalomyopathy: Fluctuating symptoms and CT. Neurology 34:1456–1460, 1984.

Yatscoff RW, Goldstein S, Freeman KB: Conservation of genes coding for proteins synthesized in human mitochondria. Somatic Cell Genet 4:633–645, 1978.

Yoza BK, Bogenhagen DF: Identification and in vitro capping of a primary transcript of human mitochondrial DNA. J Biol Chem 259:3909–3915, 1984.

Zeviani M, Nonaka I, Bonilla E, Okino E, Moggio M, Jones S, DiMauro S: Fatal infantile mitochondrial myopathy and renal dysfunction caused by cytochrome c oxidase deficiency: Immunological studies in a new patient. Ann Neurol 17:414–417, 1985.

Zuckerman SH, Solus JF, Gillespie FP, Eisenstadt JM: Retention of both parental mitochondrial DNA species in mouse-Chinese hamster somatic cell hybrids. Somatic Cell Mol Genet 10:85–92, 1984.

Zylber E, Vesco C, Penman S: Selective inhibition of the synthesis of mitochondria-associated RNA by ethidium bromide. J Mol Biol 44:195–204, 1969.

The Sources of Molecular Evolution: An Outsider's View

Samuel H. Boyer, MD

Department of Medicine, The Johns Hopkins University School of Medicine, and the Howard Hughes Medical Institute Laboratory for Biochemical Genetics, Baltimore, MD 21205

Evolution is a process whereby the genetic composition of one population comes to be different from that of another population. Although this definition seems straightforward, it was apparent at a recent workshop on evolutionary theory and molecular biology that it was not one that could be accepted by all. In particular, the fact that the definition does not include a cause, eg, natural selection, was quite distressing to some. In itself, this points up one of the principal issues that will mark any discussion in which we are asked to explain how any one distinctive characteristic evolved, be it a pattern of behavior, a morphologic feature, an amino acid sequence, or a run of nucleotides. Time and again, the issue will be: Did it evolve by Darwinian natural selection or was it, wholly or in part, the work of some other mechanism?

In what follows, the intent is to consider this issue as it bears on discoveries emanating from the capacity to examine the products of genes, and now genes themselves, at a molecular level. The focus will be first on what I believe are some of the principles or lessons to be learned from these endeavors and second on what I see as the central questions emerging in the course of such endeavors. By way of a warning, I have called this minimalist sketch an outsider's view, not just as a disclaimer but to emphasize that *any* view of evolution will necessarily be an outsider's, since a valid evolutionary synthesis must encompass disciplines as diverse as mathematical population

Birth Defects: Original Article Series, Volume 23, Number 3, pages 191–207

genetics and paleobiology and all the other biologies: behavioral, developmental, cell, and molecular. Because no one is even remotely able in all, any synthesis of current understanding will be in some way flawed.

WHEN THE MOLECULAR STUDY OF EVOLUTION BEGAN

Until about 25 years ago, the genetic characterization of populations of most species was more or less limited to the comparatively few features arising from the inspection of gross morphology, immunologic typing, and the recognition of an occasional enzyme deficiency. Only in taxa with polytenic salivary gland chromosomes, eg, Drosophila, was it possible to go much further. In the face of this comparative poverty of method, perhaps some comfort could be taken from Haldane's calculation [1957] that the cost of natural selection was far too great to support either very much variation within a population or any very rapid emergence of differences between populations. Haldane established that when selection operated on genes one at a time, the replacement of one allele by another required the loss of roughly 30 world populations of the group under selection. He reasoned that this staggering cost, which was more or less independent of the intensity of selection, could be supported only if it were spread out in time. He thus reckoned that at least 300 generations would be needed to accomplish one substitution. To illustrate, African populations paid an enormous price, in Haldane's terms, for the well-known selective advantage [Allison, 1954] conferred by the rise in frequency of the sickle cell mutation, even though this particular mutation stopped far short of complete fixation. Populations thus pressed could not coincidently afford to pay such a price for very many other entities. For this reason, it seemed to many in the 1950s that it did not much matter if the methods available for the detection of commonly occurring variations were comparatively limited: There was probably little to find.

Later, in the early 1960s, this expectation would be utterly contradicted when two sorts of technical developments began to be applied on an increasingly grand scale to the analysis of proteins. One involved the conjunction of Smithies' [1955] invention of starch gel electrophoresis with Hunter's and Markert's [1957] application of histochemical methods to gel surfaces for the detection of specific enzymes. This made it readily possible to sort proteins, especially enzymes undetectable by conventional staining, according to size and electrostatic charge. Within a few years, substantial numbers of individuals were being surveyed for variation in dozens of different proteins. More or less coincidentally, the methods introduced by Sanger [1952] for primary protein structure analysis were simplified and improved and came into widespread use. The result was that interspecies comparisons of amino acid sequences started to become available for easily purified proteins such as the hemoglobins [Ingram, 1961] and mitochondrial cytochrome c [Margoliash, 1972].

THE UNEXPECTED ABUNDANCE OF PROTEIN POLYMORPHISMS

Among the findings that emerged from the new-found methods was an unanticipated abundance of variation within populations. The rule became: scratch a protein, find a variant. Often enough, the frequency of such variants was greater than 0.01 and thus greater than could be explained by recurrent mutation alone, indicating that a genetic polymorphism was present. For example, in 1961–1962, my colleagues and I examined four human proteins for reasons unrelated to the search for common variants. We found that two of these proteins, placental alkaline phosphatase [Boyer, 1961] and blood cell glucose-6-phosphate dehydrogenase [Boyer et al, 1962], were each the site of genetic polymorphisms in which, within certain populations, least common allele frequencies were always more than 0.1. The extent of polymorphic variation was codified for man by Harris and Hopkinson [1972] and for *Drosophila* by Prakash et al [1969].

In man, 28% of 71 loci examined proved polymorphic, leading to an average heterozygosity per locus of 7%. Similar values were found in feral mice [Selander et al, 1969] and even for species as remote as horseshoe crabs [Selander et al, 1970]. In *Drosophila pseudoobscura*, 42% of loci were polymorphic, and average heterozygosity was 12%. The extent of detected polymorphism, when set against Haldane's arguments, may be too great to be explicable simply in terms of natural selection operating on genes one by one in large populations. Although some demurred [Prakash et al, 1969; Sved et al, 1967; King, 1967; Milkman, 1967], others [Kimura and Ohta, 1971] concluded that some other process had to be at work.

MOLECULAR EVIDENCE FOR DARWINIAN CONSTRAINT OF PROTEIN STRUCTURE

This unexpected richness of variation within species was, at first, antithetic to the notion developing from the interspecies comparisons of amino acid sequences, antithetic in the sense that the highly conservative character of Darwinism was at last revealed at sequence level. For example, the finding that amino acid sequences of some proteins changed far more than others during evolution was, almost from the start, attributed to selective constraints on the locale and kinds of mutations that were allowed to accumulate. At one extreme, the observation that the 102 residues found in histone IV from bovine thymus were exactly the same as found in pea seedlings [Dayhoff, 1972] was interpreted to mean that the function of this protein can be conveyed only by a unique amino acid sequence. At the other limit, the wide-ranging interspecies variability of fibrinopeptides [Dayhoff, 1972] liberated from fibrinogen during its conversion to fibrin was taken as evidence that, in a few proteins, almost any residue might be serviceable. The correctness of this view that essential function is indeed the source of evolutionary constraint was verified most notably by Perutz and his associates [1965], who analyzed a diversity of globin amino acid sequences in the context of the higher-order structures evident from crystallography. They found myriad instances in which residues essential to hemoglobin function were universally conserved. For example, the two histidines that form bonds with heme iron were invariant in all globin structures in all species. Within the interior of the hemoglobin molecule, where exclusion of water is needed for molecular integrity, amino acid interchanges between species were limited to hydrophobic residues. The nagging possibility that mutational "cold spots" might abet such limitations was later put to rest by the finding that, within any one species, such as man, mutations of hemoglobin structure are more or less randomly distributed along the length of polypeptide chains [Boyer et al, 1978]. However, when these human mutants occur at sites free of evolutionary change, they are almost invariably harmful even in heterozygous states. Thus mutations occur and recur throughout a structural gene, but most of them are harmful in most proteins and few can accumulate.

MOST AMINO ACID SUBSTITUTIONS ARE ADAPTIVELY NEUTRAL VARIANTS

The conclusion that restrictive Darwinism is a harsh arbiter that overall allows only a few percent of all mutations affecting amino acid residues to reach fixation [Boyer et al, 1978] says nothing, of course, about how fixation of the few is attained. Natural selection itself is an obvious candidate mechanism. Indeed, the protein chemists undertaking between-species comparisons were mostly good Darwinians who were hoping thereby to identify substitutions contributing to functional differences between species. Thus they, along with many geneticists, felt some parts betrayed and some parts outraged [Smith, 1970] by the arguments of Kimura [1968, 1969], King and Jukes [1969], and Boyer et al [1969] to the effect that those mutations in protein sequence that managed to attain fixation in one or another species did so not because they had been selected for but because they were adaptively neutral and had come to fixation solely by chance. Kimura's argument was based on the calculation of rates of change. By reference to the fossil record, he took as denominators the length of time that pairs of species had been divergent and then estimated the rapidity with which amino acid differences in globin structure had accumulated between the pairs of species. He found that such globin-based rates were more or less constant along various lines of descent and that, on the average, one mutation affecting amino acid sequence became fixed every 3×10^7 years. By applying this globin-derived rate to whole haploid genomes containing 4×10^9 nucleotide pairs and by assuming that all these pairs encoded proteins whose genes averaged 300 nucleotides in length, he calculated that in the aggregate one amino acid change would attain fixation *every 2 years*, ie, $3 \times 10^7/(4 \times 10^9/300) = 2$. Kimura recognized that this 2-year interval was far too brief to be compatible with Haldane's estimate that at least 300 generations were needed to accomplish the selective substitution of one allele by another. Accordingly, he argued that most amino acid changes coming to fixation were adaptively neutral.

The controversy that attended the publication of these arguments lingers today, and although many now accept Kimura's conclusion that neutralism is an important source of evolutionary change, some of his initial arguments can be faulted. This is not surprising in that the arguments themselves rested on an assortment of what were, at best, approximations. For example, it is now evident from Ohta and Kimura's [1971] own estimates of genetic load that there are unlikely to be $4 \times 10^9/300 = 10^7$ genes coding for protein structure but, in truth, perhaps no more than 30,000 genes of this sort. We now know that this same approximation can be deduced in other ways [McKusick and Ruddle, 1977]. Using this lesser number, the aggregated frequency with which mutations attain fixation falls within the scope of Haldane's estimate. Nonetheless, the fact is that neutralism makes considerable sense in terms of other findings. For example, the evolutionary rates of amino acid substitution in proteins other than hemoglobin, though unique to each kind of molecule, are themselves reasonably constant along several lines of descent [Kimura, 1983]. Although these rates may have slowed in the course of primate radiation, it is notable that *all* the rates have slowed more or less proportionately [Fitch and Langley, 1976]. The point is that constancy of evolutionary rate for any one protein is not what is expected of Darwinism. On the contrary, natural selection in the course of favoring different characters in different species should create inconstancy for the rate with which any one protein evolves. To put it in arithmetic terms, the probability that any mutation will reach fixation is proportional to the product ($N_e\mu s$) of effective population size (N_e), mutation rate (μ), and the selection coefficient (s) imposed by the mutation. Where evolutionary rates are constant, it follows that the $N_e\mu s$ product has held constant in diverse populations. This is difficult to imagine except in cases of neutrality, when it is only the N_e and the μ terms that matter. Thus, just as Kimura [1968] recognized at the outset, the occurrence of even a rough constancy is a mark in support of neutralism.

Another line of evidence in favor of neutralism is that the same *kinds* of interspecies amino acid sequence differences are seen to recur in different phyletic branches. This is illustrated in Figure 1 for β-globin interchanges encountered in a dense net of 17 primate species [Boyer et al, 1972]. As can be seen, the majority of replacements of the reconstructed archetype sequences are, in fact, repetitive and often bidirectional. For example, at the fifth position of β-globin polypeptides, glycine replaces archetype alanine at an early time in primate evolution, whereas, at a later time, alanine is restored in one line of descent. Since most of the sites involved in such substitutions are not known to be the locale of functionally important residues, and since none of the illustrated substitutions appears in man in the form of clinically aberrant heterozygotes [Boyer et al, 1978], it is reasonable to suppose that the replacements of like with like seen in Figure 1 have little in the way of adaptive consequence. In any case, it would be difficult to attribute to this dithering between similar alternatives any of the stuff of whatever it is that makes species emerge and behave and appear different from one another. The sum of these and related arguments almost forces on

RECURRENT Hbβ INTERCHANGES DURING
PRIMATE EVOLUTION (17 Species)

5	6	9	13	21	43	50	76	104
Ala→Gly		Ala→Ser		Asp→Glu		Ser→Thr		Arg→Lys
Gly→Ala		Ser→Ala		Glu→Asp		Ser→Thr		Arg→Lys
		Ser→Ala				Ser→Thr		
						Thr→Ser		
	Glu→Asp		Ala→Thr		Glu→Asp		Ala→Thr	
	Glu→Asp		Ala→Thr		Glu→Asp		Ala→Thr	
					Asp→Glu		Thr→Ala	

23 recurrent interchanges / 38 total observed mutations

Fig. 1. Recurrent interchanges that have attained evolutionary fixation at nine amino acid sequence positions in β-globins of primates. As detailed elsewhere [Boyer et al, 1972], interchanges are deduced from 1) amino acid sequence comparisons, 2) a reconstructed β-globin archetype sequence as it existed in a common progenitor of the species, and 3) the presumed pattern of descent.

us the notion that many, perhaps most, evolutionary changes in protein structure are adaptively neutral or, if not exactly so, nearly so.

The view that slightly deleterious mutants are what actually accumulate by chance is the one now espoused by Kimura [1983]. He takes this position so as to resolve the paradox that, despite equivalencies of mutation rate on a generational basis among men, mice, and flies [Kondo, 1977], taxa with short generation times do not evolve much more rapidly than ones with long generation times. To put it another way, the rate at which mutants accumulate during evolution is expected to be proportional to the rate of mutation *per generation*. This is inconsistent with the finding that, for any protein, rates of evolution *per year* are roughly constant in various phyletic branches. Both the paradox and the inconsistency can be resolved 1) when it is assumed that generation time is inversely proportional to population size and 2) when it is taken as established [Kimura, 1983] that very mildly deleterious mutants behave as if selectively neutral in small populations but are strongly selected against in large populations. When such assumptions are granted, the disturbing effect imposed by different generation times is eliminated, expected evolutionary substitution rates become the same on a *per year* basis in all taxa, and neutralism becomes quasineutralism.

The great importance of neutralism or quasineutralism lies with what it does not tell us. If we believe that the essentials of each protein structure were by and large realized in the early millenia of life and that most of the substitutions arising thereafter are often little more than noise, we must, as good Darwinians, look elsewhere for the origin of speciation and the source of the functionally important differences between species. This admittedly negative result seems to me one of the most important to arise from the analysis of protein structure. The same sort of conclusion has been reached by many others [Britten and Davidson, 1971; Valentine and Campbell, 1974; Markert et al, 1975; King and Wilson, 1975]. For example, King and Wilson [1975], after comparing a large set of proteins in man and great apes, conclude that the seat of functional differences between these species must lie elsewhere than in the substance of protein sequences.

UNRESOLVED PROBLEMS OF THE MOLECULAR EVOLUTION OF PROTEINS

One of the difficulties with reasoning of the sort just outlined is that it is easy to be misled into thinking that evolution of any one kind of characteristic proceeds by only one mechanism and not by any other. In fact, the current consensus is that evolution is a process with many origins. Natural selection, stochastic fluctuations of allele frequencies as a function of population size and composition, migration between groups, generation times, proximity to other genes, and perturbations of recombination and replication can all play a part. What is at issue for each character that has evolved is the relative weight to be assigned to each mechanism. However, what often emerges are arguments that are biased by the kinds of information at hand. This is especially evident in the case of comparisons of between-species differences in the amino acid sequence of particular proteins. Here, it is far easier to use the aggregate of information to support the neutralist idea of origin via chance than it is to point to any one amino acid substitution that has undergone selective fixation. In contrast, for any one within-species genetic polymorphism of protein structure, matters are reversed. In this case, it is sometimes possible to document the role of selection in detail. For example, augmentation of resistance to falciparum malaria attributed to sickle cell hemoglobin has been extended down to a level at which the loss of potassium peculiar to erythrocytes from sickle cell heterozygotes is seen to favor the death of potassium-dependent intraerythrocytic parasites [Freedman et al, 1979].

In principle, equivalent kinds of delineation should be applicable to other genetic polymorphisms of protein structure. Indeed, the sort of case by case approach begun by Powers [Powers and Place, 1978; Place and Powers, 1984] toward deciding in minnows (*Fundulus*) whether any one particular combination of alleles at loci for various enzymes provides demonstrable functional advantage may, in time, provide an answer to the larger question: What is the minimal extent to which protein polymorphisms, taken all together, are maintained by Darwinism? We can, of course, anticipate the direction of the answer, since the combination of Haldane's argument and the sheer abundance of protein polymorphism make it likely that a substantial fraction arises from non-Darwinian mechanisms. However, it is the actual estimate that matters, because from it can be calculated the cost of maintaining polymorphisms. Given this fractional cost and given refinements of estimates for total tolerable cost, the maximal fraction of selection available for the generation of between-species differences of any sort can be assigned. The answer, when it comes, will have potentially far-reaching consequences. For instance, if the fraction of tolerable selection assignable to between-species differences should prove to be small, the inference will be that insofar as such differences have arisen by natural selection they will depend either on a changing interplay between alleles at various polymorphic loci or on a small number of substitutions elsewhere in the genome. The former view of a changing interplay among polymorphic loci with pleiotropic effects is the sort of situation advocated by Sewall Wright [1977, 1980] in his "shifting-balance" theory of evolution and speciation. The other notion—that the functionally important distinctions between species may, in fact, depend on only a small number of changes—is one increasingly heard from developmental biologists.

MOLECULAR EVOLUTION OF DNA SEQUENCES: REAFFIRMATIONS AND ADDITIONS TO THE PRINCIPLES DEDUCED FROM PROTEIN STRUCTURE

The methods developed for the isolation of genes via recombinant DNA [Maniatis et al, 1978], for the characterization of specific gene regions by the use of endonucleases and specific probes [Southern, 1975; Smith and Bernstiel, 1976] and for the delineation of nucleotide sequence [Maxam and Gilbert, 1977; Sanger et al, 1977] have thus far had two major consequences for studies of molecular evolution. First, these approaches have led to reaffirmation and refinement of the evolutionary principles arising from the study of proteins; second, and of much greater importance, they have revealed a whole new set of evolutionary phenomena.

What had been seen by some as an overabundance of genetic polymorphisms when proteins were examined has become an undoubted superabundance at the level of nucleotide sequence. For example, 17 different DNA restriction fragment length polymorphisms have been identified in the β-globin gene cluster of man [Kazazian et al, 1983; Antonarakis et al, 1984]. Consequently, what had been an average heterozygosity for variant hemoglobins of scarcely 1% in most populations, now, on the strength of differential sensitivity to endonucleases, approaches 100% in all populations. The fact that the sites of most DNA polymorphisms in the β-globin cluster lie outside regions of assigned function suggests that they have had entirely non-Darwinian origins. The further finding that alleles at the various DNA sites are, in most populations, in genetic equilibrium with alleles at neighboring sites supports this idea. Thus, aside from the clear exception of the sickle cell gene, whose nearby presence distorts the frequency of endonuclease-site alleles [Antonarakis et al, 1984; Kan and Dozy, 1978], there is no evidence that Darwinism has played a role in generating these particular polymorphisms. Here neutralism provides the only plausible explanation.

In contrast, between-species comparisons of DNA sequences in specific gene regions have not made it any easier to recognize what is being seen by selection when it generates the important functional differences between species. This is not surprising in that the proximity to function, and thus to what Darwinism usually sees, is even less close at the gene level than it is at the protein level. What nucleotide sequence comparisons between species have done, however, is make it much easier to appreciate the evolutionary conservatism that marks much of the structure of most proteins. The reason is that DNA sequence comparisons provide "controls" for amino acid sequence comparisons. These "controls" take the form of interspecies comparisons for synonymous third nucleotide residues of expressed codons and for noncoding nucleotides found within intervening sequences and gene flanking regions, all of which usually accumulate far more differences between species than are evident from protein structure [Efstratiadis et al, 1980]. Indeed, some of these noncoding regions accumulate between-species differences with linearities in accord with the operation of an evolutionary clock [Wilson et al, 1977] and at a rate approximating the 10^{-8} substitutions per nucleotide per year expected [Boyer et al, 1978; Perler et al, 1980] for adaptively silent mutation: neutralism redux!

CONCERTED EVOLUTION

This sort of free-wheeling accumulation of mutations does not, however, lack counterbalance. Many genes coding for protein structure exist in the form of two or more juxtaposed copies between which there are recurrent exchanges. The result is that the number of mutations accumulated *per locus* becomes less than it would otherwise be. In the case of the two α-globin loci found most in mammals, exchanges recur so often that the products of the two are usually kept exactly alike within species [Boyer et al, 1974]. Their evolution is thus "concerted" [Zimmer et al, 1980]. It was supposed that this preserved similarity, which is shared to a degree by the entire α gene region and not just by coding sequences [Zimmer et al, 1980], arose by recurrent episodes of nonhomologous recombination between the two tandem loci, thereby leading to periodic gain or loss of α-globin genes on one or another chromosome. Related examples of this process are seen in human mutants, eg, the α-globin gene loss associated with several forms of α-thalassemia [Weatherall and Clegg, 1981] and the α-gene gain seen occasionally in man [Goosens et al, 1980] and often in great apes [Zimmer et al, 1980; Boyer et al, 1971]. Such unequal but reciprocal exchange is not, however, the only process responsible for concerted evolution. Shen et al [1981], in a study of nucleotide sequences within human γ-globin gene pairs, found evidence of a *nonreciprocal* process whereby one of the γ-globin copies was made partially like the other γ copy on the same chromosome. Similar events appear in each of the great apes and seem to attain evolutionary fixation approximately once every several million years [Scott et al, 1984]. The importance of these nonreciprocal "gene corrections" is that they provide, for the first time in higher taxa, a seeming example of the well-known process [Petes and Fink, 1982] of gene conversion found in yeast.

GENOMES IN FLUX: A HALLMARK OF MOLECULAR EVOLUTION

A salient feature of molecular evolution is that the number of copies of active and inactive genes identifiable with any one sort of nucleotide sequence expand and contract in the course of time. For example, the non-α-globin gene cluster in man contains two inactive pseudogenes and five active loci, each producing distinctive polypeptides [Efstratiadis et al, 1980]. In contrast, although the cluster in Old World Monkeys is remarkably similar in its overall organization [Martin et al, 1980] the δ-globin gene has been silenced. In New World monkeys, the δ gene is active, but total copies in the cluster are fewer [Barrie et al, 1981]. In contrast, copy number has expanded in goats such that, pseudogenes included, there are three tightly linked non-α-globin gene clusters with four loci each [Townes et al, 1984]. Similar features are seen when within-species and between-species comparisons are made for genes coding for immunoglobulins [Leder, 1982], histone proteins [Kedes, 1979; Coen et al, 1982a], and ribosomal RNA [Brown et al, 1972]. Clearly, with respect to numbers of genes with identifiable products, copy number and specific kinds of copies within clusters are in a state of extensive flux.

What was quite unexpected was that the evolution of sequences within many reiterated families would be concerted. In other words, the same phenomena revealed by the study of tandem pairs of α-globin genes and tandem pairs of γ-globin genes were evident among noncoding sequences. The classic illustration appears among the tandem sets of genes for ribosomal RNA, within which coding gene regions are separated by noncoding spacer regions [Brown et al, 1972]. Not unexpectedly, coding sequences for RNA are conserved in several species of frogs as they are throughout much of evolution. The surprising finding is that all copies of noncoding spacers, although distinctly different between species, are more or less alike in all the copies within each species. Equivalent phenomena appear among reiterated families of ribosomal DNA, histone DNA, and two noncoding DNAs taken from seven sib species of *Drosophila melanogaster* [Coen et al, 1982a,b]. Although one can imagine why within-species homogenization might involve sequences like globin genes or rDNA encoding for essential function, it is puzzling why it should appear among noncoding sequences. In principle, each copy of a reiterated family should be steadily diverging from the others as each accumulates mutations. Instead, all copies are being kept mostly alike within each species. The question is, how does this occur? What mechanisms lead to such unanticipated homogenization? Once again, we are face to face with the issue of Darwinian vs non-Darwinian origins.

ALTERNATIVE EXPLANATIONS FOR THE CONCERTED EVOLUTION OF REITERATED SEQUENCES

So as to account for sequence homogenization, Dover and Flavell [Dover, 1982; Dover and Flavell, 1982] advocate an ingenious mechanism they term "molecular drive," which they visualize as the result of three kinds of successive processes. The *first* consists of biased conversion or some other process of equivalent effect whereby the sequence of one member of a reiterated ensemble tends to be preferentially preserved and alternative sequences are eliminated. Of course, if this process is to lead to intrachromosomal homogenization of the sequence in question, it must occur at a rate more rapid than that with which point substitutions and small changes in nucleotide number accumulate during evolution. Although reciprocal but unequal exchanges may also contribute to homogenization [Tartof, 1974; Smith, 1974; Anderson and Roth, 1979], such exchanges develop cumbersome features when the sequences at issue are embedded, as they often are, in other sequences. *Second*, because members of a reiterated family are always found on homologous chromosomes and often dispersed among nonhomologous ones, there must also exist processes for exchanging and translocating copies between chromosomes. If such interchromosomal exchange is to lead to homogenization throughout the genome, the rates at which it occurs must exceed the rate at which neighboring copies are made similar. *Third*, the events affecting chromosomes in one individual must finally be distributed throughout a population of individuals at a faster rate than that for interchromosomal exchange. Thus molecular drive is envisioned as a set of successive processes in which the relative rates of substitution, biased conversion, translocation, and population mixing form an ascending series. If this occurs, there will then be "at any given time, in each individual, the same average ratio of new-to-old variants for a particular family" [Dover, 1982]. In short, there will be no distinctions between individuals for Darwinism to see. In the event that some function is associated with a reiterated set, the influence of any one copy should be minimal, and slight differences in either sequence or copy number should scarcely be appreciated by Darwinism.

Intrinsic to the idea of molecular drive is that one particular sequence within a family of reiterated DNA is especially "selfish" and is copied to the exclusion of others. If this sequence is also functionless, it can only become the site of interspecies differences if it is not unalterably selfish. Between-species differences might then arise either from the contributions of atypical individuals who entered the arena of speciation or from chance perturbations in preferential copying after species diverged. But what are we to say about reiterated sequences that are preserved across species barriers? Are they excessively selfish, are they unusually resistant to mutation, did they by chance escape perturbations during or after speciation, or does the very fact of their conservation imply an unrecognized function and the imposition of Darwinian constraint? Although we cannot choose among these alternatives, we can point to reiterated families whose structure is the same between species as it is within species. The *first* example concerns noncoding sequences within a multicopy fragment released from human male DNA, but not female DNA, after it is digested with *Hae*III endonuclease [Cooke, 1976]. These 3.4 kb fragment molecules contain an interspersion of two kinds of sequences: 1) those that are unique to the human Y chromosome, ie, have arisen since the ancestors of man and great apes diverged, and 2) those that have autosomal homologs in both men and apes [Kunkel et al, 1979]. A subset of these sequences with autosomal homologs exhibits thermally stable dissociation profiles ($T_m = 86°$) when hybridized either with human DNA or with DNA from any of the three great apes [Kunkel and Smith, 1982]. The inference is that such sequences have been preserved without appreciable change for at least 10 million years along several lines of descent. A *second* example of sequence conservation across taxonomic lines appears in a 5 kb subfamily of reiterated sequences found in the sea urchin *Strongylocentrotus purpuratus* [Scheller et al, 1981]. Copies of this sequence in another species of the genus exhibit only 1% divergence in the face of a 20% average interspecies difference for single-copy DNA. In yet another genus of sea urchins *(Tripneustes gratilla)*, 5 kb subfamily divergence is 5% despite a span of 320 million years of evolutionary separation from *S. purpuratus*. Thus there are in several taxa real exceptions to the usual rule that noncoding sequences diverge more rapidly than coding ones.

The important point is that we still do not know how homogenization of noncoding sequences arises, nor do we know how or when these sequences become divergent between species. It would, however, be astonishing if these events were not often the outcome of an interaction between *all* the various sources of evolution listed earlier, a list to which the sum of processes called molecular drive can now be added. It would be equally astonishing if these events did not sometimes have a single major cause. To illustrate, we can expect that some of these "noncoding" sequences contribute to a vital function and consequently are either conserved or selectively altered by Darwinism. We can also expect that some evolve more or less as Dover and Flavell [1982] imagine. Finally, we can expect that others will quickly arise by meiotic processes akin to the amplification of dihydrofolate reductase genes [Alt et al, 1978] and oncogenes [Schwab et al, 1983] in somatic cell cultures. At length, when enough examples are at hand, it may be possible to assign overall weights to Darwinian and the various non-Darwinian mechanisms.

From what has *not* been set out here, it may be apparent that the molecular origins of the important morphologic, behavioral, developmental, and functional differences between species still elude us. Only false leads have been uncovered. However, one learns from this to look elsewhere [Britten and Davidson, 1969, 1971] and, most especially, to look for an interaction between many kinds of evolutionary mechanisms. An insider [Haldane, 1959] wrote, "The history of science makes it almost certain that facts will be discovered which show that the theory of natural selection is not fully adequate to account for evolution. But the same history makes it extremely improbable that these facts will be related to the criticisms at present made of it . . . . Darwinism will, I do not doubt, be modified. Like any other successful theory it will ultimately develop its own internal contradictions."

REFERENCES

Allison AC: Protection afforded by sickle cell trait against subtertian malarial infection. Br Med J 1:290, 1954.

Alt FW, Kellems RE, Bertino JR, Schimke RT: Selective multiplication of dihydrofolate reductase genes in methotrexate-resistant variants of cultured murine cells. J Biol Chem 253:1357, 1978.

Anderson RP, Roth JR: Gene duplications in bacteria: Alteration of gene dosage by sister chromosome exchanges. Cold Spring Harbor Symp Quant Biol 43:1083, 1979.

Antonarakis SE, Boehm CD, Serjeant GR, Theisen CE, Dover GJ, Kazazian HH Jr: Origin of the β^S-globin gene in blacks. The contribution of recurrent mutation or gene conversion or both. Proc Natl Acad Sci USA 81:853, 1984.

Barrie PA, Jeffrey AJ, Scott AF. Evolution of the β-globin gene cluster in man and the primates. J Mol Biol 149:319, 1981.

Boyer SH: Alkaline phosphatase in human sera and placentae. Science 134:1002, 1961.

Boyer SH, Crosby EF, Thurman TF et al: Hemoglobins A and A_2 in the New World primates: Comparative variation and its evolutionary implications. Science 166:1428, 1969.

Boyer SH, Noyes AN, Timmons CF, Young RA: Primate hemoglobins: Polymorphisms and evolutionary patterns. J Hum Evol 1:515, 1972.

Boyer SH, Noyes AN, Vrablik GR et al: Silent hemoglobin alpha genes in apes: A potential source of thalassemia. Science 171:182, 1971.

Boyer SH, Porter IH, Weilbacher RG: Electrophoretic heterogeneity of G6PD and its relationship to enzyme deficiency in man. Proc Natl Acad Sci USA 48:1868, 1962.

Boyer SH, Scott AF, Kunkel LM, Smith KD: The proportion of all point mutations which are unacceptable: An estimate based on hemoglobin amino acid and nucleotide sequences. Can J Genet Cytol 20:111, 1978.

Boyer SH, Smith KD, Noyes AN: Immunological purification and characterization of hemoglobin chain-synthesizing polysomes: A derivative hypothesis for regulatory and evolutionary interrelationships in hemoglobin biosynthesis. Ann NY Acad Sci 241:204, 1974.

Britten RJ, Davidson EH: Gene regulation for higher cells: A theory. Science 165:349, 1969.

Britten RJ, Davidson EH: Repetitive and non-repetitive DNA sequences and a speculation on the origins of evolutionary novelty. Quart Rev Biol 46:111, 1971.

Brown DD, Wensink PC, Jordan E: A comparison of the ribosomal DNAs of *Xenopus laevis* and *Xenopus mulleri*: The evolution of tandem genes. J Mol Biol 63:57, 1972.

Coen ES, Strachan T, Dover GA: Dynamics of concerted evolution of ribosomal DNA and histone gene families in the *melanogaster* species subgroup of *Drosophila*. J Mol Biol 158:17, 1982a.

Coen ES, Thoday JM, Dover GA: Rate of turnover of structural variants in the DNA gene family of *Drosophila melanogaster*. Nature 295:564, 1982b.

Cooke H: Repeated sequence specific to human males. Nature 262:182, 1976.

Davidson EH, Britten RJ: Regulation of gene expression: Possible role of repetitive sequences. Science 204:1052, 1979.

Dayhoff MO: "Atlas of Protein Sequence and Structure." Washington, DC: National Biomedical Research Foundation, 1972, vol 5.

Dover G: Molecular drive: A cohesive mode of species evolution. Nature 299:111, 1982.

Dover GA, Flavell RB (eds): "Genome Evolution." New York: Academic Press, 1982.

Efstratiadis A, Posakony JW, Maniatis T et al: The structure and evolution of the human β-globin gene family. Cell 21:653, 1980.

Fitch WM, Langley CH: Evolutionary rates in proteins: Neutral mutations and the molecular clock. In Goodman M, Tashian RE (eds): "Molecular Anthropology." New York: Plenum Press, 1976, pp 197–219.

Freedman MJ, Roh EF, Nagel RL, Trager W: Plasmodium falciparum: Physiological interactions with the human sickle cell. Exp Parasitol 47:73, 1979.

Goosens M, Dozy AM, Embury SH et al: Triplicated α-globin loci in man. Proc Natl Acad Sci USA 77:518, 1980.

Haldane JBS: The cost of natural selection. J Genet 55:511, 1957.

Haldane JBS: Natural selection. In Bell PR (ed): "Darwin's Biological Work." Cambridge, England: Cambridge University Press, 1959, p 148.

Harris H, Hopkinson DA: Average heterozygosity per locus in man: An estimate based on the incidence of enzyme polymorphisms. Ann Hum Genet 36:9, 1972.

Hunter RL, Markert CL: Histochemical demonstration of enzymes separated by zone electrophoresis in starch gels. Science 125:1294, 1957.

Ingram VM: Gene evolution and the haemoglobins. Nature 189:704, 1961.

Kan YW, Dozy AM: Polymorphisms of DNA sequence adjacent to human β-globin structural gene: Relationship to sickle mutation. Proc Natl Acad Sci USA 75:5631, 1978.

Kazazian HH Jr, Chakravarti A, Orkin SH, Antonarakis SE: DNA polymorphisms in the human β-globin gene cluster. In Nei MM, Koehn RK (eds): "Evolution of Genes and Proteins." Sunderland, MA: Sinauer Associates, 1983, pp 137–146.

Kedes LH: Histone genes and histone messengers. Annu Rev Biochem 48:837, 1979.

Kimura M: Evolutionary rate at the molecular level. Nature 217:624, 1968.

Kimura M: The rate of molecular evolution considered from the standpoint of population genetics. Proc Natl Acad Sci USA 63:1181, 1969.

Kimura M: "The Neutral Theory of Molecular Evolution." Cambridge, England: Cambridge University Press, 1983, pp 76–97, 240–248.

Kimura M, Ohta T: Protein polymorphism as a phase of molecular evolution. Nature 229:467, 1971.

King JL: Continuously distributed factors affecting fitness. Genetics 55:483, 1967.

King JL, Jukes TH: Non-Darwinian evolution. Science 164:788, 1969.

King MC, Wilson AC: Evolution at two levels in humans and chimpanzees. Science 188:107, 1975.

Kondo S: Evolutionary considerations on DNA repair and mutagenesis. In Kimura M (ed): "Molecular Evolution and Polymorphism." Mishima, Japan: National Institute of Genetics, 1977, pp 313–331.

Kunkel LM, Smith KD: Evolution of human Y-chromosome DNA. Chromosoma 86:209, 1982.

Kunkel LM, Smith KD, Boyer SH: Organization and heterogeneity of sequences within a repeating unit of human Y chromosome deoxyribonucleic acid. Biochemistry 18:3343, 1979.

Leder P: The genetics of antibody diversity. Sci Am 246:102, 1982.

Maniatis T, Hardison RC, Lacey E et al: The isolation of structural genes from libraries of eukaryotic DNA. Cell 15:687, 1978.

Margoliash E: The molecular variations of cytochrome c as a function of the evolution of species. Harvey Lectures 66:177, 1972.

Markert CL, Shaklee JB, Whitt GS: Evolution of a gene. Science 189:102, 1975.

Martin SL, Zimmer EA, Kan YW, Wilson AC: Silent δ-globin gene in Old World monkeys. Proc Natl Acad Sci USA 77:3563, 1980.

Maxam AM, Gilbert W: A new method for sequencing DNA. Proc Natl Acad Sci USA 74:560, 1977.

McKusick VA, Ruddle FH: The status of the gene map of the human chromosomes. Science 196:390, 1977.

Milkman R: Heterosis as a major cause of heterozygosity in nature. Genetics 55:493, 1967.

Ohta T, Kimura M: Functional organization of genetic material as a product of molecular evolution. Nature 233:118, 1971.

Perler F, Efstratiadis A, Lomedico P et al: The evolution of genes: The chicken proinsulin gene. Cell 20:555, 1980.

Perutz MF, Kendrew JC, Watson HC: Structure and function of haemoglobin. II. Some relations between polypeptide chain configuration and amino acid sequence. J Mol Biol 13:669, 1965.

Petes T, Fink GR: Gene conversion between repeated genes. Nature 300:216, 1982.

Place AR, Powers DA: Purification and characterization of the lactate dehydrogenase (LDH-B_4) allozymes of *Fundulus heteroclitus*. J Biol Chem 259:1299, 1984.

Powers DA, Place AR: Biochemical genetics of *Fundulus heteroclitus* (L.) I. Temporal and spatial variation in gene frequencies of *Ldh-B*, *Mdh-A*, *Gpi-B*, and *Pgm-A*. Biochem Genet 16:593, 1978.

Prakash S, Lewontin RC, Hubby JL: A molecular approach to the study of genic heterozygosity in natural populations. IV. Patterns of genic variation in central, marginal, and isolated populations of *Drosophila pseudoobscura*. Genetics 61:841, 1969.

Sanger F: The arrangement of amino acids in proteins. Adv Protein Chem 7:1, 1952.

Sanger R, Nicklen S, Coulson AR: DNA sequencing with chain-terminating inhibitors. Proc Natl Acad Sci USA 74:5463, 1977.

Scheller RH, Anderson DM, Pasakony JW et al: Repetitive sequences of the sea urchin genome. II. Subfamily structure and evolutionary conservation. J Mol Biol 149:15, 1981.

Schwab M, Alitalo K, Varmus HE et al: A cellular oncogene (c-Ki-*ras*) is amplified, overexpressed, and located within karyotypic abnormalities in mouse adrenocortical tumour cells. Nature 303:497, 1983.

Scott AF, Heath P, Trusko S et al: The sequence of the gorilla fetal globin genes: Evidence for multiple gene conversions in human evolution. Mol Biol Evol 1:371, 1984.

Selander RK, Hunt WG, Yang SY: Protein polymorphism and genic heterozygosity in two European subspecies of the house mouse. Evolution 23:379, 1969.

Selander RK, Yang SY, Lewontin RC, Johnson WE: Genetic variation in the horseshoe crab *(Limulus polyphemus)*, a phylogenetic "relic." Evolution 24:402, 1970.

Shen S, Slightom JL, Smithies O: A history of the human fetal globin gene duplication. Cell 26:191, 1981.

Smith EL: Evolution of enzymes. In Boyer PD (ed): "The Enzymes, 3rd Ed." New York: Academic Press, 1970, vol 1, pp 267–339.

Smith GP: Unequal crossover and the evolution of multigene families. Cold Spring Harbor Symp Quant Biol 38:507, 1974.

Smith HO, Bernstiel ML: A simple method for DNA restriction site mapping. Nucleic Acids Res 3:2387, 1976.

Smithies O: Zone electrophoresis in starch gels: Group variations in the serum proteins of normal human adults. Biochem J 61:629, 1955.

Southern E: Detection of specific sequences among DNA fragments separated by gel electrophoresis. J Mol Biol 98:503, 1975.

Sved LA, Reed TE, Bodmer WF: The number of balanced polymorphisms that can be maintained in a natural population. Genetics 55:469, 1967.

Tartof KD: Unequal meiotic sister chromatid exchange and disproportionate replication as mechanisms regulating ribosomal RNA gene redundancies. Cold Spring Harbor Symp Quant Biol 38:491, 1974.

Townes TM, Fitzgerald MC, Lingrel JB: Triplication of a four-gene set during the evolution of the goat β-globin locus produced three genes now expressed differentially during development. Proc Natl Acad Sci USA 81:6589, 1984.

Valentine IW, Campbell CA: Genetic regulation and the fossil record. Am Sci 63:673, 1974.

Weatherall DJ, Clegg JB: "The Thalassemia Syndromes, 3rd Ed." Oxford, England: Blackwell Scientific Publications, 1981, pp 508–612.

Wilson AC, Carlson SS, White TJ: Biochemical evolution. Annu Rev Biochem 46:573, 1977.

Wright S: Shifting balance theories of evolution. In "Evolution and the Genetics of Populations." Chicago: The University of Chicago Press, 1977, vol 3, pp 443–473.

Wright S: Genic and organismic selection. Evolution 34:825, 1980.

Zimmer EA, Martin SL, Beverley SM, Kan YW, Wilson AC: Rapid duplication and loss of genes coding for the α chains of hemoglobin. Proc Natl Acad Sci USA 77:2158, 1980.

Human Evolution

Edward H. Birkenmeier, MD

The Jackson Laboratory, Bar Harbor, ME 04609

Human evolution is one of the more fascinating aspects of the biologic sciences. If nothing else, it stimulates the imagination and causes a tremendous amount of controversy. At one point in the Scopes Trial in Dayton, Tennessee, in 1925, Clarence Darrow asked whether it is perfectly easy to believe that the big fish swallowed Jonah. William Bryan responded by saying, "If the Bible said so; the Bible doesn't make as extreme statements as evolutionists do." Today, these issues may not seem as controversial within the scientific community, but in our state legislatures or on our college campuses, there is confusion about what evolution means and what the evidence actually is. This confusion affects the writing of textbooks and therefore has profound sociologic implications.

One of the initial questions is: "Why is there controversy to begin with?" One of the reasons is, using morphologic and behavioral criteria, we as humans have no living close relatives. We are the only surviving members of the genus *Homo* in the family Hominidae and are the last of the hominids. We think of the chimpanzee or the gorilla as our closest relatives, but they belong to another family, the Pongidae, or pongids. There are many easily discernible differences between the hominids and pongids. Hominids are exclusively ground-dwelling and bipedal. Hominids are pair bonded and have families with a home base. Females are less mobile and have continuous sexuality and multiple infant care. Hominids have food sharing, make and use tools, and during recent evolution the brain continued to enlarge. Therefore, it seems logical to place hominids and pongids into separate families.

Birth Defects: Original Article Series, Volume 23, Number 3, pages 209–219

Given this information, one might think that it would be quite easy to find the molecular or biologic mechanisms that determine these differences. Pioneering studies to discover these mechanisms were done by Sarich and Wilson [1967]. Their idea was to obtain homologous proteins from each species and then look for the amino acid substitutions in each of the proteins. They believed that the number of substitutions correlated with the evolutionary distance between the different primates or between apes and humans. For example, to compare humans and chimps, they looked at the total number of amino acid sites from several proteins and then determined the total number of differences. The surprising results summarized by King and Wilson [1975] were that chimp and human proteins were extremely similar. In fact, there were only 8.2 substitutions for 1,000 sites. This difference is only about 25 to 60 times greater than the difference between the races of man.

This same type of analysis can be done by looking at specific gene sequences such as the α- and β-globins [Goodman et al, 1983]. For both proteins, the pygmy chimp and the common chimp have the same sequence as humans. However, the gorilla β-globin at position 104 has lysine instead of arginine. Similarly, the gorilla α-globin at position 23 has aspartic acid instead of glutamic acid. The aspartic acid at position 23 is also found in orangutans, gibbons, and other primates. Therefore, it is believed to represent an older α-globin sequence. These data are used to support the suggestion that the pygmy chimp, the common chimp, and humans are more closely related to each other than they are to the gorilla. The gorilla diverged first from a common ape ancestor and then humans and chimps diverged from one another at a later time.

Another method used to study human and ape relationships is to compare mitochondrial DNA sequences. The mitochondrial genome is about 15,000 base pairs in length. When the human sequence is aligned with either the chimp, gorilla, or orangutan, sequence differences in bases are observed. These are transitions such as adenine to guanine or transversions such as cytosine to adenine. Deletions of bases are also found. The theory is that the more conserved the mitochondrial DNA sequence is between two species the more closely related the species. In a recent study Hixson and Brown [1986] examined 955 nucleotides from the 12S ribosomal RNA gene of mitochondria. They found that between the common chimp and the pygmy chimp there is a 1% difference within this sequence. However, the orangutan sequence was much more divergent. Between the orangutan and chimp (or human) there was a difference of about 9%. Human, chimp, and gorilla were essentially equidistant from one another, since there was equal sequence variation among all three. But, gorilla and chimp had an identical small deletion in the ribosomal RNA gene sequence, and this suggested that gorilla and chimp were more closely related to one another than to humans. In other words, chimp and gorilla shared a common ancestor, and this ancestor had diverged from the human lineage before the gene deletion occurred.

There are problems with the comparative studies presented above. The protein studies use only a few proteins, and these are commonly known and relatively abundant proteins. There may be 100,000 or more proteins encoded in the nucleus, and we have studied less than 1%. Maybe the known proteins such as globins do not actually contain all the information required to study evolution. The mitochondrial genome is also unique. For example, it is small and has only maternal inheritance. In addition, it is not within the nucleus, so maybe the changes we see in mitochondrial genes do not reflect the evolution of all genes.

Perhaps, if we examine the nuclear genome and look at *all* the genes, we can obtain additional information. One of the best recent studies exploring this possibility was by Sibley and Ahlquist [1984]. They studied the divergence of total nonrepetitive nuclear DNA. Their method is to isolate long, double-stranded pieces of total nuclear DNA and then sonicate the DNA. Sonication breaks the DNA into many small pieces. Next, the DNA is heated and it melts to produce small, single-stranded pieces. As the solution cools, the DNA pieces begin to rehybridize to one another. However, the sequences present in many copies find a homologous sequence much faster than a sequence that is present in only one copy per haploid nucleus.

When the rehybridization reaction is run for only a short time, all the repetitive sequences that are present in very high number find a related sequence very quickly. However, the single copy DNA is still single stranded. If the DNA is passed through a hydroxylapatite column, the double-stranded DNA binds to the column and the single-stranded DNA comes through in the wash. The purified single-stranded single copy DNA is then radiolabeled with iodine 125. Next, a thousand-fold excess of unlabeled DNA from another species is melted and hybridized to the labeled DNA. The labeled DNA finds an unlabeled homologous sequence much faster than it finds a labeled homologous sequence because of the thousand-fold excess of unlabeled DNA. The hybridized DNA is passed through another hydroxylapatite column. The labeled DNA from one species that hybridized to the related single copy sequence from the other species is then trapped on the column as double-stranded DNA. These columns are run in water jackets so that the temperature can be raised. As the temperature rises, the DNA begins to melt. The more divergent the DNA sequences are between the two species, the lower the temperature at which melting occurs. The temperature at which

50% of the label has been eluted off the column is called $T_{50}H$. Now, all possible combinations of DNA are compared. First, human DNA is rehybridized to human DNA to get a $T_{50}H$ that represents identity. Then human DNA is hybridized to chimp DNA, and a slightly lower $T_{50}H$ is obtained because there is sequence divergence between the two species. The difference between these two $T_{50}H$ values is called $\Delta T_{50}H$. For chimp and human, the $\Delta T_{50}H$ is 1.8; for gorilla and human, the $\Delta T_{50}H$ is 2.4; and for chimp and gorilla, the $\Delta T_{50}H$ is 2.1. When the orangutan is compared to either the human, chimp, or gorilla, the $\Delta T_{50}H$ is approximately 3.7. The $\Delta T_{50}H$ values are then used to estimate genetic distance. In addition, the fossil record indicates that a $\Delta T_{50}H$ of 1.0 represents approximately 4.3 million years of evolutionary separation between species. If this clock is accurate, then the pygmy and common chimp diverged about 2.4 to 3 million years ago and humans diverged from a common ancestor with the chimpanzee about 6.3 to 7.7 million years ago. The gorilla lineage separated from other apes about 8 to 10 million years ago, and the orangutan lineage separated about 13 to 16 million years ago.

Taking a different approach, George Todaro's laboratory at NIH isolated an endogenous retrovirus from the colobus monkey [Bonner et al, 1982]. The idea was that the cloned colobus retrovirus could be used as a molecular probe to detect related viruses in other primate species, including man. When the colobus virus was hybridized to Southern blots of primate genomic DNA (chimp, gorilla, colobus monkey, baboon, rhesus monkey), all of the DNAs were found to contain multiple copies of related viral sequences. However, human DNA was completely negative. Interestingly, the orangutan and gibbon were also negative. Retroviruses are unusual because they can move around within the genome and their genetic information also changes very rapidly. However, it is difficult to explain why gorilla and chimp have very similar bands on a Southern blot, while the closely related human DNA is negative. One possibility is that there were colobus-like retroviruses present in all the early primate ancestors, but somehow human and orangutan lost them during their recent evolution. Alternatively, these data could suggest that humans are actually more closely related to orangutans than to chimps and gorillas. There have been arguments made in the past that support this conclusion [Schwarz, 1984], but most scientists do not believe they are valid. Another explanation given by Benveniste and Todaro [1976] is that retroviruses infected the ancestors of chimps and gorillas in Africa at a time when the ancestors of present-day humans and orangutans were in Asia. Thus, there are many interpretations of these data; that is the difficulty with using molecular probes such as retroviruses to study evolution. However, these retroviral studies have generated intriguing hypotheses.

Given all the results discussed above, some general conclusions can be made. First of all, humans are extremely similar to chimps at both the biochemical and genetic levels. In fact, we are so close that we are essentially sib species, and from a chimp's perspective, *Homo sapiens* might be renamed *Pan sapiens*, or "wise chimp." The second conclusion is that molecular biology has yet to define the genetic mechanisms or specific genes that determine the morphologic and behavioral differences between man and chimp. Finally, it is important to remember that these data do not actually demonstrate that evolution ever occurred, and that is the controversy we started discussing. However, what they do *imply* is that about 7 or 8 million years ago humans and chimps shared a common ancestor. That does not mean that chimps evolved into humans or vice versa. What it does mean is that they had a common ancestor, and each of the species has diverged significantly from the common ancestor.

To find evidence that evolution has occurred and also to set the molecular clocks mentioned above, we need to study the fossil record. One way to begin thinking about the fossil record is to use a "family album" analogy. Imagine you have a family album of all your ancestors, pictures of your mother, her mother, so forth and so forth, page after page. There are two possible ways this album could appear. One is the creationist belief that while turning the pages, very little would change, and all of a sudden the album would end. The other possibility is the theory of evolution, that as the pages are slowly turned, one at a time, there would be gradual changes. Eventually, after several days of turning the pages, there would be organisms or individuals that you would not recognize as being human.

The fossil data that support this second possibility will be discussed in the remainder of this chapter. Actually, most of these data were on display at the Natural History Museum in New York in 1984. It was an incredible exhibit where the leading anthropologists of the world as well as the major fossil discoveries were present. The scientific meeting itself was private, but afterwards most of the fossils were available for public viewing. I was able to visit this exhibit, and it was an interesting experience. It is one thing to think about evolution and to look at pictures and drawings of fossils, but another thing actually to go and study the bones of our ancestors literally face to face. What I am going to do is take you through the exhibit.

At the beginning of the exhibit there is a small skull that is about the size of a lemon. It is from what is believed to be a primitive ape, *Aegyptopithecus*. The skull was found in Egypt. *Aegyptopithecus* was a very unspecialized, tree-dwelling primate and may prove to be one of the first early ape-like creatures that diverged from the old world monkeys about 33 million years ago. One problem with the fossil record is that there are large gaps. Thus, if we start at 33 million years ago, the next major fossils are from animals that lived about 17 million years ago. This is when Africa and Arabia touched Eurasia during continental movement. At this time there are hominoid creatures moving out of Africa and into Asia, and perhaps vice versa. The fossils we are discussing are from *Sivapithecus*. An 8-million-year-old *Sivapithecus* skull was on the cover of *Nature* a few years ago [Pilbeam, 1982], but recent findings in China and Pakistan are as old as 12 to 17 million years [Raza et al, 1983]. It was these older fossils that were used to set the molecular clocks. At one time it was believed that *Sivapithecus* was an ancestor to all later apes and hominids, but recent, more complete fossil discoveries suggest that *Sivapithecus* is much closer to the orangutan than to modern chimps. So, rather than being a common ancestor to all modern apes, it is now believed that *Sivapithecus* was a direct ancestor of modern orangutans [Hammond, 1983]. It is this reasoning that was used to set the molecular clock so that a $\Delta T_{50}H$ of 1.0 results from 4.3 million years of genetic separation.

Now we make another jump in the fossil record to about 3 or 4 million years ago. This is an important jump, because there are definitely early hominid fossils present at this time. On November 30, 1974, Donald Johanson and Tom Gray had spent the morning looking for hominid fossils in a desert in Ethiopia. The temperature had reached 110° and they were exhausted. Since it was late morning, they started back to their land rover. On the way, Johanson looked over on a barren hillside and saw a hominid arm sticking out of the ground. They investigated, and the result was the finding of a 40% complete skeleton. That night, back at camp, they were celebrating and playing the Beatles song "Lucy in the Sky with Diamonds," and that is how the skeleton got its name, Lucy. Lucy is approximately 3.5 million years old, and Johanson and Edey [1981] wrote a book about her. They believe that she represents a new species they named *Australopithecus afarensis*, or, literally, Southern ape from the Afar region of Ethiopia. The skeleton has many interesting characteristics, but the most surprising discovery was that she walked upright. One reason we know this is related to the structure of the knee joint. The angle between the tibia and the femur in the *afarensis* knee is similar to the angle in the human knee. In addition, the head of the femur is ovoid in both the *afarensis* and the human hip, while it is more circular in the chimpanzee. This is very convincing anatomic evidence that Lucy walked upright.

A second important discovery occurred in 1978. This discovery was made by a team headed by Mary Leakey. They were farther south, in the Rift Valley of Tanzania, and what they found were hominid-like footprints that were approximately 3.7 million years old [White, 1980]. The story of how these footprints were generated is very interesting. A volcano erupted in Africa producing a thick layer of ash that covered the ground. After the ash was deposited, a rainstorm wet the ash and it became the consistency of wet cement. Soon thereafter, three hominids walked across the open plain, much like we would do on a wet sandy beach today, leaving their footprints. The footprints baked in the hot sun, and the ash became very hard. The volcano erupted again, producing another layer of ash that covered and preserved the footprints until they were discovered by Mary Leakey's team. What is interesting about the footprints, other than the fact that they are hominid footprints, is that they look very similar to modern human footprints. They have a forward-pointing large toe, an arched foot, and a well-shaped heel. By casual observation there appeared to be only two sets of footprints, a small set made by a single individual and a larger set. However, the larger footprints were actually a composite made when a third smaller individual walked in the footprints of the larger hominid. Thus, three hominids made these footprints. It is interesting to speculate that these are actually *afarensis* footprints, but there is no direct proof that this is true. What is certain is that the footprints were generated about the time that *A. afarensis* was walking around in Africa.

With upright posture and walking, Lucy may seem very human; however, other data indicate that she was much different than we are. One difference is found in the upper jaw. The chimpanzee upper jaw has a box-like or rectangular shape. The human upper jaw, which is characteristic of all known hominids, has a gradual curved appearance. The *afarensis* upper jaw is intermediate between chimp and man. There are other *afarensis* dental characteristics that also appear to be transitional, and they are discussed in Johanson and Edey's book [1981]. Another interesting difference is Lucy's curved toes. Humans have straight toe bones, but in chimps and *afarensis* the bones are curved. Thus, although Lucy walked upright, she also may have climbed trees [Lewin, 1983a,b]. The main point is that Lucy is similar to modern apes in certain characteristics. Another important similarity to apes is Lucy's small brain. This was a surprise, since it was previously assumed that creatures with large brains walked upright, and creatures with small brains did not. However, the *afarensis* brain is only about 380 to 450 cc, and modern chimps are similar at 300 to 400 cc. If you saw Lucy walking around your home town, you would definitely not mistake her for a modern human.

Now we again jump ahead in the fossil record to about 2 million years ago. At this time, the fossil record is more complete as well as more complex.

There are at least three hominid species walking around simultaneously. This creates controversy, because it is not clear which one of the three, if any, is the direct forerunner of modern humans. Perhaps the fossils of our direct ancestors, which lived during this period, are yet to be found. In any case, the smallest of the three known hominids is *Australopithecus africanus*. The first *africanus* fossil skull was found in the 1920s by Raymond Dart. This is the famous skull known as the Taung child, and it is from an infant *africanus*. Since then, adult skulls have been found and the brain size is about 450 to 480 cc, which is slightly larger than modern chimps.

A second, much larger hominid was *Australopithecus robustus*. They had large skulls and huge molars that were 1 inch in diameter. A medial sagittal crest protruded from the top of the skull and is where the large temporalis muscle inserted. With the large teeth and powerful jaw muscles, these creatures probably could have ground nuts to survive.

Both of these two australopithecine species became extinct about 1 million years ago. The remaining third hominid is what Richard Leakey and many others believe may be a direct ancestor of modern man [Leakey and Lewin, 1977]. This hominid is named *Homo habilis*. *H. habilis* had an average brain size of about 700 cc and, in addition, probably used tools. However, fossils from only one new hominid species are found from 1 million years ago. These are known as Java Man and Peking Man, and they are from *Homo erectus*. The most exciting recent *H. erectus* skeleton was discovered in Africa by Alan Walker and his colleagues. Their results were published in *Nature* [Brown et al, 1985]. Their discovery was very important. Because of the completeness of the skeleton, new conclusions could be made about *H. erectus*. First of all, this skeleton is very old, about 1.6 million years. It is a male, as shown by the shape of the pelvis. The bones and teeth indicated that he was about 12 years old. However, he was already 5 feet 5 inches tall and, by modern standards, could have become a 6-foot adult. The length of his long bones is similar to North American males. However, there are also some interesting skeletal differences. First of all, the foramina of the cervical and thoracic vertebrae are significantly smaller than in modern humans, so the spinal cord must have been significantly smaller. The iliac bone has a large flare to it. There is a relatively long femoral neck, and the angle between the body and the neck is 110°, which is small by modern standards. The anatomy of the pelvis with its small pelvic opening suggests that *H. erectus* could not have given birth to infants with large brains. This led Alan Walker to suggest that one of the more important evolutionary changes that occurred between *H. erectus* and modern man was the redesigning of the pelvis. The result was that modern man has slightly less efficient walking in energetic and mechanical terms, but can give birth to babies with larger brains. The overall brain size of *H. erectus* ranges from about 775 to about 1,300 cc. This is slightly below the modern average of about 1,400 cc.

Next, we move ahead to between 250,000 and 400,000 years ago. In parts of Europe and Southern Africa, we now find early *Homo sapiens*. The skulls appear relatively modern with an increased brain size, a decreasing forehead bulge, a higher forehead, and a decrease in the brow ridges that project out over the eyes. The angular eye sockets also look very modern. One of the interesting aspects of early *H. sapiens* evolution is the appearance of a unique race of people that was quite abundant from about 100,000 to about 30,000 years ago. They were the Neanderthals. The cranial capacity of these people was the same, if not exceeding, that of modern man. However, their skulls have a unique anatomy that makes it easy to distinguish a Neanderthal skull from a modern skull. They probably used primitive tools and also hunted. Since graves have been found, we know that they buried their dead and probably developed a culture. A burial site in Iraq is particularly interesting, because the flowers in the grave indicate that the burial occurred on a June day some 60,000 years ago. What happened to the Neanderthals remains unknown. Perhaps they were eliminated by pressures from more modern *H. sapiens* moving up from Africa. There may have been interbreeding or conflict between the different groups—you can imagine your own scenario. In any case, since about 30,000 years ago modern man, *H. sapiens*, is the only surviving hominid. Cro-Magnon man is found in France from between 13,000 to 20,000 years ago. At the same time, we find very good tools, bits of clothing, settlements, communities, and paintings on cave walls. Our modern world had begun.

When you think about it, the fossil record is quite remarkable. There have been many arguments that evolution does not occur on a gradual basis but occurs in big spurts or does not even occur at all. However, the fossil record of hominid evolution is becoming complete [Rensberger, 1984]. There do not appear to be rapid periods of evolution, and evolution seems to have occurred at a relatively gradual rate. We have discussed the skulls and other bones of *A. afarensis, A. africanus, A. robustus, H. habilis, H. erectus, H. sapiens neanderthalensis*, and modern *H. sapiens*. The fossil data do not necessarily imply that all of these individuals are our direct ancestors. What they do imply is that evolution occurs as a relatively slow, gradual process and that the gaps in the fossil record are becoming smaller as research continues.

We have gone through our family album. We started with our ancient ancestors in the fossil record and then returned one step at a time back to the present. As we look at the most recent pages in the album we see ourselves, our children, and our society. It is worth asking: "What will the future hold?" I believe that many of the important facts we learn by studying human evolution have sociologic implications. Among the important conclusions to remember from this chapter are that we are all members of one species and that we are a very new species. In addition, we are the last of the hominids. The fossil record has shown that we are capable of change. A positive thought to contemplate is that the sociologic problems we face today, poverty, war, racism, you name it, are only temporary conditions created by our culture and that while evolution continues, the problems we face today will eventually become as archaic as the society that spawned them.

REFERENCES

Benveniste RE, Todaro GJ: Evolution of type C viral genes: Evidence for an Asian origin of man. Nature 261:101–108, 1976.

Bonner TI, Birkenmeier EH, Gonda MA, Mark GE, Searfoss GH, Todaro GJ: Molecular cloning of a family of retroviral sequences found in chimpanzee but not human DNA. J Virol 43:914–924, 1982.

Brown F, Harris J, Leakey R, Walker A: Early Homo erectus skeleton from west Lake Turkana, Kenya. Nature 316:788–792, 1985.

Goodman M, Braunitzer G, Stangl A, Schrank B: Evidence on human origins from hemoglobins of African apes. Nature 303:546–548, 1983.

Hammond AL: Tales of an elusive ancestor. Science 83:37–43, 1983.

Hixson JE, Brown WM: A comparison of the small ribosomal RNA genes from the mitochondrial DNA of the great apes and humans: Phylogenetic implications. Mol Biol Evol 3:1–18, 1986.

Johanson DC, Edey M: "Lucy: The Beginnings of Humankind." New York: Simon and Schuster, 1981.

King M, Wilson AC: Evolution at two levels in humans and chimpanzees. Science 188:107–116, 1975.

Leakey RE, Lewin R: "Origins." New York: E.P. Dutton, 1977.

Lewin R: Do ape-size legs mean ape-like gait? Science 221:537–538, 1983a.

Lewin R: Were Lucy's feet made for walking? Science 220:700–702, 1983b.

Pilbeam D: New hominoid skull material from the Miocene of Pakistan. Nature 295:232–234, 1982.

Raza SM, Barry JC, Pilbeam D, Rose MD, Shah SMI, Ward S: New hominoid primates from the middle Miocene Chinji Formation, Potwar Plateau, Pakistan. Nature 306:52–54, 1983.

Rensberger B: Bones of our ancestors. Science 84:29–39, 1984.

Sarich VM, Wilson AC: Rates of albumin evolution in primates. Proc Natl Acad Sci USA 58:142–148, 1967.

Schwarz JH: The evolutionary relationships of man and orangutans. Nature 308:501–505, 1984.

Sibley CG, Ahlquist JE: The phylogeny of the hominid primates, as indicated by DNA-DNA hybridization. J Mol Evol 20:2–15, 1984.

White TD: Evolutionary implications of Pliocene hominid footprints. Science 208:175–176, 1980.

BIBLIOGRAPHY

Andrews P: Hominoid evolution. Nature 295:185–186, 1982.

Brown WM, Prager EM, Wang A, Wilson AC: Mitochondrial DNA sequence of primates. Tempo and mode of evolution. J Mol Evol 18:225–239, 1982.

Bruce EJ, Ayala FJ: Humans and apes are genetically very similar. Nature 276:264–265, 1978.

Charteris J, Wall JC, Nottrodt JW: Functional reconstruction of gait from the Pliocene hominid footprints at Laetoli, Northern Tanzania. Nature 290:496–498, 1981.

Clark JD, Asfaw B, Affefa G, Harris JWK, Kurashina H, Walter RC, White TD, Williams MAJ: Paleoanthropological discoveries in the Middle Awash Valley, Ethiopia. Nature 307:423–428, 1984.

Delson E: The earliest Sivapithecus? Nature 318:107–108, 1985.

Delson E: Paleobiology and age of African Homo erectus. Nature 316:762–763, 1985.

Falk D: Cerebral cortices of East African early hominids. Science 221:1072–1074, 1983.

Ferris SD, Brown WM, Davidson WS, Wilson AC: Extensive polymorphism in the mitochondrial DNA of apes. Proc Natl Acad Sci USA 78:6319–6323, 1981.

Hasegawa M, Kishino H, Yano T: Dating of the human-ape splitting by a molecular clock of mitochondrial DNA. J Mol Evol 22:160–174, 1985.

Holloway RL: Cerebral brain endocast pattern of Australopithecus afarensis hominid. Nature 303:420–422, 1983.

Johanson DC, White TD: A systematic assessment of early African hominids. Science 203:321–330, 1979.

Jungers WL: Lucy's limbs: Skeletal allometry of locomotion in Australopithecus afarensis. Nature 297:676–678, 1982.

Leakey RE: "The Making of Mankind." New York: E.P. Dutton, 1981.

Leakey RE, Lewin R: "People of the Lake." New York: Anchor Press/Doubleday, 1978.

Lewin R: How did humans evolve big brains? Science 216:840–841, 1982.

Lewin R: Is the orangutan a living fossil? Science 222:1222–1223, 1983.

Lovejoy CO: The origins of man. Science 211:341–350, 1981.

Martin RD: Relative brain size and basal metabolic rate in terrestrial vertebrates. Nature 293:57–60, 1981.

Pilbeam D: The descent of hominoids and hominids. Sci Am March, 84–96, 1984.

Raeburn P: An uncommon chimp. Science 83(4):40–48, 1983.

Repenning CA, Fejfar O: Evidence for earlier date of Ubeidiya, Israel, hominid site. Nature 299:344–347, 1982.

Simons EL, Kay RF: Qatrania, new basal anthropoid primate from the Fayum, Oligocene of Egypt. Nature 304:624–626, 1983.

Walker A, Falk D, Smith R, Pickford M: The skull of Proconsul africanus: Reconstruction and cranial capacity. Nature 305:525–527, 1983.

Wolpoff MH: Lucy's lower limbs: Long enough for Lucy to be fully bipedal? Nature 304:59–61, 1983.

Yunis JJ, Prakash O: The origin of man: A chromosomal pictorial legacy. Science 215:1525–1529, 1982.

Yunis JJ, Sawyer JR, Dunham K: The striking resemblance of high-resolution G-banded chromosomes of man and chimpanzee. Science 208:1145–1148, 1980.

Mouse Models of Genetic Diseases

Edward H. Leiter, PhD, Wesley G. Beamer, PhD, Leonard D. Shultz, PhD, Jane E. Barker, PhD, and Priscilla W. Lane, MS

The Jackson Laboratory, Bar Harbor, ME 04609

INTRODUCTION

A popular component of the annual Short Course in Medical and Experimental Mammalian Genetics each summer at the Jackson Laboratory has been the "mouse clinic." At this "clinic," the interested student is shown representative examples of mouse mutations considered to be of biomedical significance. On display are single gene mutations that cause anemias, obesity, diabetes disorders, endocrinopathies, muscular and neurologic defects, immunodeficiency, autoimmunity, X-linked disorders, and a variety of miscellaneous defects including those involving bone, connective tissue, hair, skin, and numerous other specific organs. Although many of the students taking the course have primary interests in human medicine and genetics, the

The Jackson Laboratory is fully accredited by the American Association of Laboratory Animal Care.

Acknowledgments: The authors thank all their co-workers, past and present, at The Jackson Laboratory for identifying and conserving valuable mouse mutations. Writing of this chapter was supported by NIH grants AM17631, AM27722, and AM36175 (E.H.L.), AM17947 (W.G.B.), CA20408 (L.D.S.), AM27726 and HL29305 (J.E.B.), and NSF grant BSR-84-18828 (P.W.L.).

Birth Defects: Original Article Series, Volume 23, Number 3, pages 221–257

course teaches them an appreciation of the mouse as a more genetically defined experimental system. Consequently, during the "mouse clinic," it is not unusual for clinically oriented students to be drawn to the exhibit showing mouse mutants that share some characteristics with the human genetic disease syndromes of interest to them. For example, an endocrinologist treating patients with Prader-Willi syndrome might go to the display of single gene mutations producing obesity and dwarfism syndromes and inquire: "What do you have that produces Prader-Willi syndrome in mice?" Such questioners usually walk away with their quest unfulfilled; a given mouse model seldom provides the clinician with exactly the same symptomatology that one is accustomed to finding in the patient. The purpose of this chapter is to describe how mouse models of genetic diseases are discovered and what their value and limitations are as models for an analogous or homologous condition in humans.

GENERATION AND IDENTIFICATION OF MOUSE MUTATIONS

Inbred mice often have proved to be the laboratory animal of choice in the development of models of inherited human constitutional disorders. Chromosomal organization of linked loci and biochemical organization of enzyme pathways are very similar in mouse and man. The short generation times and relatively short life-spans (no more than 3 years) of mice have enabled biomedical scientists to study the effect of genetic mutations in the mouse through many generations, a feat not easily accomplished in humans.

Most of the single gene mutations described in the mouse have arisen spontaneously and have been identified because the genetic change has produced a phenotypic change. The Jackson Laboratory has accumulated many of these mutations as a consequence of its large breeding colonies of inbred mice, its focus on mouse genetics, and, most importantly, its dedicated personnel entrusted with husbandry. Not surprisingly, mutations producing changes in pelage (pigmentation, hair structure), eye color, and such physical characteristics as tail, foot, and ear abnormalities were among the first to be described. As inbred lines of mice became established, it became relatively easy to identify phenotypic deviants within genetically homogeneous populations of mice. The rigorous maintenance of pedigree records has ensured that the parents or any available parental sibs, as well as sibs of the prospectus, are traceable for further genetic analyses. At this writing, more than 1,100 genetic loci have been described in the mouse; these include single gene mutations affecting body size (obesity, dwarfism), skeletal structure, erythropoiesis, reproduction, immunologic function, neurologic and neuromuscular coordination, enzyme activity and/or protein electrophoretic mobility, and restriction fragment length polymorphisms.

A classification of genes according to their phenotypic effects has been published recently [Green, 1981]. In many cases, mutations at polymorphic loci have been uncovered by investigators who have been comparing different inbred strains for enzyme activities or inducibility, or for susceptibility to drugs, toxins, or infectious agents. An investigator interested in finding an allelic form of a given enzyme structural gene might undertake a "VIS" (various inbred strains) screening wherein normal representatives of a battery of unrelated inbred strains are tested for activity variants. For example, the finding of low levels of glucokinase in strain RF/J and high levels in strain C3H/HeJ allowed the delineation of the glucokinase (*Gk*) locus [Coleman, 1977]. In similar fashion, regulatory genes controlling structural genes for enzymes have been defined. Although elucidation of this genetic variation is certainly not tantamount to discovery of a potential mouse model of human disease, knowledge of which inbred strains have high or low enzyme activities or exhibit high or low responsiveness to an environmental challenge is useful in designing a screening protocol for a desired disease model.

Even when a "candidate" inbred strain of mouse is chosen as genetically appropriate for the development of a disease model, luck and serendipity appear to play major roles in the outcome. For instance, a deletion mutation causing α-thalassemia was found in one out of a group of 20 mice from a sire treated with triethylmelamine (TEM) in a mutagenesis program designed to produce chromosomal inversions [Whitney and Russell, 1980]. Subsequent screening of progeny from a much larger group of TEM-treated mice for more mutations at this locus was unsuccessful. Similarly, attempts to produce genetically diabetic mice by selective breeding of normal mice with fasting plasma glucose levels at the high end of the normal distribution have not been successful; the NOD (nonobese diabetic) mouse strain, developing spontaneous autoimmune-associated insulin-dependent diabetes, was discovered during inbreeding to select for something entirely different—cataract formation [Makino et al, 1980].

Sometimes a mouse mutation will arise whose significance in understanding the biology of man is not immediately apparent. In one case, a mouse model of microcytic anemia [Russell et al, 1970] preceded by several years the reporting in the human clinical literature of its human counterpart [Staven et al, 1973]. Alternatively, even when the human disease is well described before the mouse mutation is discovered, single gene mutations producing phenotypic deviants usually have multiple effects on the organism (pleiotropism) such that it often takes years of investigation before the biomedical potential of a given mutation is fully recognized; eg, the coat color mutation "beige" has become recognized as a model for Chediak-Higashi syndrome in man because of similar abnormalities in pigment and lysosomal granules [Oliver and Essner, 1973]. Given that most of the mouse models of human disease currently available have occurred spontaneously, or have been fortuitous discoveries in mutagenesis or breeding programs not specifically targeted toward induction of the particular model, does this mean that biomedical scientists in search of a mouse model for a human constitutive disorder must sit back patiently and passively while nature takes its course?

Mutations of desired types can be selected from random mutations produced by chemical mutagens or irradiation if an intelligent screening program can be devised and if the investigators are willing to screen large numbers of mice. For example, five mutations at the catalase-1 (*Cas-1*) locus producing acatalasemic and hypocatalasemic mice were discovered in 12,306 progeny of irradiated sires [Feinstein et al, 1966]. One of the most potent chemical mutagens in the mouse is ethylnitrosurea (ENU); this reagent produces a very high frequency of mutations in spermatogonia of treated males that are subsequently mated. In one such study, nine electrophoretically detectable mutations were recovered from a screening of 22,512 loci [Johnson and Lewis, 1981].

Molecular geneticists have provided valuable new tools for producing "man-made" mutations by means of the experimental insertion of foreign DNA. In some cases, this is achieved by infecting embryos with a retrovirus. For example, an embryonic recessive lethal mutation was induced by infecting embryos with a Moloney murine leukemia virus [Schnieke et al, 1983]. The retrovirus was shown to integrate at a site called the *Mov-13* locus and, by so doing, to block completely the downstream transcription of the $\alpha 1$(I) collagen gene. More often, the recipient is a mouse zygote into which a cloned DNA is microinjected (commonly into the male pronucleus). If the introduced DNA integrates stably into the genome of the zygote, the live progeny are termed transgenic. A review of experiments with transgenic mice has recently appeared [Gordon and Ruddle, 1985]. Of interest is an inherited limb deformity created in transgenic mice by inserting a fusion construct of a mouse mammary tumor virus long terminal repeat and the c-myc oncogene [Woychik et al, 1985]. This mutation proved to be a remutation at the locus *ld*, limb deformity on chromosome 2, only two centimorgans (cM) away from another locus where mutation produced developmental limb deformity (Strong's luxoid; *Lst*). An elegant demonstration that some naturally occurring mutations may also reflect retroviral transpositions into or out of the genome was provided in a study of the coat color mutation dilute (*d*). The *d* allele was associated with an ecotropic murine leukemia virus DNA sequence (provirus) from DBA/2J mice; when revertants (back to wild-type) at this locus were analyzed, a concomitant loss of the proviral sequence was found [Jenkins et al, 1981]. Indeed, inbred strains of mice have been identified that, when intercrossed, show a high frequency of proviral insertions such that hybrid progeny exhibit stable new ecotropic proviral insertion sites not present in either parental strain [Jenkins and Copeland, 1985]. Such hybrid mice may prove useful for generating new mutations resulting from proviral insertions and for cloning neighboring genomic sequences.

Retroviral gene sequences are particularly suited as transmission vectors for cloned DNA because of the ease with which they integrate into mouse genomic DNA; the packaging-defective retroviruses have proved exceptionally valuable in this regard [Mann et al, 1983]. However, a major problem still facing molecular geneticists working with transgenic mouse models has been the difficulty of achieving the appropriate tissue-specific expression of the inserted gene. To circumvent this problem, genes of interest have been fused to promoter-regulatory gene sequences flanking the retroviral genome or specific cellular promoters such as the metallothionein promoter. This latter sequence was utilized to effect expression of a human growth hormone (GH) gene in mice [Palmiter et al, 1982]. Supplementation of the drinking water of the transgenic mice with $ZnSO_4$ induced high levels of promoter function and of GH mRNA transcription in the liver, an anomalous site for GH biosynthesis (the appropriate site being the somatotrope cells of the pituitary gland). Promoter sequences are being cloned that will allow expression of foreign DNA in the appropriate target cell. For example, the pancreatic β cell can be targeted by making fusion constructs to an upstream promoter sequence of one of the insulin genes; a recombinant DNA sequence for the SV40 large T-antigen linked to the rat insulin-promoter produced insulinomas in transgenic mice [Hanahan, 1985].

MATCHING MOUSE MODELS TO HUMAN CONSTITUTIONAL DISEASE: CYSTIC FIBROSIS AND DIABETES AS CASES IN POINT

Although the advantages of having a genetically well-defined mouse model of a human disease are obvious, the limitations of the mouse models are not so readily recognized. To paraphrase a famous quotation, some men may be "mice" with regard to their timorous behavior under duress, but the reverse will never hold—mice will not be men. Thus, although geneticists often emphasize the similarities between the chromosomal organization of linked genes, and the similarities in the biochemical pathways controlled by these genes, the major differences in the biology of the genera *Homo* and *Mus* should not be forgotten. Nevertheless, the inherent limitations of research of human subjects have intensified the search for adequate animal models. A case in point is cystic fibrosis (CF), reputedly the most common lethal hereditary disease transmitted as a mendelian recessive trait in the United States [Di Sant'Agnese, 1973]. The disease afflicts children and is expressed as a glandular exocrinopathy associated with generalized mucoviscidosis. It is clinically characterized by the triad of chronic pulmonary disease, exocrine pancreatic deficiency, and elevated sweat electrolytes. How could a desperately needed mouse model for CF be developed? A genetic locus on human chromosome 7 responsible for almost all CF (designated as the CF gene) has recently been discovered [White, 1986].

Reasoning that since CF is a disease of children, and manifests itself early by reduced weight gain and steatorrhea (feces containing partially digested fat), Pivetta and Green [1973] screened 15,462 culls (runts or otherwise substandard mice obtained at weaning) from among 13 different inbred strains in the large colonies of the Animal Resources Unit of The Jackson Laboratory. They found three CBA/J mice with clinical and histopathologic features of exocrine pancreatic insufficiency; on the basis of preliminary breeding tests, the disease was inferred to be produced by an autosomal recessive mutation they designated exocrine pancreatic insufficiency (*epi*). Was this the long-awaited model for CF in man? Subsequent analysis indicated that susceptibility to spontaneous exocrine pancreatic necrosis was a feature of the particular subline of CBA/J rather than a single autosomal recessive mutation [Leiter and Cunliffe-Beamer, 1977]; moreover, pathophysiologic analysis indicated strain susceptibility to autoactivation of zymogens inside the pancreatic exocrine cell rather than mucoviscidosis as the cause of the lesion [Eppig and Leiter, 1977; Leiter et al, 1977]. Since mucoviscidosis, chronic pulmonary disease, and sweat electrolyte imbalance were lacking in the mouse syndrome, it was clear that the model was incomplete. Nevertheless, it was of considerable biomedical interest in that the exocrine pancreatic insufficiency was well managed with pancreatic enzyme-replacement therapy [Leiter and Cunliffe-Beamer, 1977], thereby affording gastroenterologists an excellent model for study of exocrine pancreatic insufficiency and its management [Kwong et al, 1978]. Those components of human CF missing in the exocrine pancreatic insufficiency syndrome have been separately associated with the expression of other mouse gene mutations. These include cribiform degeneration (*cri*) showing electrolyte disturbance in saliva [Kaiser et al, 1974] and susceptibility to pneumonitis [Pivetta et al, 1980]. However, mutation at a single gene locus affecting the same multiplicity of pathophysiologic changes in the mouse as observed in CF remains to be identified.

The mouse models of obesity and diabetes syndromes further underscore the problems associated with direct comparison of mice and men. Single gene mutations at six different loci in the mouse have been identified that produce insulin-resistant obesity syndromes [Coleman, 1978a]. In general, such obese mice exhibit a "thrifty" metabolism in which more caloric energy is converted to fat. It has been suggested that such mutations are of selective advantage in feral populations where food supplies are limited, but that such genes are deleterious under laboratory conditions where enriched foods are constantly available [Coleman, 1978b]. Human counterparts for these obese mice would be certain Indians from the American Southwest (the Pimas) or Melanesian tribes such as the Nauru Islanders, who have recently departed from traditional food sources and partaken of the 20th century "Coke and bun" diets. This change in life-style has been associated with high (50%) incidences of obesity and diabetes. In such human populations, obesity genes that may have been valuable in the "primitive" society would be rendered deleterious by "progress" [Neel, 1962].

To the diabetologist, the intriguing aspect of the murine obesity genes is their interaction with other modifying genes in the background genome of various inbred mouse strains. Obese (*ob*) and diabetes (*db*) are two unlinked recessive obesity genes; when either mutation is homozygous in C57BL/6J mice, an obesity/insulin-resistance state uncomplicated by a permanent diabetes ensues. In contrast, either mutation, when homozygous on the C57BL/KsJ background, expresses as obesity accompanied by a severe diabetes syndrome [Coleman, 1978a]. This dependence on the genetic background for the type of disease syndrome expressed strongly underscores the difficulty in elucidating the genetics of diabetes in the genetically heterogeneous human population. Because of the marked obesity, insulin resistance, and hyperinsulinemia associated with the development of diabetes in mice homozygous for these recessive obesity genes, most clinicians have tended to view such mice as models of noninsulin-dependent (type II) diabetes mellitus in man.

The model for type I, or insulin-dependent diabetes mellitus with autoimmune involvement is considered to be the nonobese diabetic (NOD) inbred strain [Makino et al, 1980]. Development of insulin-dependent diabetes in the absence of obesity in NOD mice is under polygenic control [Makino et al, 1985]. Whereas the major histocompatibility complex (MHC) is not an important modifier of the obesity-diabetes syndrome in the single gene obesity models [Leiter, 1984], diabetogenesis in NOD mice is strongly associated with a recessive MHC-linked susceptibility trait [Hattori et al, 1986], as is type I diabetes in man [Nerup et al, 1984].

Superficially, it would seem that the C57BL/KsJ-*db*/*db* diabetes mouse and NOD mouse each serve as a model for one of the two major forms of diabetes in man. However, a careful analysis of diabetes in these mouse models leads to a blurring of distinctions. As with MHC association of susceptibility to type I diabetes, autoimmunity is considered to be a feature of type I diabetes, yet it is not limited to the NOD model. The C57BL/KsJ-*db*/*db* mouse reportedly develops both humoral and cellular anti-islet immunity [Meade et al, 1981; Debray-Sachs et al, 1983]. Similarly, viral associations in diabetes pathogenesis have also fallen within the confines of type I, and not type II, diabetes, yet susceptibility to the diabetogenic action of the obesity gene is strongly associated with the expression of endogenous retroviruses in the pancreatic β cells [Leiter and Kuff, 1984], an expression shared by the NOD mouse [Fujita et al, 1984]. Although insulin-dependent diabetes is generally thought to be more "juvenile-onset" than noninsulin-dependent diabetes, the C57BL/KsJ-*db*/*db* mouse develops hyperglycemia and insulin-resistant diabetes within 2 weeks of weaning (age 6 weeks), whereas the peak incidence of insulin-dependent diabetes in NOD mice may not be realized until more than 22 weeks of age. Moreover, there are gender effects on diabetes susceptibility that are not common to man; *db*/*db* males are more susceptible to insulin-resistant diabetes than females [Leiter et al, 1981], whereas NOD females are more susceptible to the autoimmune insulin-dependent diabetes than males [Makino et al, 1980]. Given the sexual dimorphism reflected by greater male susceptibility to the diabetogenic action of the *db* gene, it is not surprising that such mutants respond very well to estrogen therapy [Leiter, 1981; Coleman et al, 1984]. Estrogen treatment has allowed dissociation of the primary expression of the *db* mutation (producing obesity and insulin resistance) from secondary consequences such as β-cell hypersecretion of insulin [Prochazka et al, 1986]. There is no evidence for such pronounced sexual dimorphism or sensitivity to estrogen therapy in either type I or type II human diabetics. Clearly, the mouse disease models have to be accepted on their own terms; they have species-unique features as well as features shared with the counterpart syndrome in humans. Accordingly, they cannot be expected to be exact replicas of human conditions.

THE PROBLEM OF DEFINING MUTANT GENE ACTION: STUDIES WITH MUTANT GENES AFFECTING ENDOCRINE GLANDS

The preceding section has established that use of human clinical descriptive terms to describe mouse genetic diseases can lead to premature assumptions about the similarity of disease processes in the two species. The incredibly diverse genetic constitutions of humans have a significant impact on the phenotypic expression of a mutant gene. Homogeneity of loci within an inbred strain of mice, on the other hand, maximizes a consistent expression of a mutant gene, thus facilitating experimental investigations. To determine how genetic background modulates expression of a specific mutant gene in the mouse, the investigator can place the mutation on different inbred strain backgrounds. The differential expression of two obesity genes (*ob* and *db*) described above when placed on two different inbred strain backgrounds illustrates the complexity of these interactions. In other words, a specific mouse mutant gene acting within an inbred background to disturb a specific biochemical process will give rise to a phenotype that may be identical to or differ in subtle to substantial ways from the paradigm description of the human genetic disorder.

Because of the remarkable conservation of genes on the X chromosome throughout evolution of species, one tends to find excellent homologies between genetic diseases in mouse and man caused by X-linked mutations [Ohno, 1969]. A well-known example of such an X-linkage homology is testicular feminization (TFM in humans, *Tfm* in mice). Affected males in both species present an external female phenotype; internally, testes are present but are very small, and other male reproductive tract tissues (vas deferens, epididymis) are lacking. *Tfm*/Y males are insensitive to androgens because of a defect in cytosolic androgen-binding protein [Bullock, 1974; Wilson et al, 1983]. Another homologous X-linked endocrinopathy in mouse and man is murine dominant hypophosphatemia (*Hyp*) and human dominant familial hypophosphatemic rickets (HPDR). In both species, hemizygous males and heterozygous females have low serum phosphate levels, rachitic bone lesions, defective phosphate transport, and subtly disordered vitamin D_3 metabolism [Eicher et al, 1976; Tennenhouse, 1983]. Although not endocrinopathies, several other clear X-linked homologies are noteworthy. Mutations at the bare patches (*Bpa*) locus in mice and the chrondrodysplasia punctata (CDPP) locus in man both express as skeletal dysplasias, cataracts, and ichthyosis [Happle et al, 1983]. Similarly, mutations at the murine mottled (*Mo*) locus and human Menkes kinky hair (MKH) locus both result in defects in copper transport [Prins and Van Den Hamer, 1979].

Detailed analyses of a genetic disorder in humans and mice lead to more precise definition of both genetic disorders through the uncovering of similarities and differences. When such results emphasize similarities between humans and mice, fundamental principles of mammalian genetics are established. Alternatively, when research results point toward more differences than similarities, then our understanding of species-specific biology increases. In either case, a benefit often forthcoming is the opening of new therapeutic approaches for genetic disorders.

Mouse mutant genes affecting endocrine gland function will most often result in deficiency conditions. The consequent atrophied state of target tissue underscores a major function of hormones—the support of specific tissue and bodily growth. It is of interest that, as yet, there are no clear examples in mice of overproduction of hormones directly attributable to mutant gene action. When confronted by a mutation with a possible endocrine deficiency, the initial challenge, of course, is to define the tissue site of mutant gene action. For example, suppose that a mutant phenotype presents with a small body size (albeit normal in form), infertility, and demonstrable autosomal recessive inheritance. In such a case, where in the cascade of amplifying signals represented by the endocrine system could this mutant gene act? Even though the "mouse clinician" may have some experimental advantages over the human clinician, the basic tools of endocrinology—systemic analyses of serum hormone levels, glandular responses to challenge stimuli, and tissue biopsies—still provide the first lines of discriminatory data whose interpretation leads to the next level of analysis. Such research into these endocrine gland metabolic disturbances has revealed the complexity of mutant gene-induced changes one would anticipate from modern knowledge of cell biology. Thus these mutations can affect the primary structure of DNA; transcription or translation processes; hormone synthesis, storage, or release systems; hormone receptors and postreceptor signal systems, or metabolizing enzymes, etc. The following examples of mouse mutant gene-induced endocrine dysfunction leading to genetic disease are offered as illustrative models.

The functional "meeting ground" of the nervous and endocrine systems is the hypothalamus, located at the base of the brain. This specialized brain region produces releasing or inhibiting hormones that govern synthesis and release of pituitary trophic hormones. The best documented inherited trait affecting hypothalamic neuroendocrine function is that found in hypogonadal (*hpg*) mice [Cattanach et al, 1977]. This autosomal recessive gene reduces gonadotropin-releasing hormone (GnRH) content of the hypothalamus to essentially undetectable levels. The biochemical basis for the effect of this mutant gene was uncovered through use of a human placental cDNA sequence for pro-GnRH that was identical to rat brain pro-GnRH sequences [Nikolics et al, 1985]. Utilizing this human cDNA placental sequence for pro-GnRH, it has been reported recently that the *hpg* mutation is actually a partial deletion of the GnRH gene [Mason et al, 1986]. The hypogonadism in these mutant mice is amenable to a novel form of therapy. Restoration of full reproduction has been accomplished in adult female *hpg/hpg* mice through surgical implantation of fetal hypothalamic tissue into the preoptic region of the hypothalamus [Gibson et al, 1984]. In humans, there is a remarkably similar syndrome known as isolated hypogonadotropic hypogonadism that involves deficient GnRH secretion [Valk et al, 1980]. In contrast to *hpg/hpg* mice, direct therapy of such patients has been limited to use of exogenously administered GnRH. Increasing availability of GnRH analogs and alternate methods of administration (other than IV injection) offer a bright therapeutic future. Although it remains for future research to establish whether some of these patients are characterized by a deletion of some or all of the GnRH gene, the *hpg* mutation has provided valuable insight into severe GnRH deficiency and opportunity for analyses of new experimental therapies, perhaps involving genetic engineering.

The pituitary gland has been dubbed the "master gland" by its aficionados and is composed of three parts that collectively secrete a combined sum of at least nine different products with broad impact on the organism. With respect to pituitary hypofunction in the mouse, there are three known autosomal recessive genes: Snell's dwarf (*dw*), Ames dwarf (*df*), and little (*lit*). The *dw* and *df* genes are nonallelic genes that manifest similar phenotypic characteristics [Schaible and Gowen, 1961; Bartke, 1965]. Both cause severe, ateliotic dwarfism that results from marked deficiency or absence of pituitary growth hormone (GH), prolactin, and thyroid-stimulating hormone (TSH). Studies of both *dw/dw* and *df/df* pituitaries for the presence of the GH gene, appropriate GH mRNA, and pituitary GH content have shown that the GH gene is present [Phillips et al, 1982], but neither mRNA for GH [Cheng et al, 1983] nor pituitary GH hormone is detectable [Eicher and Beamer, 1980]. Thus the GH deficiency in both *dw* and *df* results from independent mutant genes blocking transcription of mRNA from the GH gene.

The underlying basis of the prolactin and TSH deficiencies is not known, nor has a relationship been established between the GH transcription block and the deficiencies of these latter two hormones. The first endocrine therapy with *dw/dw* mice was reported by Smith and MacDowell [1930], who found that daily implantation of fresh rat anterior pituitary tissue restored normal growth. This pioneering endocrine therapy was refined and corroborated with purified hormone preparations [Bartke, 1965]. In humans, the most common form of familial panhypopituitarism (deficiency of GH and at least one other trophic hormone) is inherited as an autosomal recessive trait [Rimoin and Schimke, 1971]. Such individuals are dwarfed in stature and may display signs of additional pituitary hormone defects. Therapy with purified human GH undertaken in affected children is usually successful in significantly increasing their growth rate. Reports on the integrity of the GH structural gene and its mRNA transcript in familial panhypopituitary dwarfism have not yet appeared. Thus the relationship between the two mouse mutant genes and the human genetic disease at the molecular level is not clear.

Single gene-induced pituitary hormone deficiencies can also involve just one trophic hormone. A classic example may be found in little (*lit/lit*) mice that are characterized by moderate ateliotic dwarfism with marked deficiency of pituitary and serum GH [Eicher and Beamer, 1976]. Studies at the molecular level reveal that, even though *lit/lit* pituitaries possess the GH gene, only small amounts of GH mRNA and of pituitary GH are being synthesized [Cheng et al, 1983]. The presence of normal-appearing somatotropes coupled with reduced amounts of GH mRNA in *lit/lit* pituitary glands strongly suggests a reduction in the normal hypothalamic stimulation of the pituitary. Such stimulation is normally provided by GH-releasing hormone (GHRH) from the hypothalamus that acts on pituitary somatotropes to increase the synthesis of GH as well as govern its release into the blood. With the recent availability of GHRH, two different laboratories have shown that *lit/lit* pituitary somatotropes are totally resistant to the action of this releasing hormone [Clark and Robinson, 1985; Jannson et al, 1986]. Therefore, the *lit* mutant gene involves the GHRH receptor on the somatotrope, but it remains to be determined if the defect involves primary alteration of the plasma membrane receptor molecule or the postreceptor signal system. In humans, autosomally recessive isolated GH deficiency (IGHD) can be either type A with severe GH deficiency or type B with moderate GH deficiency. Type A patients respond to human GH therapy with formation of anti-GH antibody that prevents growth repair, whereas GH therapy is quite successful in boosting growth velocity in type B IGHD patients. A group of patients with type A IGHD recently has been shown to have a deletion of the normal GH gene and thus cannot make normal GH [Phillips et al, 1981]. GH therapy in these patients is analogous to administering a foreign antigen—hence the formation of GH-blocking antibodies. It is clear that the *lit/lit* mice are closely analogous to type B IGHD and not to type A IGHD.

The elegant progress made in research with the *dw*, *df*, and *lit* mutant genes contrasts sharply with research on the pygmy (*pg*) mutation, in which even the tissue site of gene action has eluded identification. The homozygous *pg/pg* mice are approximately half normal size, marginally fertile (then only if placed on an outbred background), and achieve maximal body growth by 6 weeks of age. Even though all initial phenotypic indicators point to a pituitary-related problem, the *pg/pg* mutants have morphologically normal pituitaries, adrenals, and thyroids [King, 1955]. In pursuit of possible endocrine dysfunction related to the *pg* gene, Rimoin and Richmond [1972] provided suggestive evidence of GH resistance (when compared with similarly treated *dw/dw* mice), yet serum levels of immunoreactive GH [Sinha et al, 1979], thyroid hormones and corticosterone (Beamer, unpublished observations), and total biologically active somatomedins [Nissley et al, 1980] are normal in *pg/pg* mice. Although the name "pygmy" was originally selected to denote undersized mutant mice distinguishable from *dw/dw*, a more precise analysis of serum somatomedins (insulin-like growth factors IGF-I and IGF-II) in *pg/pg* mice could be promising in view of the reported isolated IGF-I deficiency in human pygmies and patients with the pygmy trait [Merrimee et al, 1982]. These humans are diminutive in stature, have normal serum GH levels, are reproductively intact, and have been considered to be GH-resistant [Rimoin and Schimke, 1971]. More recently, ethnic pygmies and patients with the pygmy trait have been shown to differ from GH-deficient patients in that the latter responded to GH treatment with increases in both serum levels of IGF-I and IGF-II, whereas pygmies responded only with increased serum IGF-II. Demonstration of a specific IGF-I deficiency in *pg/pg* serum would establish the appropriateness of this mutant mouse as a model of the pygmy trait. On the other hand, demonstrating that serum IGF-I levels are normal in the *pg/pg* mice would suggest that these mice represent yet another genetic growth disorder, perhaps not involving endocrine functions at all. Clearly, it is premature to speculate about the site of *pg* action.

General body metabolism and growth are profoundly influenced by another gland, the thyroid. Its activities can be both dependent and independent of the pituitary; thus mutant genes disturbing thyroid function can have widely ranging effects. Primary disturbance of thyroid function has been found to result from the action of two different autosomal mutant genes in mice. Hypothyroid (*hyt/hyt*) mice have severe hypothyroidism and thyroidal agenesis caused by resistance to the action of pituitary TSH [Beamer and Cresswell, 1982]. Since the *hyt/hyt* thyroid will secrete thyroxine in vitro in response to forskolin or cholera toxin (Beamer, unpublished observations), it appears that this mutant gene acts either on the thyrofollicular cell TSH receptor or a postreceptor signalling step prior to activation of the plasma membrane-bound adenylate cyclase catalytic subunit. The *hyt/hyt* mice are also characterized by a number of behavioral anomalies [Beamer and Targ, 1980]. Although the impact of replacement therapy on behavioral anomalies is not known, such therapy from 3 weeks of age onward is effective in restoring *hyt/hyt* body growth to normal [Beamer et al, 1981] but is not effective in repairing *hyt*-induced cerebral myelination deficits [Noguchi et al, 1985]. In humans, thyroidal agenesis is the major autosomal recessively inherited thyroid disorder of the newborn [Fisher et al, 1979]. Therapy of these severely hypothyroid infants must be initiated within the first few months of life for maximum relief from the effects of hypothyroidism. Although little is known about the inborn error of metabolism in this human genetic disorder, resistance to the action of TSH is a possible explanation. The *hyt/hyt* mice represent an excellent research model for this human genetic disease.

The second mouse mutant gene affecting thyroid function is named congenital goiter (*cog*). The *cog/cog* mice develop a hypertrophied thyroid gland induced by excess TSH released from the pituitary in response to deficient serum thyroid hormone levels [Beamer et al, 1986]. In the search for the mutant gene-induced metabolic error in the thyroid, defects in iodide trapping or in incorporation of iodide into organic molecules have been eliminated. Since these mutant glands do contain small amounts of immunoreactive thyroglobulin, it is possible that the *cog* mutant allele affects the intracellular control of thyroglobulin synthesis. Disorder of thyroglobulin synthesis is one of four classes of defects in thyroid function leading to human goitrogenesis [Stanbury and Dumont, 1983]. Within these disorders of synthesis, there are undoubtedly several different genetic defects whose etiologies remain to be elucidated. Further research with *cog/cog* mice will define the mechanism for one such defect.

MOUSE MODELS FOR GENETICALLY DETERMINED IMMUNOLOGIC DYSFUNCTION

Single gene mutations affecting the immune system of mice have made important contributions to our understanding of the complex interrelationships between various components of this system. Several of these mutant genes will be highlighted to illustrate selective defects at the level of T lymphocytes (the "nude" mutation), B lymphocytes (X-linked immunodeficiency), or both ("severe combined immunodeficiency disease," "motheaten," and "viable motheaten"). These mutations will be discussed to underscore the wide range of immunologic dysfunction in these mutant mice and to show how these "experiments of nature" are being utilized to increase our understanding of the mammalian immune system in normal and pathologic states.

Deficient T-Lymphocyte Development in Nude Mice

The congenitally hairless and athymic "nude" (*nu*) mouse is the most widely studied immunologic mutant and has been the subject of numerous workshops. Nude mice have been used extensively in studies of the role of T cells in immune responses to a great variety of antigens. Moreover, the finding that human neoplasms can grow in nude mice has resulted in many investigations on the mechanisms of transplantable human tumor growth, including metastasis and the response to experimental therapeutic measures. Absence of the thymus in nude mice is apparently traceable to a developmental failure of the thymic anlage that arises from the third pharyngeal pouch. The thymic rudiment remains small and cystic, whereas other derivatives of the third pharyngeal pouch are unaffected. Such normal development of other organs arising from the third pharyngeal pouch of nude mice is in sharp contrast to humans affected with DiGeorge syndrome, which is characterized by aplasia of both the thymus and parathyroids [Rosen et al, 1984]. Although the majority of such patients appear to have suffered intrauterine injury early in gestation, there are several cases of affected sibs on record, suggesting an underlying genetic basis. An association between DiGeorge syndrome and an unbalanced chromosomal rearrangement entailing partial trisomy for chromosome 20 and partial monosomy for chromosome 22 suggests that this syndrome can be caused by deletion of a gene, probably in band 22q11 [De La Chapelle et al, 1981]. The developing thymus in normal mice incorporates ectodermal as well as endodermal tissue by 9–10 days of gestation. By 14 days, thymic epithelial cells contain tonofilaments and desmosomes characteristic of the adult gland. Moreover, at this stage, the lymphocytic population of the thymus has begun [Mandel, 1970]. It has been suggested that the underlying basis for the failure of stromal elements in nude mice to interact with normal lymphocytic precursors might be an abnormal distribution or expression of Ia antigens by epithelial components. Although normal thymic epithelial cells express such antigens by 13 days of gestation, there is no detectable Ia expression in the 13-day fetal thymic rudiment of nude mice [Jenkinson et al, 1981]. Since the thymus plays a critical role in determining the maturation of functional T cells, it is not surprising that this mutant is T cell deficient. However, recent work has shown that by 6 months of age low numbers of functional T cells appear in lymphoid tissues of nude mice [MacDonald, 1984].

Because the original mutation to nude occurred in a noninbred strain, much of the early work with this mutation was subject to variation associated with heterogeneity in the genetic background. Although the nude gene has been backcrossed onto a variety of inbred strains, the possibility of "contaminating" genes (ie, genes other than the *nu* gene from the donor strain that are not removed by subsequent backcrosses to the recipient strain) must always be considered during experimental design and data analysis. More recently, a second spontaneous mutation at the nude locus called streaker (nu^{str}) occurred in the inbred AKR/J strain of mice [Schultz et al, 1982]. Homozygotes (nu^{str}/nu^{str}) resemble nude mice, the difference between the two alleles being no greater than reported variations among nude mice. Although nearly all AKR/J − +/+ mice acquire thymic lymphomas by 1 year of age, AKR/J-nu^{str}/nu^{str} mice, as expected, are refractory to the development of these neoplasms. Nevertheless, thymic lymphomas can be induced experimentally in this mutant when preceded by grafting thymic implants under the kidney capsule. Autosomal recessive mutations causing thymic aplasia have also been reported in the rat and guinea pig. Although the phenotypic effects of these mutations are not as well characterized as those at the *nu* locus in mice, investigations to date show that homozygosity for the mutations in rats and guinea pigs causes defects in the skin and immune system similar to those reported in nude mice [Festing, 1982; Reed and O'Donoghue, 1982].

Impaired B-Lymphocyte Development in Mice

Although there is no known mouse mutant with a selective block in the development of all B-cell lineage lymphocytes, hemizygous males (*xid/Y*) and homozygous females (*xid/xid*) for the mutation X-linked immunodeficiency appear to lack a specific population of mature B cells. The *xid* mutation was originally discovered in CBA/N mice at the National Institutes of Health. This mutation causes a defective antibody response to certain thymic-independent antigens and to thymic-dependent antigens, decreased serum IgM and IgG3 levels, and a defective proliferative response to B-cell mitogens [Scher, 1982]. Studies of B-cell surface marker expression and immune responsiveness have indicated that B cells in CBA/N mice are immature, resembling the B cells of young (1–3 week old) normal mice. However, recent studies have demonstrated that B cells from these mutant mice are idiosyncratic, and they do not represent normal B cells [Mosier, 1985]. Recently, Cohen et al [1985] sought *xid* mRNA by the "subtraction" cloning method. They prepared a lymphocyte-specific cDNA clone that recognizes an X-linked gene family with 15–20 members. These genes, named X-linked lymphocyte-regulated (XLR), apparently include the locus defined by the *xid* mutation. Since the X chromosome of eutherian mammals has been conserved throughout evolution, it appears likely that this gene family is present on the X chromosome of man and may be important in regulating immune functions that are disturbed in certain X-linked immunodeficiency diseases. Although *xid* is not an exact homolog of any known human immunodeficiency disease, this mutation shares certain characteristics with Wiskott-Aldrich syndrome. Patients with this X-linked disease have decreased levels of serum IgM and are unable to respond to polysaccharide antigens. However, unlike *xid* mice, patients affected with Wiskott-Aldrich syndrome have defects in T cells and platelets [Rosen et al, 1984).

Severe Combined Immunodeficiency Disease

Severe combined immunodeficiency disease (SCID) includes a broad spectrum of syndromes characterized by defective function of both T and B lymphocytes. Inherited deficiencies of purine metabolic enzymes underlie two immunodeficiency diseases of man. Patients with adenosine deaminase (ADA) deficiency account for approximately 25% of cases of the autosomal recessive form of SCID. Purine nucleotide phosphorylase (PNP) deficiency is a rarer disorder and causes a selective defect in cellular immunity. The biochemical basis for immunodeficiency in these diseases appears to be the selective trapping of the ADA substrate deoxyadenosine and the PNP substrate deoxyguanosine by thymic lymphocytes. These lymphocytes accumulate high levels of deoxyribonucleoside triphosphates that cause an inhibition of ribonucleotide and DNA synthesis [Boss and Seegmiller, 1982].

Although mice exhibiting ADA and PNP deficiency might also be anticipated to exhibit immunodeficiency resulting from the selective toxicity of deoxyribonucleosides accumulating inside lymphocytes, it has been reported that human and mouse cells differ in their metabolism of purines. In human thymic lymphocytes, high levels of deoxycytidine kinase, for which deoxyadenosine and deoxyguanosine are known substrates, and low levels of dephosphorylating activity are responsible for the build-up of deoxyribonucleoside triphosphates in the absence of ADA or PNP. However, mice differ from humans in the tissue distribution of deoxyadenosine phosphorylating activity [Carson et al, 1980]. Therefore, it is possible that an inherited deficiency in purine metabolism in mice might not result in a severe combined immunodeficiency disease as in humans.

Mutations at two loci in the mouse produce severe immunodeficiency disease without obvious defects in enzymes controlling nucleic acid metabolism. A recently discovered autosomal recessive mutation termed "severe combined immunodeficiency" (*scid*) severely impairs lymphopoiesis. Homozygotes (*scid/scid*) have greatly reduced numbers of lymphocytes, are defective in functions mediated by T and B cells, and have profound hypogammaglobulinemia [Bosma et al, 1983; Dorshkind et al, 1984]. These mutant mice have an extremely small thymus; approximately 10% of homozygotes develop thymic lymphoma by 9 months of age. The lymphopoietic microenvironment of *scid/scid* mice is conducive to lymphocyte differentiation as shown in the development of functional lymphocytes after reconstitution with normal bone marrow. Bone marrow cells from *scid/scid* mice rescue lethally irradiated normal syngeneic recipients but fail to produce differentiated lymphocytes. Thus the *scid* mutation selectively impairs the differentiation of lymphopoietic stem cells. It appears to be a homolog of human SCID characterized by normal adenosine deaminase levels and a dearth of functional lymphocytes [Rosen et al, 1984].

Immunodeficiency in man is often accompanied by autoimmune disease. Rheumatoid arthritis is a frequent complication of X-linked agammaglobulinemia. In addition, genetic deficiency of several components of the complement system in man is accompanied by systemic lupus erythematosus [Amman, 1977; Tourraine, 1976]. Genetic deficiency of the second component of complement (C2) is the most common genetically determined complement-deficiency state. Approximately 50% of affected patients studied have had autoimmune diseases [reviewed in Cole et al, 1985]. Similarly, homozygosity for the autosomal recessive mutation called "motheaten" (*me*) causes both severe immunodeficiency and systemic autoimmune disease. Homozygotes (*me/me*) can be recognized within 2 days after birth by the appearance of abscesses in the skin. Many of these mice die before weaning and the mean life-span is 3 weeks [Green and Schultz, 1975]. Motheaten mice do not respond to thymic-dependent or thymic-independent antigens. They have severely impaired proliferative responses to both T-cell and B-cell mitogens and also lack cytotoxic T-cell as well as natural killer (NK)-cell activity [Sidman et al, 1978; Davidson et al, 1979; Clark et al, 1981; Shultz and Green, 1976]. These mice produce abnormally high levels of serum immunoglobulins, express multiple autoantibodies, and show deposition of immune complexes in the kidneys, skin, and lungs [Shultz and Green, 1976]. The short life-span of this mutant is associated with the accumulation of macrophages in pulmonary alveoli. Investigations with this mutant are restricted because even the longest-lived homozygotes die from pneumonitis by 7 weeks of age. Recently, a spontaneous remutation at the motheaten locus occurred. Since homozygotes for this new allele show a mean life-span of 9 weeks compared to 3 weeks for *me/me* mice, it was designated "viable motheaten" (me^{v}) [Shultz et al, 1984]. Homozygotes (me^{v}/me^{v}) are severely immunodeficient and develop hyperimmunoglobulinemia with multiple autoantibodies. However, a delay in the development of autoimmune pneumonitis facilitates investigation with this mutant. Recent investigation on the biochemical basis of hyperimmunoglobulinemia in me^{v}/me^{v} mice has revealed a dramatic overproduction of two B-cell maturation-promoting lymphokines. These lymphokines drive normal resting B cells to differentiate into antibody secreting cells and appear to underlie the observed polyclonal B-cell activation in *me*^{v}/*me*^{v} mice [Sidman et al, 1984]. Although these mutant mice are not homologs for a recognized, specific human immune dysfunction, determination of mechanisms underlying their lack of immune function, and the development of systemic autoimmune disease early in life, may help to elucidate the bases of severe immunodeficiency and autoimmunity diseases in man.

In addition to the immunologic mutations described above, there are many other single genes known to cause defects in the immune system. Detailed descriptions of these mutant mice can be found in a comprehensive treatise on the known genetic variants of the mouse [Green, 1981] and in a review that focuses on the immune system in many of these mutant mice [Shultz and Roths, in press]. It should be stressed that these single genes do not act alone but function in concert with the background genome. Although most of the published studies based on research with these mutant mice have concentrated on the whole animal and changes at cellular levels, emphasis is now shifting to investigations at the molecular level.

INHERITED ANEMIAS IN MICE

Mouse red blood cell dyscrasias can be arbitrarily separated into several classes that include the globinopathies (thalassemias and hemoglobinopathies), the spectrin-deficient anemias, the microcytic anemias (except deletion thalassemias), and the macrocytic anemias. Until recently, inherited anemias in mice were diseases with no definitive biochemical defect and no known human counterparts. Mice with hemoglobinopathies did not exist, presumably because carriers were too severely affected to survive. The erythrocytes of mice with hemolytic anemias, unlike the erythrocytes in some humans with hemolytic diseases, showed no decreased activity of glycolytic and pentose phosphate pathway enzymes and had normal levels of methemoglobin [Hutton and Bernstein, 1973]. Only one of the four known mutations causing a microcytic anemia showed signs of a thalassemia-like imbalance of α- and β-globin synthesis, and that imbalance was secondary to a malfunction in iron metabolism [Cole et al, 1974]. The defects in mice with two of three known types of macrocytic, normochromic anemias were traced, one to an intrinsic and one to an environmentally determined malfunction of hemopoietic stem cells [Russell, 1979].

Newer techniques for quantifying erythroid stem cell precursors [Gregory, 1976] for the rapid analysis of hemoglobins [Whitney, 1978] and membrane proteins [Fairbanks et al, 1971; Laemmli, 1970], and for cloning cDNA from isolated cells types have expedited our search for new hereditary variants and our capacity to define the defects in existing mutations. A similar revolution, occurring in the analysis of human hereditary blood dyscrasias, has indicated that human models do, indeed, exist for several of the mouse mutations. As the lesions in the genome are determined, more similarities between human and mouse diseases will undoubtedly be uncovered. In mice, as in humans with hereditary blood dyscrasias, research is directed toward defining the cellular and molecular defect, the genetic lesion, and the efficacy of various therapeutic measures, including bone marrow transplantation. The expectation that mice with inherited anemias will be good models for examining the potential of foreign gene transfer has recently been tested in β-thalassemic mice [Constantini et al, 1986]. With the exception of thalassemic mice, gene transfer either into the male pronucleus of a fertilized egg or into the progenitors of myeloid and lymphoid cells must await identification of the defective gene and generation of recombinant DNA for the normal allele. When gene therapy is perfected, it will probably require actual repair of the genetic lesion through either site-specific recombination or gene conversion. Only by such extreme measures will the gene be subject to the appropriate regulatory signals. In the meantime, the dissection of blood dyscrasias at the cellular, biochemical, and molecular levels in mice with a common genetic background augments our knowledge of erythroid cell ontogeny, structure, and function. This information will be useful in designing protocols for alleviating disease symptoms in both mouse and man. A representative sampling of the known mutations, their defects, and their relationships to known human dyscrasias are described below.

Thalassemias and Hemoglobinopathies

There are now three known deletions that cause α-thalassemia [Whitney and Russell, 1980; Russell et al, 1976], one, $Hba^{th\text{-}J}/+$, maintained at The Jackson Laboratory and two, $Hba^{b-2(th)}/+$ (formerly 352HA) and $Hba^{b-3(th)}/+$ (formerly 27HB), maintained by Dr. Raymond A. Popp at Oak Ridge National Laboratory. Homozygotes die in utero soon after implantation [Popp et al, 1980]. Heterozygotes are discernible from their normal littermates by decreased osmotic fragility of their red blood cells in 0.45% NaCl in H_2O. Neither the two adult α-globin chains [Popp et al, 1979] nor the embryonic α-like-globin chains [Popp et al, 1980; Whitney and Russell, 1980] are expressed from the affected chromosome. An embryonic β-like-globin forms a tetramer similar to hemoglobin Bart in the yolk sac-derived red blood cells of the heterozygotes [Popp et al, 1980; Whitney and Russell, 1980]. Adult erythrocytes have bizarre shapes [Popp et al, 1979] and are probably short-lived. The affected mice are good models for stem cell transfer experiments in that their stem cells compete poorly with transplanted normal marrow cells [Barker and McFarland, 1985]. As such, they may also be good recipients of their own marrow cells transfected with normal α-globin genes. One drawback to such experiments is the large size of the deletions that thus far preclude site-specific replacement with the normal haplotype in cells from the mutant mice.

A mouse model for human β-thalassemia, $Hbb^{d3(th)}$ (formerly $Hbb^{(th)}$), was described in 1983 by Skow et al. The major adult β-globin sequence, including regulatory sequences and all coding blocks, is deleted, but the minor adult β-like-globin remains intact. Increased synthesis from the minor adult β-like-globin, which has some of the characteristics of a human fetal β-like-globin [Whitney, 1977], compensates in part for the deletion in the homozygotes. The affected α- and β-thalassemic mice, since they accumulate iron in their tissues, are excellent models for experimental chemotherapeutic measures that prevent the accumulation of iron and development of hemosiderosis, since they accumulate iron in their tissues [Van Wyck et al, 1984] (also, M. Garrick, personal communication).

A hemoglobinopathy, Hbb^{a4}, involving a single amino acid alteration from Tyr → Cys at position 145 in the major adult mouse β-globin has been described by Peters et al [1985]. It is a homolog of hemoglobin Rainier in man, and the homozygote is characterized by polycythemia, rubor, increased oxygen affinity, and decreased heme-heme interaction. Other recently described globin-chain mutations appear to have no deleterious effects on the carriers. These include mutations in the β chains designated Hbb^{d2} (formerly Hbb^{e} [Murota et al, 1982]) and Hbb^{s2} [Lewis et al, 1985] and a mutation in an α chain, Hba^{g2} (formerly Hba^{y9}) [Popp et al, 1983]. Hbb^{s2} alters the valine at position 60 in an adult β-globin to a glutamic acid, and Hba^{g2} changes one of the adult α-globin chains from histidine to leucine at position 89. Electrophoretically neutral substitutions in the α-globin chains carried by different mouse strains were described previously [Popp, 1963, 1967; Russell and McFarland, 1974] and the list has recently been extended to include a new variant and various new combinations of the known α chains in wild mice [Whitney et al, 1985].

Spectrin-Deficient Anemias

The mutations that affect the stability of the red blood cell membrane now include four alleles at the spherocytosis, *sph*, locus [Unger et al, 1983], which is the site of the structural gene for erythroid α-spectrin. Three of these, *sph*, sph^{2BC}, and sph^{2J}, decrease α-spectrin synthesis, and the fourth, hemolytic anemia, *sph*^{ha}, alters α-spectrin membrane binding [Bodine et al, 1984]. In another mutant mouse known as jaundiced, *ja/ja*, there is no detectable β-spectrin synthesis, suggesting that *ja* disrupts β-spectrin transcription, processing, or translation. The mutation called normoblastosis, *nb*, affects ankyrin, the membrane-binding protein for tetramers of α- and β-spectrin. Molecular probes are now available for erythroid α- and β-spectrin, for ankyrin [Moon et al, 1985; Cioe and Curtis, 1985], and for the brain-specific form of the α-spectrin [Birkenmeier et al, 1985]. The molecular probes will be of inestimable value in further defining the genetic lesions in the spectrin-deficient mice and their human counterparts and in mapping the structural gene loci [Huebner et al, 1985].

Microcytic Anemias

The genesis of defects in iron metabolism is being studied in mice with four different mutations that cause microcytic anemias. Male mice with sex-linked anemia, *sla*/Y, have a specific defect of iron absorption by the intestine; this is completely corrected if the lesion is bypassed by parenteral injection of iron [Bannerman, 1981]. The primary cellular defect in mice with microcytic anemia, *mk/mk*, is more difficult to pinpoint, since there is a general impairment of iron entry into cells [Bannerman, 1981] including the intestinal mucosa and erythroid precursors. Anemia is transferred to lethally irradiated normal recipients by transplanting bone marrow from *mk/mk* mice [Harrison, 1972] but not by transplanting *sla*/Y marrow cells [Bennett et al, 1968]. Dissecting out the role iron metabolism plays in flexed, *f/f*, mice is even more difficult. The anemia is transitory, affecting primarily fetal erythropoiesis, but the mutation also causes prolonged recovery from induced anemic crises in the adult [Fowler et al, 1967]. β-Globin synthesis is decreased when compared to α-globin synthesis in fetal erythrocytes [Chui et al, 1977]. Iron accumulates and siderocytes are common in prenatal erythrocytes [Grüneberg, 1942]. The question as to whether intracellular iron mobilization, the hemoglobinopathy, or some hitherto unsuspected cell anomaly is the primary defect is still unanswered. Mice homozygous for the mutation hemoglobin deficient, *hbd*, seem to have basic problems with iron mobilization within the erythroid cells [Bannerman et al, 1986].

Macrocytic Anemias

Mice with macrocytic anemias have relatively severe defects. In this group are mice with mutations at various loci, including dominant white spotting, *W*, and its multiple alleles on chromosome 5; Steel, *Sl*, and its multiple alleles on chromosome 10; and the recessive gene Hertwig anemia, *an*, on chromosome 4. Mice carrying a double dose of one of the *W* alleles have a hemopoietic stem-cell deficit, but whether this involves the lack of either a totipotent stem cell or a progenitor cell with more restricted potential is not yet clear (Barker and Russell, in preparation). Sl^{d}/Sl^{d} mice have an environmental defect that decreases the numbers of erythroid [McCulloch et al, 1965] and mast cell [Kitamura and Go, 1979] committed progenitors. The fact that mice with Hertwig anemia lose cells during successive stages of erythroid differentiation [Barker and Bernstein, 1983] led to experiments showing that there was an unequal distribution of chromosomes in at least 5% to 15% of the dividing cells [Eppig and Barker, 1984]. The defect in *an/an* mice is probably common to all rapidly dividing cells and is not limited to precursors of the hemopoietic system.

EXAMPLES OF OTHER CLASSES OF MOUSE MUTATIONS

Neurologic and Muscular Mutants

At present, there are well over 100 mutations in the mouse specifically affecting the central and peripheral nervous systems, and this number is steadily increasing. Many recessive neurologic mutants are juvenile lethals, and many others are infertile, conditions that necessitate breeding from carriers. Most of these neurologic mutants show some kind of abnormal behavior and generally some histologic abnormality. A rough classification of the neurologic mutants discussed here would include those with cerebellar abnormalities, myelination abnormalities, and neural crest disorders. Cerebellar mutants have several characteristics that make them particularly useful tools. They are easily distinguished from normal sibs and often live to adulthood, and the lesion(s) is generally limited to the cerebellum. Many of these mutants are given names depicting the behavioral symptoms such as reeler, *rl*, staggerer, *sg*, and weaver, *wv*. Others have names indicative of the known pathology, such as Purkinje cell degeneration, *pcd*. One of the basic questions to be answered in any genetic neurologic disorder is which cell population is affected by the mutation. Much research on cerebellar mutants of the mouse has addressed this question, and much has been learned about the development of spatial organization in the embryonic and adult brain.

Homozygotes of the four mutations mentioned above, *rl*, *sg*, *wv*, and *pcd*, have abnormal cerebellar architecture and ataxia with a reeling-type gait [Green, 1981]. In reeler (*rl/rl*) mice, the Purkinje cells, which are the sole output neurons of the cerebellar cortex, are not in their proper alignment. DeLong and Sidman [1970] suggested that the disorder could be due to a failure of cell-cell recognition. Recently, Edelman and Chuong [1982] discovered that the staggerer (*sg/sg*) mouse, with its large reduction in Purkinje cells and a degeneration of most of the granule cells, had a defective development of the neural cell adhesion molecule (N-CAM). They found that *sg/sg* mice process the N-CAM molecule aberrantly and have suggested that some of the features of synapse formation and cell survival in the cerebellum of the *sg/sg* may result from this anomaly. Weaver (*wv/wv*) and Purkinje cell degeneration (*pcd/pcd*) mice have been intensively investigated. In weaver mice, the cerebellum is almost devoid of granule cells that fail to migrate to the internal granular layer and degenerate in their proliferative zone. In addition *wv/wv* mice also show a deficit in dopamine content in the dorsal striatum but not in the ventral striatum [Roffler-Tarlov and Graybiel, 1984]. In *pcd/pcd* mice, almost all cerebellar Purkinje cells degenerate after the completion of normal cerebellar development. Recently, Roffler-Tarlov et al [1984] used this genetically programmed death of Purkinje cells in the *pcd/pcd* mouse to determine that the absence of the Purkinje cells, target cells for norandrenergic projection, did not affect the number of noradrenergic terminals in the mutant cerebellar cortex.

Mutant mice with myelin degeneration or deficiency are also recognized by their locomotor abnormalities, particularly by a jerking, unsteady gait. Myelin forms the sheath of the nerve fibers and facilitates conduction of nerve impulses. Central nervous system myelin contains two basic proteins, proteolipid protein and myelin basic protein (PLP and MBP, respectively). Genetically determined demyelinating diseases have been associated with deficiencies in one or both of these constituent proteins. Structural genes for these proteins have recently been identified. Molecular mapping of the gene for myelin PLP using a complementary DNA for bovine PLP has allowed X-chromosome assignment in both mouse and man [Willard and Riordan, 1985]. Jimpy, *jp*, is a recessive X-linked mutation producing myelin deficiency in the mouse and may have its homolog in a rare human X-linked disorder of myelin synthesis, the Pelizaeus-Merzbacher disease. Comparison of the gene maps of the human and mouse X chromosome indicated that the *jp* locus in the mouse probably represents a mutation in the myelin PLP structural gene [Willard and Riordan, 1985]. In shiverer (*shi/shi*) mice, the central and peripheral nervous systems are depleted of MPB. Since the structural gene for this protein and *shi* both map to chromosome 18, and since the *shi* mutation produces a partial deletion in the gene for MBP, *shi* probably represents a mutation in the structural locus for MBP [Roach et al, 1985; Sidman et al, 1985]. The defect produced by the *shi* mutation is intrinsic to the myelin-forming cells, oligodendrocytes of the central nervous system and Schwann cells of the peripheral nervous system [Billings-Gagliardi et al, 1984; Shine and Sidman, 1984].

Twitcher, *twi*, is a relatively new recessive autosomal mutation with demyelination in the central and peripheral nervous systems caused by deficient activity of galactosylceramidase [Kobayashi et al, 1980]. It is considered to be homologous to the human globoid cell leukodystrophy, Krabbe disease. The defect in twitcher is intrinsic to Schwann cells [Scaravilli and Jacobs, 1981]. No definitive therapy for this human heritable storage disease, in which the toxic metabolite psychosine accumulates in globoid macrophages, has been provided. Therapeutic measures for enzymatic deficiency diseases have involved enzyme replacement either by organ transplant or by administration of purified enzyme. Such measures have not provided significant disease control in the human disease [Yeager et al, 1984]. However, survival time was prolonged in *twi/twi* mice with bone marrow transplants, but no improvement in the neurodegenerative process in the CNS was found [Yeager et al, 1984]. Perhaps specific gene replacement can be the next experimental therapy now that a homologous model in the mouse is available.

Two mutations that probably act through the neural crest are expressed as a deficiency of pigment cells and of enteric ganglion cells in the terminal gut resulting in congenital megacolon (or Hirschsprung disease in man). The recessive mutations used as models of Hirschsprung disease, which is one of the most common causes of intestinal obstruction in the newborn child, are lethal spotting, *ls*, and piebald lethal, s^1. These mutant mice are recognized by white spotting in the *ls* homozygote and an almost entirely white coat in the s^1 homozygote. In both mutants, megacolon may become evident at any time from age 10 days to age 10 months [Lane, 1966]. It has been shown by Webster [1973] that the absence of enteric ganglia in the distal gut of the s^1/s^1 mouse is the result of a slower than normal migration of vagal neuroblasts. Consequently, the migration of the neuroblasts and the elongation of the gut are out of phase, so the migrating neuroblasts never reach the end of the gut. Recent studies with s^1/s^1 mice have shown that cholinergic innervation is absent in the aganglionic colon [Ueki et al, 1985] and that tissue concentrations of neuropeptides are decreased [Kaufman et al, 1985]. Rothman and Gershon [1984] reported that the terminal 2 mm of the *ls/ls* gut appears to be intrinsically abnormal and cannot be colonized by enteric neurons. It was concluded that more than a deficiency in the migration of precursor cells is involved in the pathogenesis of aganglionosis.

McKusick [1986] lists numerous forms of human muscular dystrophy. Several forms are inherited as autosomal recessive traits, one of which resembles the X-linked Duchenne-type dystrophy with onset before 6 years of age, wheelchair confinement by age 12, and death usually before age 20. The two major X-linked forms of dystrophy are Duchenne and Becker. In Becker, the onset is later and survival time is longer than in Duchenne. Recent linkage analysis with restriction fragment length polymorphisms suggests that the Duchenne and Becker X-linked types could be allelic [Murray et al, 1982; Kingston et al, 1983; Bulfield et al, 1984]. Although Duchenne muscular dystrophy is one of the most common and serious human X-linked disorders, the primary biochemical defect still remains unknown [Murray et al, 1982]. The autosomal murine mutation, dystrophia muscularis, *dy*, its allele, dy^{2J}, and another mutation in the mouse, myodystrophy, *myd* [Lane et al, 1976], have provided the scientific community with animal models of progressive atrophy of skeletal muscle that resemble features of the human muscular dystrophies. However, in the murine dystrophies, both *dy* and *myd* homozygotes have defective myelination of the dorsal and ventral roots, probably resulting from a defect intrinsic to the Schwann cells [Green, 1981]. Until recently, there has been no murine model of the X-linked Duchenne muscular dystrophy. In 1984, X-linked muscular dystrophy, *mdx*, in the mouse was reported by Bulfield et al. Homozygotes were shown to be viable, have high serum levels of muscle enzymes, and exhibit muscle lesions similar to human muscular dystrophy. Comparative gene mapping has placed both the mouse *mdx* gene and the human Duchenne gene locus at the same relative position on the X chromosome [Bulfield et al, 1984]. However, the clinical symptoms of mild tremor and uncoordinated gait found in the affected mouse are later in onset and less severe than those found in the affected human.

Among the numerous other myopathies of the mouse are myotonia, *mto*, which closely resembles the human condition of myotonia cogenita [Heller et al, 1982], and motor end-plate disease, *med*, which exhibits progressive weakness of skeletal muscles in proximal limbs [Green, 1981]. These mutant genes are autosomal recessive, and neither mutant phenotype is known to be homologous to any specific human muscle disease, but each provides an unusual tool for experimental study.

Mutations With Bone and Cartilage Defects

Again, close to 100 mutations in the mouse affect the skeleton. Mice exhibiting conditions of osteopetrosis, chondrodysplasia, and congenital limb deformities (including polydactyly, syndactyly, and hypodactyly) as well as those with degenerative joint disease are available as animal models. Since it is beyond the scope of this chapter to cover all mutants in the mouse, this discussion will be limited to those mutations producing osteopetrosis and chondrodysplasia.

Osteopetrosis is a metabolic bone disease with reduction in bone resorption that results from a defect in the production and function of the osteoclast cell [Marks et al, 1984]. Human osteopetrosis or Albers-Schonberg syndrome is also referred to as "marble bone disease." There are four genetically distinct osteopetrotic mutants in the mouse; grey-lethal, *gl*, microphthalmia, *mi*, osteopetrosis, *op*, and osteosclerotic, *oc*. Each is an autosomal recessive and is located on a different chromosome. In all these mutants, the bones do not grow normally, and there is no marrow cavity [Green, 1981]. The grey lethal (*gl/gl*) mouse dies between 20 and 30 days of age; the teeth never erupt, and the grey coat color results from the clumping of phaeomelanin granules. Transplantation of bone marrow or spleen cells from normal sibs cures the disease of *gl/gl* mice [Walker, 1975]. The microphthalmia (*mi/mi*) mouse dies at weaning; it is devoid of pigment; the incisors fail to erupt, and, like *gl/gl* mice, bone resorption is restored by marrow and spleen transplants [Green, 1981; Walker, 1975]. The osteopetrotic (*op/op*) mouse survives if provided adequate soft food. The teeth do not erupt; the rate of bone formation is initially elevated but drops by age 6 weeks, and a marrow cavity starts to form in the long bones. However, unlike *gl/gl* and *mi/mi* mice, the *op/op* mouse is not cured by bone and spleen transplants [Marks et al, 1984]. The osteosclerotic (*oc/oc*) mouse dies by 20 days of age; its teeth do not erupt, and it has markedly thickened epiphyseal cartilage growth plates characteristic of rickets [Marks et al, 1985]. No transplantation studies with *oc* have yet been reported.

The therapeutic treatment of children with congenital osteopetrosis by bone marrow transplantation followed the early work in *gl/gl* and *mi/mi* mice by Walker [1975]. Some human transplants have been successful [Coccia et al, 1980]; others have not [Sorell et al, 1981], indicating that human osteopetrosis, like that in the mouse, may result from mutation at several different genetic loci.

The human osteochondrodysplasias are a very large heterogeneous group of disorders affecting the size and shape of the limbs, trunk, and skull. They frequently result in disproportionate dwarfing and are classified primarily on the basis of their clinical, radiologic, and genetic features [Rimoin and Lachman, 1983]. There are several mutations in the mouse that show phenotypic resemblance to some of these human dysplasias. Both achondroplasia, *cn*, and brachymorphic, *bm*, are autosomal recessive mutations that, in the homozygous state, exhibit features similar to those of spondyloepiphyseal dysplasias in children [Wikstrom; 1984].

The achondroplasia (*cn/cn*) mouse is a severely disproportionate dwarf with shortened limb bones, short thick tail, and a domed skull [Lane and

Dickie, 1968]. The disturbance in endochondral growth seems to be due primarily to underdevelopment of the hypertrophic cartilage zone. However, no abnormalities of the proteoglycans, one of the major components of cartilage, or collagen content in the epiphyses have been observed [Kleinman et al, 1977; Wikstrom, 1984].

The brachymorphic (*bm/bm*) mouse is also a disproportionate dwarf with disturbed endochondral bone formation. However, unlike the cartilage of the *cn/cn* mouse, that of *bm/bm* mice contains undersulfated chondroitin sulfate—the major glycosaminoglycan of cartilage [Wikstrom, 1984]. According to Wikstrom [1984], the study of human skeletal dysplasias is very difficult because it is almost impossible to study different stages of the disease. Furthermore, it is equally difficult to obtain proper control materials. Probably the best solution to these problems is the use of the appropriate mouse models.

CONCLUSIONS

This chapter has discussed selected genetic mutations producing inborn errors of metabolism in mice, with the view that knowledge gained from an analysis of the disease in mice might be applicable to an understanding of the ontogeny of analogous or homologous conditions in man and to the development of effective therapies. Numerous additional genetic loci known to control susceptibility or resistance of various inbred strains of mice to a variety of infectious agents, drugs, and toxins were not discussed. Species differences between man and mouse notwithstanding, the many kinds of genetically defined mice have clearly established this small mammal as the premier experimental tool for gene therapy experiments. The use of animals in scientific research is currently a topic of debate by a public concerned that animals not be exposed to needless pain or suffering. Indeed, the question has been raised as to whether it is ethical to seek and to propagate deleterious mutations in animals to serve the "greater good" of mankind. Knowledge gained from the study of the molecular and biochemical genetics of the mouse has already provided much insight into human disease processes. When a single gene mouse mutation of biomedical significance is well characterized from the organismic and cellular levels all the way to the defect at the nucleotide level of the DNA, that mutant mouse becomes invaluable as a vehicle for development of gene, drug, or transplantation therapy. The technologies so developed will undoubtedly revolutionize the diagnosis and treatment of human constitutional disorders in the century to come.

REFERENCES

Amman AA: Immunodeficiency disorders and autoimmunity. In Talal N (ed): "Autoimmunity—Genetic, Immunologic, Virologic, and Clinical Aspects." New York: Academic Press, 1977, pp 479–512.

Bannerman RM: Of mice, men and microcytes. J Pediatr 98:760–762, 1981.

Bannerman RM, Garrick LM, Rusnak-Smalley P, Hoke JE, Edwards JA: Hemoglobin deficit: An inherited hypochromic anemia in the mouse. Proc Soc Exp Biol Med 182:52–57, 1986.

Barker JE, Bernstein SE: Hertwig's anemia: Characterization of the stem cell defect. Blood 61:765–769, 1983.

Barker JE, McFarland EC: The hematopoietic stem cells of α-thalassemic mice. Blood 66:595–601, 1985.

Barker JE, Russell ES: Investigations in vitro of marrow erythrocyte progenitors from mice with six different *W* locus genotypes. Blood (submitted for publication).

Bartke A: The response of two types of dwarf mice to growth hormone, thyrotropin, and thyroxine. Gen Comp Endocrinol 5:418–426, 1965.

Beamer WG, Creswell LA: Defective thyroid ontogenesis in fetal hypothyroid (*hyt/hyt*) mice. Anat Rec 202:387–393, 1982.

Beamer WG, Maltais LJ, De Baets MH, Eicher EM: Inherited congenital goiter in mice. Endocrinology 120:838–840, 1987.

Beamer WG, Targ E: Behavioral Impact of Genetic Hypothyroidism. Bar Harbor, ME: Annu Rep Jackson Laboratory, 1980, p 87.

Bennett M, Pinkerton PH, Cudkowicz G, Bannerman RM: Hemopoietic progenitor cells in marrow and spleen of mice with hereditary iron deficiency anemia. Blood 32:908–921, 1968.

Billings-Gagliardi S, Hall AL, Stanhope GBS, Alichuler RJ, Sidman RL, Wolf MK: Cultures of shiverer mutant cerebellum injected with normal oligodendrocytes make both normal and shiverer myelin. Proc Natl Acad Sci USA 81:2558–2561, 1984.

Birkenmeier CS, Bodine DM, Repasky EA, Helfman DM, Hughes SH, Barker JE: Remarkable homology among the internal repeats of erythroid and nonerythroid spectrin. Proc Natl Acad Sci USA 82:5671–5675, 1985.

Bodine DM, Birkenmeier CS, Barker JE: Spectrin deficient inherited hemolytic anemias in the mouse: Characterization by spectrin synthesis and mRNA activity in reticulocytes. Cell 37:721–729, 1984.

Bosma GC, Custer RP, Bosma MJ: A severe combined immunodeficiency mutation in the mouse. Nature 301:527–530, 1983.

Boss GR, Seegmiller JE: Genetic defects in human purine and pyrimidine metabolism. Annu Rev Genet 16:297–328, 1982.

Bulfield G, Siller WG, Wright PAL, Moore KJ: X chromosome-linked muscular dystrophy *mdx* in the mouse. Proc Natl Acad Sci USA 1189–1192, 1984.

Bullock LP, Bardin CW: Androgen receptor in mouse kidney: A study of male, female, and androgen-insensitive (*Tfm*/Y) mice. Endocrinology 94:746–756, 1974.

Carson DA, Kaye J, Wasson DB: Differences in deoxyadenosine metabolism in human and mouse lymphocytes. J Immunol 124:8–12, 1980.

Cattanach BM, Iddon A, Charlton HM, Chiappa SA, Fink G: Gonadotropin releasing hormone deficiency in a mutant mouse with hypogonadism. Nature 269:338–340, 1977.

Cheng TC, Beamer WG, Phillips JA III, Bartke A, Mallonee RL, Dowling C: Etiology of growth hormone deficiency in Little, Ames, and Snell dwarf mice. Endocrinology 113:1669–1677, 1983.

Chui DHK, Sweeney GD, Patterson M, Russell ES: Hemoglobin synthesis in siderocytes of flexed-tailed mutant (*f/f*) mice. Blood 50:165–177, 1977.

Cioe L, Curtis P: Detection and characterization of a mouse α-spectrin cDNA clone by its expression in *Escherichia coli*. Proc Natl Acad Sci USA 82:1367–1371, 1985.

Clark EA, Shultz LD, Pollack SB: Mutations in mice that influence natural killer (NK) cell activity. Immunogenetics 12:601–613, 1981.

Clark RG, Robinson ICAF: Effects of a fragment of human growth hormone-releasing factor in normal and "little" mice. J Endocrinol 106:1–5, 1985.

Coccia PF, Krivit W, Cervenka J, et al: Successful bone marrow transplanation for infantile malignant osteopetrosis. N Engl J Med 302:701–708, 1980.

Cohen DI, Hedrick SM, Nielsen EA, D'Eustachio R, Ruddle F, Steinberg AD, Paul WE, Davis MM: Isolation of a cDNA clone corresponding to an X-linked gene family (XLR) closely linked to the murine immunodeficiency disorder (*xid*). Nature 314:369–372, 1985.

Cole FS, Whitehead AS, Phil D, Auerbach HS, Lint T, Zeitz HJ, Kilbridge P, Colten HR: The molecular basis for genetic deficiency of the second component of human complement. N Engl J Med 313:11–16, 1985.

Cole RJ, Garlick J, Tarbutt RG: Disturbed haem and globin synthesis in reticulocytes of prenatal flexed-tail (*f/f*) anemic mice. Genetics 61:631–642, 1974.

Coleman DL: Genetic control of glucokinase activity in mice. Biochem Genet 15:297–305, 1977.

Coleman DL: Obese and diabetes: Two mutant genes causing diabetes-obesity syndromes in mice. Diabetologia 14:141–148, 1978a.

Coleman DL: Diabetes and obesity: Thrifty mutants? Nutr Rev 36:129–132, 1978b.

Coleman DL, Leiter EH, Applezweig N: Therapeutic effects of dehydroepiandrosterone metabolites in diabetes mutant mice (C57BL/KsJ-*db/db*). Endocrinology 115:239–243, 1984.

Constantini F, Chada K, Magram J: Correction of murine β-thalassemia by gene transfer into the germ line. Science 233:1192–1194, 1986.

Davidson WF, Morse HC, Sharrow SO, Chused TM: Phenotypic and functional effects of the motheaten gene on murine B lymphocytes. J Immunol 122:884–891, 1979.

Debray-Sachs M, Dardenne M, Sai P, Savino W, Quiniou MC, Boillot D, Gepts W, Assan R: Anti-islet immunity and thymic dysfunction in the mutant diabetic C57BL/KsJ *db/db* mouse. Diabetes 32:1048–1054, 1983.

De La Chapelle A, Herva A, Koivisto M, Aula P: A deletion in chromosome 22 can cause DiGeorge syndrome. Hum Genet 57:253–256, 1981.

DeLong GR, Sidman RL: Alignment defect of reaggregating cells in culture of developing brains of reeler mutant mice. Dev Biol 22:584–600, 1970.

Di Sant'Agnese PA: Cystic fibrosis (mucoviscidosis). Am Fam Physician 7:102–111, 1973.

Dorshkind K, Keller GM, Phillips RA, Miller RG, Bosma GC, O'Toole M, Bosma MJ: Functional status of cells from lymphoid and myeloid tissues in mice with severe combined immunodeficiency disease. J Immunol 132:1804–1808, 1984.

Edelman GM, Chuong CM: Embryonic to adult conversion of neural cell adhesion molecules in normal and staggerer mice. Proc Natl Acad Sci USA 79:7036–7040, 1982.

Eicher EM, Beamer WG: Inherited ateliotic dwarfism in mice: Characteristics of the mutation, little, on chromosome 6. J Hered 67:57–91, 1976.

Eicher EM, Beamer WG: New mouse *dw* allele: Genetic location and effects on lifespan and growth hormone levels. J Hered 71:157–190, 1980.

Eicher EM, Southard JL, Scriver GR, Glorieux FH: Hypophosphatemia: Mouse model for human familial hypophosphatemic (vitamin D resistant) rickets. Proc Natl Acad Sci USA 73:4667–4671, 1976.

Eppig JT, Barker JE: Chromosome abnormalities in mice with Hertwig's anemia. Blood 64:727–732, 1984.

Eppig JT, Leiter EH: Exocrine pancreatic insufficiency (EPI) syndrome and CBA/J mice. Ultrastructural study. Am J Pathol 86:17–30, 1977.

Fairbanks G, Steck TL, Wallach DFH: Electrophoretic analysis of other major polypeptides of the human erythroid membrane. Biochemistry 10:2606–2616, 1971.

Feinstein RN, Howard JB, Braun JT, Seaholm JE: Acatalasemic and hypocatalasemic mouse mutants. Genetics 53:923–933, 1966.

Festing MW: Characteristics of nude rats. In Reed ND (ed): "Proceedings of the Third International Workshop in Immune-Deficient Animals." New York: Fisher, 1982, pp 41–49.

Fisher DA, Klein AH: Thyroid development and disorders of thyroid function in the newborn. N Engl J Med 304:702–712, 1981.

Fowler JH, Till JE, McCulloch EA, Simonovitch L: The cellular basis for the defect in haemopoiesis in flexed-tailed mice. II. The specificity of the defect for erythropoiesis. Br J Haematol 13:256–264, 1967.

Fujita H, Fujino H, Nonaka K, Tarui S, Tochino Y: Retrovirus-like particles in pancreatic B-cells of NOD (non-obese diabetic) mice. Biomed Res 5:67–70, 1984.

Gibson MJ, Krieger DT, Charlton HM, Zimmerman EA, Silverman A, Parlow MJ: Mating and pregnancy can occur in genetically hypogonadal mice with preoptic area brain grafts. Science 225:949–951, 1984.

Gordon JW, Ruddle FH: DNA-mediated genetic transformation of mouse embryos and bone marrow—A review. Gene 33:121–136, 1985.

Green MC: Catalogue of mutant genes and polymorphic loci. In Green MC (ed): "Genetic Variants and Strains of the Laboratory Mouse." Stuttgart: Gustave Fischer Verlag, 1981, pp 8–278.

Green MC, Shultz LD: Motheaten, an immunodeficient mutant of the mouse. I. Genetics and pathology. J Hered 66:250–258, 1975.

Gregory CJ: Erythropoietin sensitivity as a differentiation marker in the hemopoietic system: Studies of three erythropoietic responses in culture. J Cell Physiol 89:289–302, 1976.

Grüneberg H: The anaemia of flexed-tailed mice (*mus musculus L.*). II. Siderocytes. J Genet 44:246–271, 1942.

Hanahan D: Heritable formation of pancreatic B-cell tumours in transgenic mice expressing recombinant insulin/simian virus 40 oncogenes. Nature 315:115–122, 1985.

Happle RR, Phillips RJ, Roessner A, Junemann G: Homologous genes for X-linked chondrodysplasia punctata in man and mouse. Hum Genet 63:24–27, 1983.

Harrison DE: Marrow transplanation and iron therapy in mouse hereditary microcytic anemia. Blood 40:893–901, 1972.

Hattori M, Buse JB, Jackson RA, Glimcher L, Dorf ME, Minami M, Makino S, Kuzuya K, Imura H, Strauss WM, Seidman JG, Eisenbarth GS: The NOD mouse: Recessive diabetogenic gene in the major histocompatibility complex. Science 231:733–735, 1986.

Heller AH, Eicher EM, Hallet M, Sidman RL: Myotonia, a new inherited muscle disease in mice. J Neurosci 2:924–933, 1982.

Huebner K, Palumbo AP, Isobe M, Kozak CA, Monaco S, Rovera G, Croce CM, Curtis PJ: The α-spectrin gene is on chromosome 1 in mouse and man. Proc Natl Acad Sci USA 82:3790–3793, 1985.

Hutton JJ, Bernstein SE: Metabolic properties of erythrocytes of normal and genetically anemic mice. Biochem Genet 10:297–307, 1973.

Jansson JO, Downs TR, Beamer WG, Frohman LA: Receptor-associated resistance to growth hormone-releasing hormone in growth hormone deficient dwarf "little" mice. Science 232:511–512, 1986.

Jenkins NA, Copeland NG: High frequency germline acquisition of ecotropic MuLV proviruses in SWR/J-RF/J hybrid mice. Cell 43:811–819, 1985.

Jenkins NA, Copeland NG, Taylor BA, Lee BK: Dilute coat color mutation of DBA/2J mice is associated with the site of integration of an ecotropic MuLV genome. Nature 293:370–374, 1981.

Jenkinson EJ, Van Ewijk W, Owen JJT: Major histocompatibility complex antigen expression on the epithelium of the developing thymus in normal and nude mice. J Exp Med 153:280–292, 1981.

Johnson FM, Lewis SE: Electrophoretically detected germinal mutations induced in the mouse by ethylnitrosourea. Proc Natl Acad Sci USA 78:3138–3141, 1981.

Kaiser D, Pivetta O, Rennert OM: Autosomal recessively inherited electrolyte excretory defect in the parotid of the "cribiform degeneration" mouse mutant - Possible analogy to cystic fibrosis. Life Sci 15:803–810, 1974.

Kaufman BH, Koch TR, Szurszewski JH, Go VL: Distribution and quantification of immunoreactive gut neuropeptides in piebald mice and their normal littermates. J Surg Res 38:479–483, 1985.

King JWB: Observations on the mutant "pygmy" in the house mouse. J Genet 53:487–497, 1955.

Kingston HM, Thomas NST, Pearson PL, Sarfarazi M, Harper PS: Genetic linkage between Becker muscular dystrophy and a polymorphic DNA sequence on the short arm of the X-chromosome. J Med Genet 20:755–758, 1983.

Kitamura Y, Go S: Decreased production of mast cells in Sl/Sl^{d} mice. Blood 53:492–497, 1979.

Kleinman HK, Pennypacker JP, Brown KS: Proteoglycan and collagen of "achondroplastic" (*cn/cn*) neonatal mouse cartilage. Growth 41:171–177, 1977.

Kobayashi T, Yamanaka T, Jacobs JM, Teixeira F, Suzuki K: The twitcher mouse; an enzymatically authentic model of human globoid cell leukodystrophy (Krabbe's disease). Brain Res 202:479–483, 1980.

Kwong WKL, Seetharam B, Alpers DH: Effect of exocrine pancreatic insufficiency on small intestine in the mouse. Gastroenterology 70:1277–1282, 1974.

Laemmli UK: Cleavage of the structural proteins during the assembly of the head of bacteriophage T4. Nature 227:680–685, 1970.

Lane PW: Association of megacolon with two recessive spotting genes in the mouse. J Hered 57:29–31, 1966.

Lane PW, Beamer TC, Meyers DD: Myodystrophy, a new myopathy on chromosome 8 of the mouse. J Hered 67:15–138, 1976.

Lane PW, Dickie MM: Three recessive mutations producing disproportionate dwarfing in mice. J Hered 59:300–308, 1968.

Leiter EH: The influence of genetic background on the expression of mutations at the diabetes locus in the mouse. IV. Male lethal syndrome in CBA/Lt mice. Diabetes 30:1035–1044, 1981.

Leiter EH: Genetic control of pathogenesis of diabetes in C3H mice. Influence of the major histocompatibility complex. Diabetes 33:1068–1071, 1984.

Leiter EH, Coleman DL, Hummel KP: The influence of genetic background on the expression of mutations at the diabetes locus in the mouse. III. Effect of H-2 haplotype and sex. Diabetes 30:1029–1034, 1981.

Leiter EH, Cunliffe-Beamer T: Exocrine pancreatic insufficiency (EPI) syndrome in CBA/J mice. III. Pathological and genetic analysis. Gastroenterology 73:260–266, 1977.

Leiter EH, Dempsey EC, Eppig JJ: Exocrine pancreatic insufficiency (EPI) syndrome in CBA/J mice. II. Biochemical studies. Am J Pathol 86:31–46, 1977.

Leiter EH, Kuff EL: Intracisternal type A particles in murine pancreatic B cells. Am J Pathol 114:46–55, 1984.

Lewis SE, Johnson FM, Skow LC, Popp D, Barnett LB, Popp RA: A mutation in the β-globin gene detected in the progeny of a female mouse treated with ethylnitrosourea. Proc Natl Acad Sci USA 82:5829–5831, 1985.

MacDonald HR: Phenotypic and functional characteristics of "T-like" cells in nude mice. In Sordat B (ed): "Immune-Deficient Animals." Basel: Karger, 1984, pp 2–6.

Makino S, Kunimoto K, Muraoka Y, Mizushima Y, Katagiri K, Tochino Y: Breeding of a non-obese, diabetic strain of mice. Exp Anim 29:1–13, 1980.

Makino S, Muraoka Y, Kishomoto Y, Hayashi Y: Genetic analysis for insulitis in NOD mice. Exp Anim 34:45–432, 1985.

Mandel T: Differentiation of epithelial cells in the mouse thymus. Z Zellforsch 106:498–515, 1970.

Mann R, Mulligan RC, Baltimore D: Construction of a retrovirus packaging mutant and its use to produce helper-free defective retrovirus. Cell 33:153–159, 1983.

Marks SC Jr, Seifert MF, Lane PW: Osteosclerosis, a recessive skeletal mutation on chromosome 19 in the mouse. J Hered 76:171–176, 1985.

Marks SC Jr, Seifert MF, McGuire JL: Congenitally osteopetrotic mice *op/op* are not cured by transplants of spleen or bone marrow cells from normal litter mates. Metab Bone Dis Relat Res 5:183–186, 1984.

Mason AJ, Hayflick J, Zoeller RT, Young WS, Phillips HS, Nikolies K, Seeburg PH: A deletion truncating the gonadotropin-releasing hormone gene is responsible for hypogonadism in the *hpg* mouse. Science 234:1366–1371, 1986.

McCulloch EA, Simonovitch L, Till JE, Russell ES, Bernstein SE: The cellular basis of the genetically determined hemopoietic defect in anemic mice of genotype Sl/Sl^{d}. Blood 26:399–410, 1965.

McKusick VA: "Mendelian Inheritance in Man, 7th ed." Baltimore: John Hopkins University Press, 1986.

Meade CJ, Brandon DR, Smith W, Simmons RG, Harris S, Sowter C: The relationship between hyperglycemia and renal immune complex deposition in mice with inherited diabetes. Clin Exp Immunol 43:109–120, 1981.

Merrimee TH, Zapf J, Froesch ER: Insulin-like growth factors (IGFs) in pygmies and subjects with the pygmy trait: Characterization of the metabolic actions of IFG-I and IGF-II in man. J Clin Endocrinol Metab 55:1081–1085, 1982.

Moon RT, Ngai J, Wold BJ, Lazarides E: Tissue-specific expression of distinct spectrin and ankyrin transcripts in erythroid and non-erythroid cells. J Cell Biol 100:152–160, 1985.

Mosier DE: Are *xid* lymphocytes representative of any normal B-cell population? J Mol Cell Immunol 2:170, 1985.

Murota T, Shibuya T, Tutikawa K: Genetic analysis of an N-ethyl-N-nitrosourea-induced mutation at the hemoglobin β-chain locus in mice. Mutat Res 104:317–321, 1982.

Murray JM, Davies KE, Harper PS, Meredith L, Mueller CR, Williamson R: Linkage relationships of a cloned DNA sequence on the short arm of the X-chromosome to Duchenne muscular dystrophy. Nature 300:69–71, 1982.

Neel JV: Diabetes mellitus, a thrifty genotype rendered detrimental by "progress." Am J Hum Genet 14:353–362, 1962.

Nerup J, Christy M, Platz P, Ryder LP, Svejgaard A: Aspects of the genetics of insulin dependent diabetes mellitus. In Andreani D, DiMario U, Federlin KF, Heding LG (eds): "Immunology in Diabetes." London: Kimpton Medical Publications, 1984, pp 63–70.

Nikolics K, Mason A, Phillips H, Szonyi E, Ramachandran J, Seeburg PS: Bacterial expression and purification of the C-terminal fragment of PRO-GnRH. Abstract 782, The Endocrine Society Annual Meeting, 1985, p 196.

Nissley SP, Knazek RA, Wolff GL: Somatomedin activity in sera of genetically small mice. Horm Metab Res 12:158–164, 1980.

Noguchi T, Sugisaki T, Satoh I, Kudo M: Partial restoration of cerebral myelination of the congenitally hypothyroid mouse by parenteral or breast milk administration of thyroxine. J Neurochem 45:1419–1426, 1985.

Ohno S: Evolution of sex chromosomes in mammals. Annu Rev Genet 3:495–52, 1969.

Oliver C, Essner E: Distribution of anomalous lysosomes in the beige mouse. A homologue of Chediak-Higashi syndrome. J Histochem Cytochem 21:218–228, 1973.

Palmiter RD, Brinster RL, Hammer RE et al: Dramatic growth of mice that develop from eggs microinjected with metallothionein-growth hormone fusion genes. Nature 300:611–615, 1982.

Peters J, Andrews SJ, Loutit JF, Clegg JB: A mouse β-globin mutant that is an exact model of hemoglobin Rainier in man. Genetics 110:709–721, 1985.

Phillips JA III, Beamer WG, Bartke A: Analysis of growth hormone genes in mice with genetic defects of growth hormone expression. J Endocrinol 92:405–407, 1982.

Phillips JA III, Hjelle BL, Seeburg PH, Zackman M: Molecular basis for familial isolated growth hormone deficiency. Proc Natl Acad Sci USA 78:6372–6375, 1981.

Pivetta OH, Cassino RJJ, Sordelli DO: Pulmonary cell response to bacterial challenge in mutant mice with some hereditary alterations resembling cystic fibrosis. Life Sci 26:1349–1357, 1980.

Pivetta OH, Green EL: Exocrine pancreatic insufficiency: A new recessive mutation in mice. J Hered 64:301–302, 1973.

Popp RA: Hemoglobin loci: Mice classified for their Hb and Sol alleles. Science 140:893, 1963.

Popp RA: Hemoglobins of mice: Sequence and possible ambiguity at one position of the alpha chain. J Mol Biol 27:9, 1967.

Popp RA, Bailiff EG, Skow LC, Johnson FM, Lewis SE: Analysis of a mouse α-globin gene mutation induced by ethylnitrosourea. Genetics 105:157–167, 1983.

Popp RA, Brodsleaw BS, Skow LC: Effects of alpha thalassemia on mouse development. Differentiation 17:205–210, 1980.

Popp RA, Stratton LP, Hawley KP, Effron K: Hemoglobin of mice with radiation-induced mutations at the hemoglobin loci. J Mol Biol 127:141–148, 1979.

Prins HW, Van Den Hamer CJA: Primary biochemical defect in copper metabolism in mice with a recessive X-linked mutation analogous to Menkes' disease in man. J Inorg Biochem 10:19–27, 1979.

Prochazka M, Premdas FH, Leiter EH, Lipson LG: Estrone treatment dissociates primary versus secondary consequences of "diabetes"(*db*) gene expression in mice. Diabetes 35:725–728, 1986.

Reed C, O'Donoghue JL: The hairless athymic guinea pig. In Reed ND (ed): "Proceedings of the Third International Workshop in Immune-Deficient Animals." New York: Fisher, 1982, pp 51–57.

Rimoin DL, Lachman RS: The chondrodysplasias. In Emery AEH, Rimoin DL (eds): "Principles and Practice of Medical Genetics." New York: Churchill Livingstone, 1983, Vol 2, pp 703–735.

Rimoin DL, Richmond L: The pygmy (*pg*) mutant of the mouse—A model of the human pygmy. J Clin Endocrinol Metab 35:467–468, 1972.

Rimoin DL, Schimke RN: "Genetic Disorders of the Endocrine Glands." St. Louis: CV Mosby Co, 1971, pp 11–65.

Roach A, Takahashi M, Provtcheva D, Ruddle F, Hood L: Chromosomal mapping of mouse myelin basic protein gene and a structure and transcription of the partially deleted gene in shiverer mutant mice. Cell 42:149–155, 1985.

Roffler-Tarlov S, Graybiel AM: Weaver mutation has differential effects on the dopamine containing innervation of the limbic and nonlimbic striatum. Nature 307:62–66, 1984.

Roffler-Tarlov S, Landis SC, Zigmond MN: Effects of Purkinje cell degeneration on the noradrenergic projection to mouse cerebellar cortex. Brain Res 298:303–311, 1984.

Rosen FS, Cooper MD, Wedgwood RJ: The primary immunodeficiencies. Part II. N Engl J Med 311:300–310, 1984.

Rothman TB, Gershon MD: Regionally defective colonization of the terminal bowel by the precursors of enteric neurons in lethal spotted mutant mice. Neuroscience 12:1293–1311, 1984.

Russell ES: Hereditary anemias of the mouse: A review for geneticists. Adv Genet 20:357–459, 1979.

Russell ES, McFarland EC: Genetics of mouse hemoglobins. Ann NY Acad Sci 241:25–37, 1974.

Russell ES, Nash DJ, Bernstein SE, Kent EL, McFarland EC, Mathews SM, Norwood MS: Characterization and genetic studies of microcytic anemia in the house mouse. Blood 35:838–850, 1970.

Russell LB, Russell WL, Popp RA, Vaughan C, Jacobson KB: Radiation-induced mutations at mouse hemoglobin loci. Proc Natl Acad Sci USA 73:2843–2846, 1976.

Scaravilli F, Jacobs JM: Peripheral nerve grafts in hereditary leukodystrophic mutant mice (twicher). Nature 290:56–58, 1981.

Schaible R, Gowen JW: A new dwarf mouse. Genetics 46:896, 1961.

Scher I: CBA/N immune defective mice: Evidence for the failure of a B-cell subpopulation to be expressed. Immunol Rev 64:117–136, 1982.

Schnieke A, Harbers K, Jaenisch R: Embryonic lethal mutation in mice induced by retrovirus insertion into the α(I) collagen gene. Nature 304:315–320, 1983.

Selmanowitz VJ: Ectodermal dysplasias including epitheliogenesis imperfecta, ichthyosis, and follicular/glandular anomalies. In Andrews EJ, Ward BC, Altman NH (eds): "Spontaneous Animal Models of Human Disease." New York: Academic Press, Inc, 1979, Vol II, pp 3–10.

Shine HD, Sidman RL: Immunoreactive myelin basic proteins are not detected when shiverer mutant Schwann cells and fibroblasts are cocultured with normal neurons. J Cell Biol 98:1291–1295, 1984.

Shultz LD, Bedigian HG, Heiniger HJ, Eicher EM: The congenitally athymic streaker mouse. In Reed ND (ed): "Proceedings of the Third International Workshop in Immune-Deficient Animals." New York: Fisher, 1982, pp 33–40.

Shultz LD, Coman DR, Bailey CL, Beamer WG, Sidman CL: "Viable motheaten," a new allele at the motheaten locus. I. Pathology. Am J Pathol 116:179–192, 1984.

Shultz LD, Green MC: Motheaten, an immunodeficient mutant of the mouse. II. Depressed immune competence and elevated serum immunoglobulins. J Immunol 116:936–943, 1976.

Shultz LD, Roths JB: Euthymic murine models for immunologic dysfunction. In Rygaard J (ed): "Immune Deficient Animals in Biochemical Research." Basel: Karger (in press).

Sidman CL, Marshall JD, Mastiello NC, Roths JB, Shultz LD: Novel B-cell maturation factor from spontaneously autoimmune viable motheaten mice. Proc Natl Acad Sci USA 81:7199–7202, 1984.

Sidman CL, Shultz LD, Unanue ER: The mouse mutant "motheaten." II. Functional studies of the immune system. J Immunol 121:2399–2404, 1978.

Sidman RL, Conover CS, Carson JH: Shiverer gene maps near the distal end of chromosome 18 in the house mouse. Cytogenet Cell Genet 39:241–245, 1985.

Sinha YN, Wolff GL, Baxter SR, Domon OE: Serum and pituitary concentrations of growth hormone and prolactin in pygmy mice. Proc Soc Exp Biol Med 162:221–223, 1979.

Skow LC, Burkhart BA, Johnson FM, Popp RA, Popp DM, Goldberg SZ, Anderson WF, Barnett LB, Lewis SE: A mouse model for β-thalassemia. Cell 34:1043–1052, 1983.

Smith PE, MacDowell EG: An hereditary anterior-pituitary deficiency in the mouse. Anat Rec 46:249–257, 1930.

Sorell M, et al: Marrow transplantation for juvenile osteopetrosis. Am J Med 70:1280–1287, 1981.

Stanbury JB, Dumont JE: Familial goiter and related disorders. In Stanbury JB, Wyngaarden JB, Fredrickson DS, Goldstein JL, Brown MS (eds): "The Metabolic Basis of Inherited Disease. 5th ed." New York: McGraw Hill, 1983, pp 231–269.

Staven PE, Saltvedt E, Elgjo K, Rootwelt K: Congenital hypochromic-microcytic anaemia with iron overload of the liver and hyperferraemia. Scand J Haematol 10:153–160, 1973.

Tennenhouse HS: Abnormal renal mitochondria 25-hydroxyvitamin D_3-1-hydroxylase activity in the vitamin D and calcium deficient X-linked *Hyp* mouse. Endocrinology 113:816–818, 1983.

Touraine JL: Autoimmunity in children and animals with a primary immunodeficiency. Paediatrician 5:18–29, 1976.

Ueki S, Okamato E, Kuwata K, Toyasaka A, Nagai K: Qualitative and quantitative analysis of muscarinic acetylcholine receptors in the piebald lethal mouse model of Hirschsprung's disease. Gastroenterology 88:1834–1841, 1985.

Unger AE, Harris MJ, Bernstein SE, Falcone JC, Lux SE: Hemolytic anemia in the mouse. J Hered 74:88–92, 1983.

Valk TW, Corley KP, Kelch RP, Marshall JC: Hypogonadotropic hypogonadism: Hormonal responses to low dose pulsatile administration GnRH. J Clin Endocrinol Metab 51:730–738, 1980.

Van Wyck DB, Popp RA, Foxley J, Witte MH, Witte CL, Crosby WH: Spontaneous overload in α-thalassemic mice. Blood 64:263–266, 1984.

Walker DG: Control of bone resorption by hematopoietic tissue. The induction and reversal of congenital osteopetrosis in mice through use of bone marrow and splenic transplants. J Exp Med 142:651–663, 1975.

Webster W: Embryogenesis of the enteric ganglia in normal mice and in mice that develop congenital aganglionic megacolon. J Embryol Exp Morphol 30:573–585, 1973.

White R: The search for the cystic fibrosis gene. Science 234:1054–1055, 1986.

Whitney JB III: Differential control of the synthesis of two hemoglobin β-chains in normal mice. Cell 12:863–871, 1977.

Whitney JB III: Simplified typing of mouse hemoglobin (*Hbb*) phenotypes using cystamine. Biochem Genet 16:667–672, 1978.

Whitney JB III, Cobb RR, Popp RA, O'Rourke TW: Detection of neutral amino acid substitutions in proteins. Proc Natl Acad Sci USA 82:7646–7650, 1985.

Whitney JB III, Russell ES: Linkage of genes for adult α-globin and embryonic α-like globin chains. Proc Natl Acad Sci USA 77:1087–1090, 1980.

Wikstrom B: Growth cartilage in some inherited skeletal dysplasias in the mouse: A morphological and biochemical study. In: "Kongl Carolinska Medico Chirurgiska Institutet." Stockholm: ReproPrint AB, 1984, pp 1–135.

Willard HF, Riordan JR: Assignment of the gene for myelin proteolipid protein to the X-chromosome: Implications for X-linked myelin disorders. Science 230:940–942, 1985.

Wilson JD, Griffin JE, Leshin M, MacDonald PC: The androgen resistance syndrome: 5 α-reductase deficiency, testicular feminization, and related disorders. In Stanbury JB, Wyngaarden JB, Fredrickson DS, Goldstein JL, Brown MS (eds): "The Metabolic Basis of Inherited Disease, 5th ed." New York: McGraw Hill, 1983, pp 1001–1026.

Woychick RP, Stewart TA, Davis LG, D'Eustachio P, Leder P: An inherited limb deformity created by insertion mutagenesis in a transgenic mouse. Nature 318:36–40, 1985.

Yeager AM, Brennan S, Tiffany C, Moser HW, Santos GW: Prolonged survival and remyelination after hematopoietic cell transplantation in the twitcher mouse. Science 225:1052–1054, 1984.

Gene Diagnosis: Detection of Genetic Disorders by DNA Analysis

John A. Phillips III, MD

Division of Genetics, Vanderbilt University School of Medicine, Nashville, TN 37232

INTRODUCTION

DNA analysis using molecular genetic techniques has identified the basic gene defects that cause a variety of genetic disorders. Using restriction endonucleases, Southern blots, and DNA hybridization techniques, gene diagnosis of many types of hemoglobinopathies, hemophilia A and B, various urea cycle defects, Duchenne muscular dystrophy, cystic fibrosis, and a growing number of other genetic disorders is now possible. Future developments promise to increase the variety of disorders that can be diagnosed and make knowledge of these techniques pertinent to almost every medical subspecialty.

BACKGROUND

Classic Genetic vs DNA Approach

In classic genetic studies, the approach often used is to identify a phenotype and then determine the protein alteration that is characteristic of this

This study was supported in part by NIH grants AM35592 and AM01434. The author thanks Ms. Judy Copeland for expert preparation of the manuscript.

Birth Defects: Original Article Series, Volume 23, Number 3, pages 259–295

phenotype. This approach has been very useful in elucidating many biochemical disorders, but it has proved to be a difficult method of study for many other mendelian disorders, especially those with autosomal dominant modes of inheritance. The next logical step after examining the protein or enzyme alteration is to examine the abnormal gene that encodes the aberrant protein causing the phenotype. In the case of autosomal dominant disorders, this usually has been impossible because of the difficulty in detecting the basic biochemical derangement. However, with the increasing availability of techniques for DNA analysis, it has become practical to study directly the genes likely to be deranged in such disorders. Thus the classic genetic approach of examining the intermediate biochemical alteration can be bypassed and the gene under question examined directly. The DNA approach allows a very quick screening method to determine if a candidate gene such as the pro-α2(I) collagen gene is genetically linked to a disorder such as Marfan syndrome. This can be done simply by examining DNA samples from a large family with multiple individuals affected with Marfan syndrome to see if the phenotype in that family cosegregates with the pro-α2(I) collagen gene. Using this approach, DNA from a family having sufficient members who display any abnormal phenotype can be examined to determine if genetic linkage exists between the segregation patterns of the disorder and the DNA sequences under study.

DNA and Gene Structure

In examining specific genes for mutations to determine if they cause a given mendelian disorder, one has the problem of the immense amount of DNA contained in each cell. Since about 3 billion base pairs (bp) of DNA is the haploid complement of each cell, and eukaryotic genes often are from 1 to 4 kilobases (kb) in size, it follows that a given human gene would comprise roughly one-millionth of the human haploid complement. This complement of DNA is arranged on the 22 pairs of autosomes, the two sex chromosomes (XX or XY, depending on the sex), and the mitochondrial chromosome. When a chromosome is examined after release of the DNA from the protein scaffold of the chromosome, a virtual cloud of DNA loops radiates out from the scaffold [1]. Under higher magnification, the loops are seen to be comprised of two strands. These two strands run in opposite directions (antiparallel) and form the double helix described by Watson and Crick [2,3]. The two strands of DNA are held together in the helical form by hydrogen bonding between the adenine and thymine (A,T) and the guanine and cytosine (G,C) nucleotides on opposite strands. On the coding strand, units of three nucleotides (a triplet) constitute a codon, the unit of the genetic code that encodes a specific amino acid. The codons thus form the blueprint for the protein encoded and are arranged in sequence to form the gene.

Additional complexity in the human genome was discovered when the nucleotide sequences of various genes were determined [4,5]. As the DNA sequence on the coding strand was read, the sequences first encountered encoded the protein, and these sequences were referred to as an exon. However, unexpectedly, as more of the given gene was sequenced, it was found that the series of nucleotides following the exon did not encode for the next amino acids in the protein sequence. These noncoding DNA sequences were called intervening sequences (IVS) or introns [5] (Fig. 1). Further sequencing into the gene revealed a second and a third and possibly more exons and introns. Thus human genes were found to be split, and coding sequences or exons were separated by intervening sequences or introns. From this complexity, it should be apparent that mutations causing human diseases can occur within exons, and, if so, they can result in protein sequence alterations in the translated product of the gene. However, mutations also occur in introns or sequences of DNA that flank the gene and are not contained in an exon (see Kazazian, this volume). These mutations can

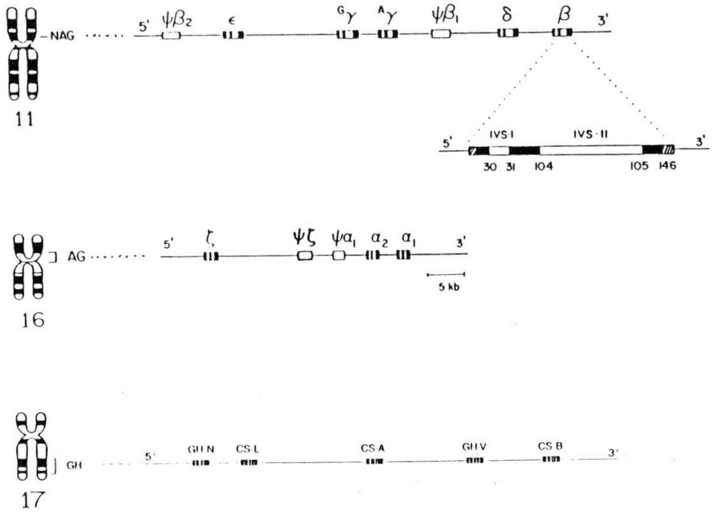

Fig 1. α-Globin (AG), β-globin (NAG), and growth hormone (GH) gene clusters. For every gene shown, the black boxes represent coding regions (exons), small white boxes represent introns, and hatched boxes are the 5′ and 3′ untranslated regions. The large white boxes represent pseudogenes that are not expressed because of DNA sequence alterations. The numbers below the area of the coding sequences represent the corresponding amino acid encoded by that sequence.

affect various steps necessary for normal expression of the gene but may not be detectable by studies of its protein product. In contrast, using recombinant DNA techniques, mutations that affect the transcription or splicing steps of gene expression often are detected easily.

Another order of complexity of the human genome became apparent when it was discovered that many human genes occur in families or clusters of related or homologous loci. Good examples of this are the human α-globin gene cluster on chromosome 16 that contains the paired α-globin loci and the ζ locus; the human β-globin gene cluster on chromosome 11 that contains the β, δ, γ, and ϵ loci; and the growth hormone gene cluster on chromosome 17 that has five loci that are evolutionarily related (Fig. 1) [5–8]. In such clusters or gene families there may be remnants of genes called pseudogenes that have homology to active genes but because of DNA sequence alterations are now nonfunctional, that is, not expressed (see $\psi\alpha$ and $\psi\beta$, Fig. 1) [5]. Thus, in analyzing human DNA to detect mutations in a given gene, the presence of a cluster of related sequences can make the process of detecting a mutation in one member of the gene family difficult. Although these complexities pose problems in the detection of genetic disorders by DNA analysis, a variety of molecular genetic methods can be used to detect the mutations that cause a growing list of genetic disorders.

METHODS OF DETECTING MUTANT GENES BY DNA ANALYSIS: PROTOTYPES

The prototypic methods used to detect mutant genes are restriction endonuclease analysis, cloning and DNA sequencing, and use of mutation-specific oligonucleotide probes.

Restriction Endonuclease Analysis

Restriction endonuclease analysis can be used to detect 1) gene deletions and 2) certain point mutations and as a very useful tool in linkage analysis to 3) determine if cosegregation exists between a given DNA sequence and a mendelian phenotype. Restriction endonucleases were first isolated about 15 years ago by Arber, Smith and Nathans, who were awarded the 1978 Nobel Prize for Medicine and Physiology for this discovery [9–11]. Today, more than 200 restriction endonucleases have been described, and many of these are commercially available [12].

Restriction endonucleases are bacterial enzymes that recognize and cleave double-stranded DNA at specific recognition sequences of four or more nucleotides that are nearly always palindromic. This property enables the long double strands of DNA to be cleaved into fragments of reproducible size. If the endonuclease used is one like *Eco*RI, it will cut the DNA in a manner that leaves the ends of the two strands slightly offset from each other (Table 1) [10–12]. This overhang of one strand over the other produces what are referred to as "sticky ends." These sticky ends are very useful in promoting pairing between complementary ends of other fragments also cleaved with *Eco*RI so as to enable the formation of recombinant molecules (Fig. 2). Other endonucleases, for example, *Hae*III, cut the ends of the two strands of the DNA molecule off equally (Table 1), and this is referred to as a "blunt end" cleavage. The ability to give sticky or blunt ends is an inherent property of a given restriction endonuclease, and we will return to their utility in the discussion of cloning.

The first step in restriction endonuclease detection of a mutation is isolation of genomic DNA from the nuclei of leukocytes, tissue, or cultured fibroblasts. The method of DNA extraction is, essentially, digestion with proteinase K followed by phenol, then chloroform extractions and precipitation with ethanol [13].

After the DNA is isolated, a small aliquot (1–5 μg) is digested with a restriction endonuclease (Table 1). This digestion yields DNA fragments of reproducible size as determined by the number and location of recognition sequences for that restriction endonuclease within the DNA. Following cleavage by the selected restriction endonuclease, agarose gel electrophoresis is used to separate the DNA fragments by size (Fig. 3) [14]. Because of the large number of DNA fragments that result, a smear, rather than discrete bands, is seen after electrophoresis, with large fragments at the top and small fragments at the bottom of the gel. Since it is impossible to recognize the one or two fragments containing the gene sequences of interest in this smear, the sensitive method of molecular hybridization is used. To carry out hybridization, the DNA fragments are rendered single-stranded (denatured) by alkali and transferred from the agarose gel to a filter membrane as described by Southern [14]. The DNA fragments are transferred from the agarose gel to the filter by diffusion of a salt solution through the gel and filter membrane into a stack of blotting paper (Fig. 3) or by electrophoretic transfer from the gel to the filter membrane. The DNA is firmly attached to the membrane by baking or chemical treatment. The filter-bound DNA is then hybridized to a solution containing a probe (^{32}P-radiolabeled single-stranded DNA or RNA copies of known DNA sequences). These radiolabeled molecules can anneal to their complementary sequences in the filter-bound fragments to form double-stranded DNA. One strand of these molecules is from the patient's DNA; the other strand is the radiolabeled copy that is being used as a probe.

TABLE 1. Selected Restriction Endonucleases and Their Specificities†

Endonuclease	Sequence cleaved
*Bam*HI	G↓GATCC CCTAG↑G
*BGl*II	A↓GATCT TCTAG↑A
*Eco*RI	G↓AATTC CTTAA↑G
*Hae*III	GG↓CC CC↑GG
*Hinc*II	GTPy↓PuAC CAPu↑PyTG
*Hind*III	A↓AGCTT TTCGA↑A
*Hinf*I	G↓ANTC CTNA↑G
*Hpa*I	CTT↓AAC CAA↑TTG
*Mbo*II	TCTTCNNNNNNNN↓ AGAAGNNNNNNN↑N
*Mst*II	CC↓TNAGG GGANT↑CC
*Pst*I	CTGCA↓G G↑ACGTC
*Sst*I	GAGCT↓C C↑TCGAG
*Taq*I	T↓CGA AGC↑T
*Xba*I	T↓CTAGA AGATC↑T

†The sequence recognized and cleaved by a restriction endonuclease is usually palindromic; ie, it reads the same in opposite directions on the two DNA strands. Abbreviations: A, adenine; C, cytosine; G, guanine; T, thymine; Py, pyrimidine (C or T); Pu, purine (A or G); N = A, G, T, or C. Arrows represent sites of cleavage.

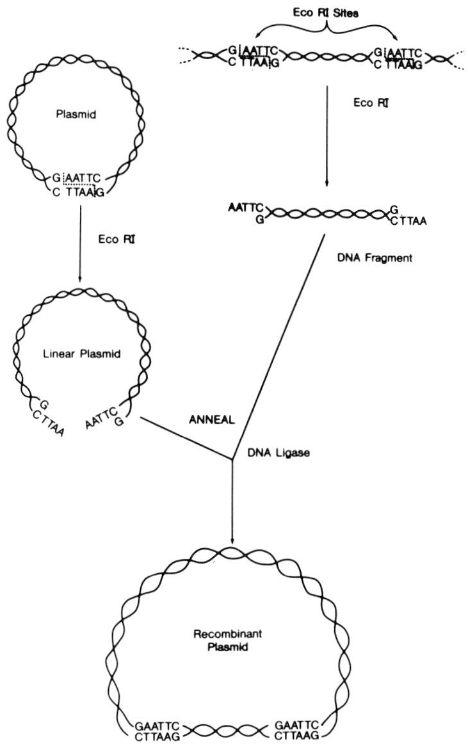

Fig. 2. Recombinant DNA molecule. Cleavage of plasmid DNA linearizes it, enabling it to be ligated with another DNA fragment that was also cleaved by the restriction endonuclease *Eco*RI. When the linear plasmid and DNA fragment are mixed, the "sticky ends" anneal and form hydrogen bonds, resulting in a recombinant molecule.

After hybridization, the filter is washed to remove nonannealed probe and is subjected to autoradiography. Using these methods, DNA fragments that have annealed with the radioactive probe appear as bands on the developed film. By comparing the location of these bands with the location of markers of known molecular weight, the size of fragments containing hybridizing DNA is determined (Fig. 3).

Gene Deletions: Familial Growth Hormone Deficiency and α^0-Thalassemia

An example of the use of restriction endonuclease analysis to detect gene deletions is familial isolated growth hormone deficiency (IGHD type 1A). This rare disorder is autosomal recessive, and affected children have severe

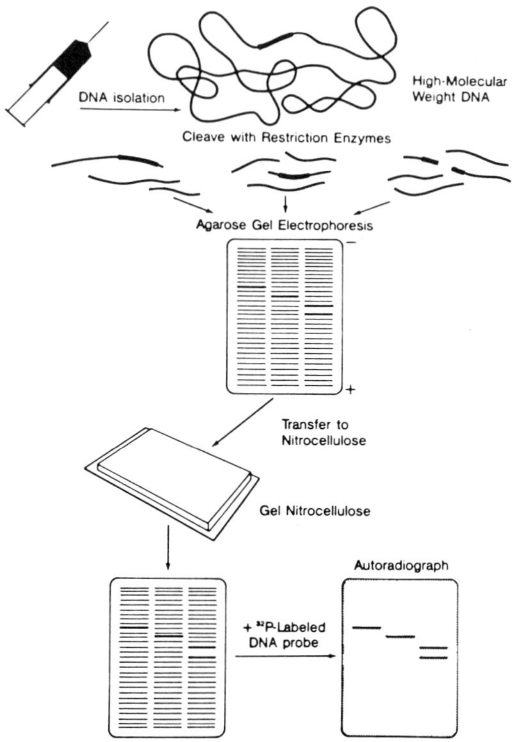

Fig. 3. Restriction endonuclease analysis of genomic DNA. Genomic DNA is digested with one or more restriction endonucleases, and the resulting DNA fragments are separated by size through gel electrophoresis and blotted to nitrocellulose. The nitrocellulose-bound DNA fragments are then hybridized to a radiolabeled gene probe, and the resulting radioactive double-stranded DNA molecules are visualized by autoradiography.

growth retardation. Unfortunately, many respond only briefly to exogenous growth hormone (GH) because of their subsequent production of anti-GH antibodies [13–15].

Restriction endonuclease analysis is used to diagnose such cases through studies of the structural gene for growth hormone (GH-N, see Fig. 1). The GH-N gene is flanked by consistent *Bam*HI recognition sites that are 5′ (upstream) and 3′ (downstream). The distance between these *Bam*HI sites is 3,800 bp or 3.8 kb (Table 1, Fig. 4) [16]. Thus, of all the cleavage products of the GH gene cluster, the 3.8 kb fragment is specific for the GH-N gene. Following electrophoresis, Southern blotting, molecular hybridization with radiolabeled single-stranded GH sequences, and autoradiography, the pattern

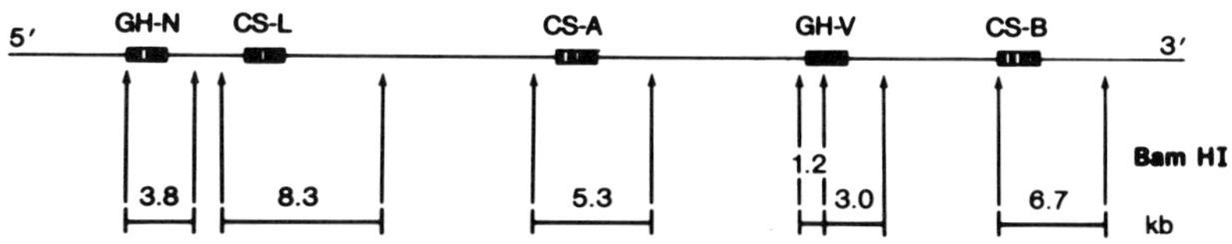

Fig. 4. Partial restriction endonuclease map of the human GH gene cluster. The sizes of the *Bam*HI-derived genomic DNA fragments (below) containing the various components of the human GH gene cluster (above) are shown. Note that the structural gene for GH (GH-N) is contained in the 3.8 kb *Bam*HI-derived fragment.

shown in Figure 5 is seen [13]. Multiple bands are due to the different related genes that comprise the GH gene cluster (Figs. 1, 4, 5). Note that the sample from the subject with IGHD type IA lacks the 3.8 kb fragment that contains the GH-N gene. Furthermore, the amount of hybridization seen for the 3.8 kb fragment in the DNA of parents is intermediate between that seen in DNA from the control and the affected child. These findings demonstrate that the affected child is homozygous for GH-N gene deletions and that his parents are heterozygous and his sister homozygous normal. Since such deletions preclude production of any GH, these children are often immunologically intolerant to exogenous GH analogously to certain hemophiliacs who develop factor VIII:C inhibitors after treatment with factor VIII:C [17].

Gene deletions are the usual basis of α^0-thalassemia, in which the absence of α-globin genes and chain production results in the formation of γ^4 tetramers because no α chains are available to form $\alpha_2\gamma_2$ tetramers or fetal hemoglobin [18–20]. The γ^4 tetramers are defective in the transport of oxygen to the fetal tissues, and the usual outcome is demise in utero or death immediately after delivery. The resulting anoxic tissue damage and edema are referred to as hydrops.

Point Mutations: β^S, β^{0Arab}, and Certain Factor VIII:C Mutations

The detection of the sickle cell mutation is an example of the detection of point mutations using restriction endonuclease analysis. This mutation causes sickle cell anemia, which is associated with irreversible sickling of red cells because of the polymerization of the abnormal hemoglobin. This deformation of the red cells results in infarction of tissues because of interruption of the blood supply, hemolysis, and anemia and susceptibility to infections. Couples at risk for a child with sickle cell anemia are usually carriers who have a 25% recurrence risk. In carriers by affected matings the risk of recurrence is 50%. The basis of sickle cell anemia is a single nucleotide substitution (A→T) in the sixth codon of the β-globin chain [21]. The method of detecting

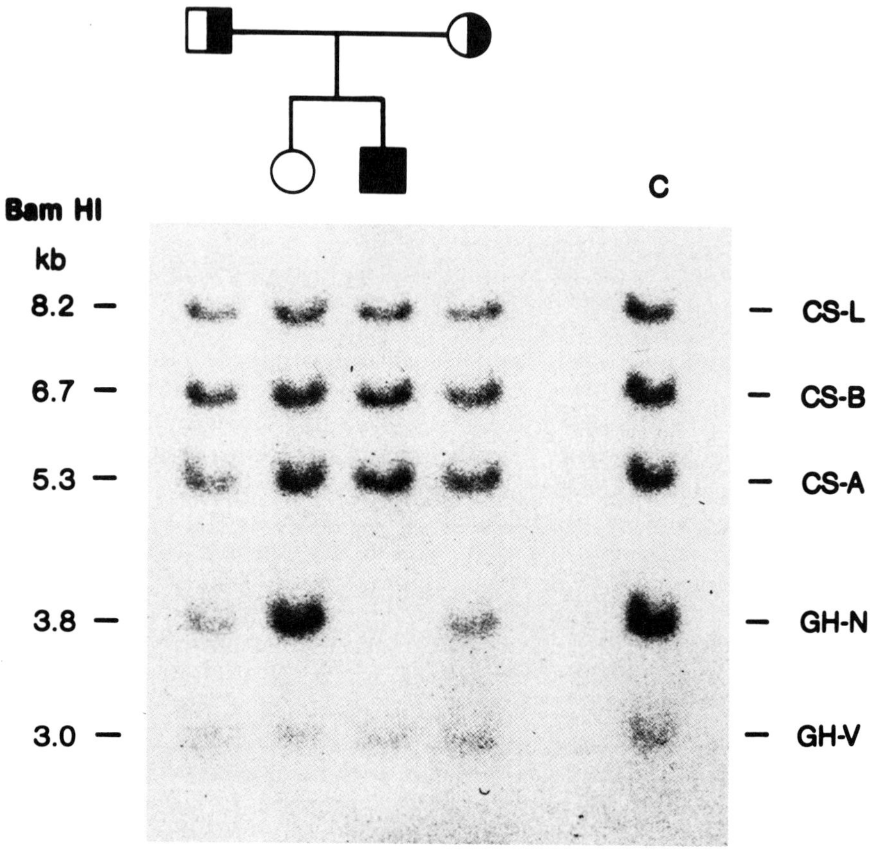

Fig. 5. Autoradiogram patterns of DNA from a family with a child (solid symbol) affected with isolated growth hormone deficiency (IGHD) type IA and a control after digestion with *Bam*HI restriction endonuclease (see Fig. 4) and hybridization to the GH cDNA probe. Fragment sizes in kb are shown on the left [13] (see text).

this substitution has to be sufficiently sensitive to recognize one difference in the 3 billion nucleotides present. This is achieved by use of the restriction endonuclease *Mst*II (Table 1). In the case of the normal (β^A-) globin gene, CCT is the fifth codon and GAG the sixth codon. The A→T substitution in the sickle (β^s) gene changes the sixth codon from GAG to GTG, which substitutes valine for glutamic acid and destroys this sequence as a recognition site for *Mst*II. When the β^A-globin gene is cleaved with *Mst*II, a recognition site 5′ to the gene is normally cut and another cleavage occurs at codons 5 and 6, generating a 1.15 kb fragment (Fig. 6) [21–23]. In the case

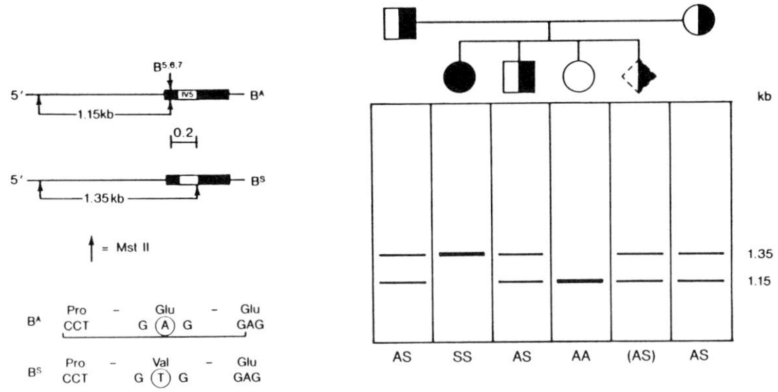

Fig. 6. Detection of the sickle mutation by restriction endonuclease analysis. An A to T substitution in the sixth codon of the β^A-globin gene yields the β^S allele. This change eliminates an *Mst*II recognition site, which results in the 5′ region of the β^S allele generating a 1.35 kb fragment rather than a 1.15 kb fragment. For details, see text.

of the β^s gene, the cleavage at codons 5 and 6 does not occur because the A→T substitution has destroyed the *Mst*II recognition site and the next site 3′ is cleaved, generating a 1.35 kb fragment. For the AS parents, their β^A-gene sequences are recognized as a 1.15 kb fragment, whereas the sequences from their β^s-genes generate 1.35 kb fragments. When DNA from their offspring is examined, if only the 1.15 kb fragment is seen, the child would be predicted to be an AA homozygote. If 1.15 and 1.35 kb fragments are seen, the child would be predicted to be a carrier (AS) like each of the parents. Finally, if only a 1.35 kb fragment is seen, the fetus would be diagnosed as being homozygous SS, or affected.

The use of restriction endonuclease analysis to detect point mutations has also been applied to the β^{0Arab} mutation. Codons 121 and 122 of the β^A-globin gene are GAATTC, which is an *Eco*RI recognition site (Table 1). However, the β^{0Arab} gene has a G→A substitution in codon 121 that destroys the *Eco*RI recognition site. After digestion with *Eco*RI, the β^{0Arab} gene is not cleaved at 121–122, and a larger 8.4 kb fragment occurs that is easily differentiated from the 5.2 and 3.2 kb fragments cleaved from the β^A gene (Fig. 7) [24].

Recently, Gitschier et al [25] have detected altered patterns of factor VIII:C genes of different hemophiliacs following *Taq*I digestion. Such site alterations are likely to be the mutations responsible for factor VIII:C deficiency. This prediction can be made because two of the three stop or termination codons can be produced in the underlined codons by a single

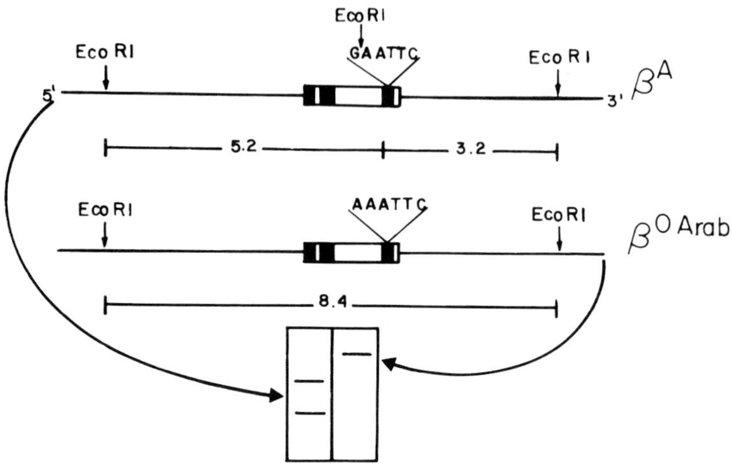

Fig. 7. Detection of the β^{0Arab} mutation by restriction endonuclease analysis. The mutation eliminates an *Eco*RI restriction site and generates an 8.4 kb fragment that contains the β^{0Arab} gene.

nucleotide substitution in the C position of the *Taq*I recognition sequence TCGA→T*TGA* and TCGA→*TAGA*A (Table 1). Since five of the seven *Taq*I sites in the factor VIII:C gene are in exons, the introduction of termination codons at these points would produce an incomplete factor VIII:C protein molecule [25,26].

Linkage Analysis: β-Thalassemia and Hemophilia A

The third prototypic method of detecting mutant genes using restriction endonucleases is linkage analysis. When two genes are linked and the linkage phase (in coupling when the disease and marker are on same chromosome or in repulsion when the disease and marker are on opposite chromosomes of the pair) can be determined within a given family, the status (presence or absence) of the disease gene in certain relatives can be inferred from the status of the marker. Increasingly, gene diagnosis by linkage analysis makes use of the high frequency of restriction fragment length polymorphisms (RFLPs). These RFLPs result from the many DNA polymorphisms or base differences that are normally seen between homologous chromosomes. Sequence data suggest that about one in every 100–200 nucleotides in regions not encoding a protein is polymorphic without any clinical consequences [27, 28]. These differences are usually single nucleotide substitutions, and occasionally they affect the sequence so as to create or eliminate a restriction endonuclease recognition site. The addition or loss of such sites results in different DNA fragment sizes after digestion. Since many substitutions exist, there are many chances to affect recognition sites. The large numbers of resulting RFLPs provide useful markers for linkage analysis.

A good example of the use of the resulting RFLPs in linkage analysis is that of β-thalassemia. The β-globin gene cluster spans about 50 kb, and many different polymorphic restriction sites have been identified throughout this cluster (Fig. 1) [29–31]. For example, in Figure 8, a polymorphic *Hinc*II restriction site is shown that is just upstream to the ε-globin gene on the 5′ end of the β-globin gene cluster. After digestion with *Hinc*II, two common patterns occur when the DNA fragments are studied. Since there is a constant *Hinc*II site 3′ to the ε-globin gene, if the site just 5′ to the gene is present, then a 3.7 kb fragment results that contains the gene (Fig. 8). If the site just 5′ to the ε-globin gene is absent, then an 8 kb fragment results. The chromosomes that produce the 3.7 and 8 kb fragments are both assumed to be normal. However, the difference in resulting fragment sizes enables one to differentiate easily between these two chromosomes. The advantage of being able to differentiate the two chromosomes is that one can then follow the inheritance of the β-globin gene, which is a very tightly linked genetic marker for the β-thalassemia gene that is on the other end of the cluster. The usefulness of these markers in linkage analysis is illustrated in Figure 8, which shows the pedigree and autoradiogram patterns of a family at risk for β-thalassemia major. In this family, the parents, who are carriers for β-thalassemia, are heterozygous for the RFLP detected by *Hinc*II, and both

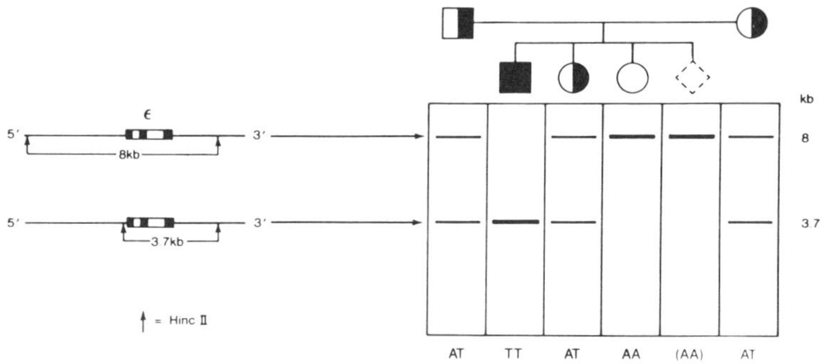

Fig. 8. Prenatal exclusion of β-thalassemia using restriction fragment length polymorphisms (RFLPs) and linkage analysis. Digestions of the ε-globin genes of both β^{thal} trait parents indicate that each is heterozygous for the *Hinc*II RFLP shown. From the autoradiogram on the right, it can be seen that the β^{thal} allele is in coupling with the 3.7 kb fragment of both parents. The restriction patterns and corresponding β-globin genotypes for various offspring are shown on the right. For details, see text.

have 3.7 and 8 kb fragments after their DNA is digested with *Hinc*II, blotted, and probed with an ϵ-globin gene probe. The 8 kb fragment results from absence of the polymorphic *Hinc*II site just 5′ to the ϵ-globin gene, whereas the 3.7 kb fragment results from their ϵ-globin gene on the chromosome 11 homolog having the polymorphic site. DNA from a previously affected child can be used to establish the linkage phase between the RFLP being examined and the locus under linkage study. In this case, the child is homozygous for the 3.7 kb fragment, indicating that the β^{thal} gene is in coupling phase with the 3.7 kb fragment in both parents. Given this, one can predict the outcome of all subsequent pregnancies that are studied using this method. If the fetus is homozygous for the 8 kb fragment, it should be a β^A homozygote having inherited the β^A gene of both parents. If the outcome is an 8/3.7 kb pattern, the fetus should be heterozygous (β^A/β^{thal}), like the parents. However, if the fetus inherits a 3.7 kb fragment pattern from each parent, the diagnosis would be β^{thal} homozygote (Fig. 8).

Hemophilia A is a second example of the use of RFLPs and linkage analysis in gene diagnosis. The factor VIII:C gene is closely linked to an X chromosome-specific clone named DX13 (also known as DNA segment DXS-15) that has been localized to Xq28 along with the factor VIII:C locus [32]. A polymorphic *Bgl*II restriction site is detected by the DX13 clone, and two common patterns are seen when the resulting DNA fragments are studied (Fig. 9). The 5.8 kb fragment occurs when the polymorphic recognition site is absent and the 2.8 kb fragment when it is present. The difference in fragment sizes enables one to differentiate easily the two chromosome patterns of a heterozygous female (see Fig. 9). Since about 50% of females have the 5.8/2.8 pattern, one can follow the inheritance of the RFLP detected by DX13 and infer the inheritance of the tightly linked factor VIII:C gene. The usefulness of this marker in linkage analysis is seen when the pedigree and autoradiogram patterns of a family at risk for hemophilia A are examined (Fig. 9). In this case, the grandmother is heterozygous for the RFLP detected by *Bgl*II/DX13 as well as being a carrier for hemophilia A. The pattern of her affected son can be used to establish the linkage phase between the RFLP and the mutant factor VIII:C gene. In this case, the affected son inherited her 2.8 kb fragment and mutant factor VIII:C gene, indicating that her normal factor VIII:C gene is in coupling phase with the 5.8 kb fragment (Fig. 9). Note that her daughters both inherited her 2.8 kb fragment and are carriers, because this fragment is in coupling with the mutant factor VIII:C allele. For these carriers, the hemophilic status of their sons and carrier status of their daughters can be determined as shown for the third generation. This approach has been used for both carrier detection as well as prenatal diagnosis of hemophilia A [32,33].

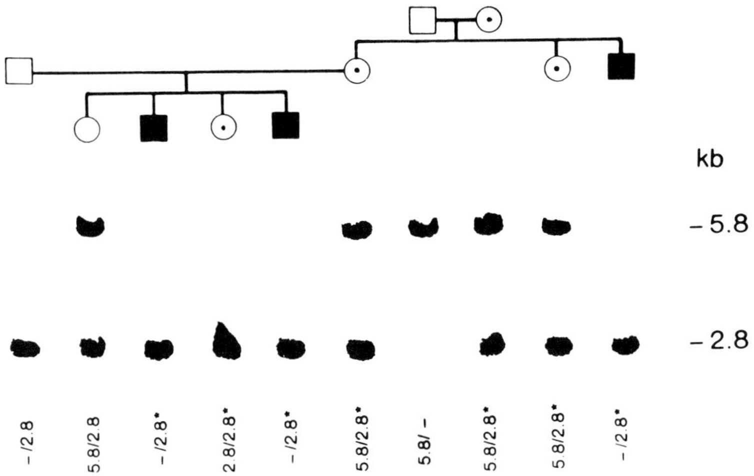

Fig. 9. Autoradiogram patterns of males with hemophilia A (solid squares) or carrier females (circles with dots) following digestion with *Bgl*II restriction endonuclease and hybridization with the DX13 probe derived by Harper et al [32]. Note that the 2.8* allele segregates with the mutant factor VIII:C gene in this family (see text).

Three general points about linkage analysis must be emphasized. First, the linkage phases shown in Figures 8 and 9 (β-thal genes in coupling with 3.7 kb *Hinc*II fragments and mutant factor VIII:C genes in coupling with DX13 sequences that cleave to 2.8 kb fragments following *Bgl*II digestion) are not observed in all families at risk for β-thalassemia or for hemophilia A. The reason for the differences is that many different mutations probably exist for both disorders and the relationship or coupling phase between the size of the marker and the abnormal allele may be different in different families. This possible difference in linkage phase from family to family must always be considered to avoid erroneous conclusions from linkage analysis.

Second, one can use such RFLPs to detect the inheritance of any mutation within a family through linkage analysis without actually knowing the nature of the mutation. However, to do this, one often needs samples from certain close relatives, such as an affected or normal sib or, alternatively, spouses or grandparents. These are necessary to establish the linkage phase between the given mutation and the RFLP being used as a marker.

Third, the distance between the RFLP being used as a marker and the mutant gene is critical. The greater the distance between the two, the greater the chance that an erroneous diagnosis will be made. This is due to the increasing probability of recombination with increasing genetic distance between the RFLP and the gene under study. A recombination (crossing-over) between the gene and RFLP during meiosis will cause a reversal of the previous linkage phase, resulting in an error in the inferred genotype of offspring. The map distance between two genes that corresponds to a 1% chance of recombination is called a centimorgan (cM). In the case of the ϵ-globin gene and the β^{thal} allele, the physical distance is about 50 kb (Figs. 1, 8). This would seem to be a rather large distance, but the corresponding genetic distance is in fact quite small. Estimates of the physical distance required to give 1% recombination suggest that about 1,000,000 bp or 1,000 kb would be associated with 1% recombination [34]. Thus the 50 kb distance separating the RFLP and mutant allele in the above study would represent about one-twentieth of 1% recombination, suggesting an extremely low error rate resulting from recombination. In the case of the factor VIII:C mutation and DX13 (Fig. 9), the physical distance is not precisely known. However, in the first 24 informative offspring studied, no recombinations were detected, suggesting that the error rate should be low (95% confidence limits of 0–12 cM) [32]. Although the factor VIII:C gene and DX13 may be very close, one of 82 meioses reported is a recombinant, and an exact estimate of the risk for errors resulting from recombination must await further studies [26,35]. Another X chromosome-specific clone that is potentially more reliable than DX13 in linkage studies of hemophilia A is ST14, also known as DNA segment DXS52 [36]. Use of ST14 allows detection of hemophilia A in over 90% to 92% of families at risk, with only one recombination detected in 138 offspring studied [35].

An additional application of linkage analysis is the use of RFLPs in differentiating mutant alleles. The association of different RFLPs with different mutant alleles has been very well documented in the case of the β-thalassemias [29,30]. These studies demonstrate the usefulness of examining each of the different sites to establish the presence or absence of the respective sites on the chromosome. The symbols usually used are a + to indicate the presence of a given polymorphic restriction site and a − to indicate its absence, and the combination of these sites is often referred to as a haplotype. Thus, when families are examined, affected children may have their β^{thal} alleles on β-globin gene clusters that have two different haplotypes, suggesting that the mutations may be different. The different haplotypes can be used to select for different β^{thal} alleles prior to sequence analysis. The approach of sequencing one of each of the different haplotypes observed proved to be a very effective way of avoiding duplication of cloning and sequencing the same mutant allele multiple times [29,30]. Also, haplotype studies have demonstrated that in various populations the individuals affected with β-

thalassemia are in most cases genetic compounds rather than true homozygotes [29]. The term "genetic compound" refers to the fact that they have two different mutant alleles at the same locus in contrast to true homozygotes, who have the same mutation on both alleles.

Cloning and Sequencing Mutant Genes

To determine the specific nucleotide alterations that characterize a mutation, cloning and DNA sequencing are usually done. The information gained can be improved by carefully selecting the mutants that are to be cloned and sequenced. One method mentioned in the previous paragraph is determining the haplotype around the mutation. As a general rule, within a given population the mutant alleles on different haplotypes have different mutations. Although there are exceptions to this rule, it has proved to be a useful tool in the selection of certain mutant alleles for sequence analysis [29,30].

The next step in cloning and sequencing is to establish a genomic library [37]. To do this, the DNA sample is cleaved using a restriction endonuclease, and the fragment length that is desired is enriched by gradient centrifugation or by electrophoresis in an agarose gel. Fragments of the size that contain the allele to be sequenced are then cloned into a vector, and the complete collection of such cloned DNA fragments is known as a genomic library. In principle, each DNA fragment to be cloned is introduced into a simple organism (bacteria), and during the process the inserted piece of DNA replicates along with the host DNA. There are three vector systems frequently used: 1) a bacterial plasmid, such as PBR322, 2) a bacterial virus or phage, usually phage λ, and 3) a synthetic hybrid of a plasmid and a phage that is referred to as a "cosmid" [37,38].

Following the selection of the vector to be used, the collection of random fragments of genomic DNA that have been cleaved is mixed with vector DNA that has been cleaved with the same restriction endonuclease (Fig. 2). Physical conditions are adjusted to give maximum likelihood ligation to form recombinant DNA molecules. Next, a suitable strain of bacterium is exposed to the DNA, usually in the presence of high calcium concentration to enable the DNA to enter into the bacteria. Selection for bacteria carrying the desired recombinant vector is done using antibiotic selection. In the case of recombinant PBR322, there are two antibiotic-resistance genes, and usually the genomic DNA is cloned into one of these. Bacteria that have taken up the recombinant will be resistant to the antibiotic for which the resistance gene of PBR322 remains intact. They will be sensitive to the alternative antibiotic whose resistance gene contains the cloning site. Using appropriate antibiotic selection, the colonies containing the various genomic fragments are grown on agar plates, and then the colonies are overlaid with a filter to make a replica plate. The DNA is denatured on the filter to separate the strands, and then the filter is incubated with a radiolabeled probe containing sequences homologous to the mutant gene that is going to be sequenced. Following autoradiography, colonies are identified that contain the specific DNA sequence that is complementary to the probe. These colonies can then be picked from the original plate and amplified and their DNA isolated for DNA sequencing.

DNA sequencing is usually done by the dideoxy method [38,39]. In principle, the dideoxy method makes use of a single-stranded phage called M13 [40]. Once the fragment to be sequenced has been subcloned into M13, then DNA polymerase is used to synthesize a complementary radioactive copy of the insert. During the DNA polymerase reaction 2′, 3′ dideoxynucleotides of A, C, G, or T are added to four separate reactions containing one of the deoxyribonucleotides radiolabeled with ^{32}P or ^{35}S (Fig. 10). When one of the dideoxynucleotides is incorporated, the 3′ end of the reaction is no longer a substrate for chain elongation, and the growing DNA chain is terminated. Thus in each of these four separate reactions there is a population of DNA molecules that is radioactive with a common 5′ end, but each of which is of varying length because of the incorporation of a specific 3′ end. Next, parallel-size fractionation of the four reaction products is done and autoradiography and sequence deduction are performed (Fig. 10). Comparison of the mutant and wild-type sequence allows deduction of the mutation. Sequencing not only identifies the mutation but can enable derivation of short oligonucleotide sequences that can be used as probes to detect the presence of that specific mutation in genomic DNA samples from other individuals.

Use of Oligonucleotide Probes: α1-Antitrypsin and β-Thalassemia

Oligonucleotide probes are small pieces of DNA, usually on the order of 14 to 30 nucleotides in length, which are used to detect directly the presence of mutations in genomic DNA. An example is the oligonucleotide probe that was derived by Kidd et al [41] to detect the mutation commonly responsible for α1-antitrypsin (α1-AT) deficiency. Although over 30 different α1-AT alleles have been detected, most normal individuals are MM homozygotes. Individuals who are ZZ homozygotes have severe α1-AT deficiency, but most other phenotypes including MZ do not. About 1/2,000 whites are ZZ homozygotes, and 5% are carriers of the Z gene (allele) [42].

Kidd et al [41] found that the Z mutation was a G→A substitution in codon 342, which results in a lysine substitution for glutamic acid. A 19 base oligonucleotide probe was synthesized that was specific for the M allele (Fig.

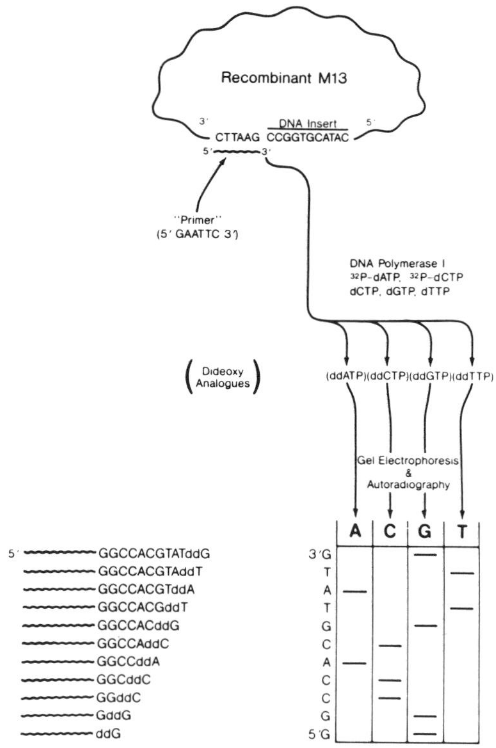

Fig. 10. DNA sequencing by the dideoxy method. The DNA insert to be sequenced is subcloned into M13, and a "primer" is annealed to sequences adjacent to the insert. Following the DNA polymerase reaction in the presence of dideoxy analogs, the four reaction products are subjected to gel electrophoresis and autoradiography allowing sequence deduction.

11). This M-specific oligonucleotide differed from the corresponding sequence of the Z gene in that the latter had an A rather than a G at the central nucleotide. The radiolabeled M-specific oligonucleotide was used as a probe for genomic DNA as shown in Figure 11. Under low-temperature hybridization and washing conditions, the M-specific probe hybridized to DNA from both MM and ZZ individuals. However, if either the hybridization or washing temperature was increased (see Fig. 11), the one base mismatch decreased hybridization to ZZ sequences. When both temperatures were increased, striking differences in hybridization intensity occurred that allowed differentiation of MM and ZZ genotypes. The MZ genotype can be identified by its intermediate signal to both the M- and Z-specific oligonucleotide probes.

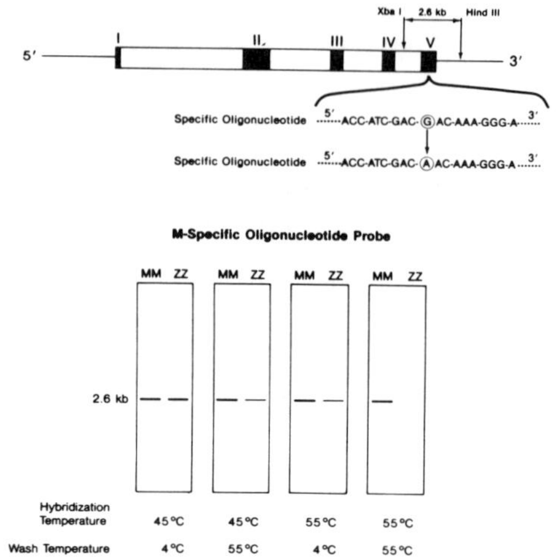

Fig. 11. Detection of a single base substitution using an oligonucleotide probe. The sequence of a portion of exon V of the normal (M) and Z alleles of α1-antitrypsin is shown above. The M-specific 19 base oligonucleotide was radiolabeled and used as a probe to study DNA from MM and ZZ individuals after *Xba*I and *Hind*III digestion and electrophoresis (shown below). Different stabilities of the M-specific oligonucleotide with the M and Z alleles allow identification of the MM and ZZ genotypes. (Adapted from Kidd et al [41]).

Mutation-specific oligonucleotide probes also have been of great utility in detecting certain β-thalassemia mutations [43]. Thus specific probes, each of which allows detection of single base differences, can be used to differentiate mutant alleles in the heterozygous or homozygous state.

APPLICATIONS OF DNA STUDIES TO GENE DIAGNOSES

Selected examples of cloned human sequences are listed in Table 2 that are potential tools for determining the etiology of various genetic disorders. The disorders and the applications of various probes are constantly growing in number. In the following section, examples of applications of these sequences to identify mutations at their respective loci will be discussed.

CHROMOSOME-SPECIFIC PROBES

Y Chromosome-Specific Probes: Fetal Sex Determination

The usefulness of Y chromosome-specific sequences has been demonstrated in studies of fetal DNA to determine the sex. Using chorionic villus sampling (CVS) at 9 weeks of gestation, an adequate amount of chorionic villi can be obtained to isolate sufficient DNA to analyze for the presence of Y-specific sequences [44]. Following digestion with *Hae*III and use of a probe for sequences on the Y chromosome, specific sequences occur in a 3.4 kb fragment that is present in DNA from males but absent in DNA from females (Fig. 12). This technique has been used to determine the sex of fetuses at risk for Lesch-Nyhan syndrome, hemophilia A, and other X-linked disorders and cytogenetically abnormal Y chromosomes [45–47]. Subsequently, a simpler technique has been developed in which DNA isolated from CVS or as little as 0.2 ml of amniotic fluid can be assayed without restriction endonuclease digestions. In this case, DNA is obtained by rapid alkali digestion of the cells and after neutralization dotted onto filters. Following hybridization with a Y-specific probe that anneals 1,000 times more strongly to male genomic DNA than to female DNA, the fetal sex can be determined [48,49].

X Chromosome-Specific Probes

Duchenne muscular dystrophy (DMD). DMD is an X-linked disease with progressive muscle wasting that leads to death in the late teens or early 20s [50]. In cytogenetic studies of certain females who were affected with DMD, chromosome deletions of Xp21 were found, suggesting that this is the location of the DMD locus [51,52]. Using this clue, Murray et al [53] isolated X chromosome-specific clones, one of which (RC8) mapped to Xp21–22.3. A *Taq*I RFLP was detected with RC8, but in linkage studies the map distance between RC8 and the DMD locus was shown to be about 15 cM. Thus an error rate of approximately 15% would occur in diagnoses inferred by using only RC8. Subsequently, Davies et al [54] isolated a second clone (L1.28) that is also 15 cM from DMD but is on the opposite side from RC8.

To provide additional markers that are tightly linked to the DMD locus, Kunkel et al [55] used the phenol-enhanced reassociation technique (PERT) to clone DNA sequences absent in a male who had DMD, retinitis pigmentosa, and chronic granulomatous disease [56]. One isolated clone (pERT87) was deleted in five of the DNAs isolated from 57 unrelated males with DMD. In addition, RFLPs identified by subclones from the same region of Xp21 showed no recombination with DMD in the 34 meioses studied [57]. Thus pERT87 and subclones derived from it should provide the markers necessary for accurate prenatal detection of most pregnancies at risk for DMD when a previous case is available for study to confirm the linkage phase or to detect the occurrence of deletions.

X-linked retinitis pigmentosa (XLRP). The same L1.28 clone used above for DMD has been used to study a specific X-linked form of retinitis

TABLE 2. Selected Cloned DNAs That Are Potential Tools for Diagnosis of Various Genetic Disorders†

DNA sequence	Chromosome assignment	Genetic disorder
Adenosine deaminase (ADA)	20q13–qter	ADA deficiency
α-Globin	16p12–pter	α-Thalassemia
α1Antitrypsin (α-1-AT)	14q24–q32	α1-AT deficiency
Antithrombin-III (AT-III)	1q23	AT-III deficiency
Apolipoprotein CII (APOC2)	19p13–q13	Myotonic dystrophy
Argininosuccinate synthetase	9q34	Citrullinemia
β-Globin	11p12	β-Thalassemia, sickle cell and other β-globin mutations
Carbamylphosphate synthetase I (CPS I)	2p	CPS I deficiency
Collagen		
pro-α1(I)	17q21–q22	Osteogenesis imperfecta
pro-α2(I)	7q21–q22	
Cystathionine β-synthetase	21	Homocystinuria
DNA segment		
ST14	Xq28	Adrenoleukodystrophy
3′ HVR	16p12–pter	Adult polycystic kidney disease
DOCR1-917	7cen–q22	Cystic fibrosis
met oncogene	7q21–q31	Cystic fibrosis
pj3.11	7cen–q22	Cystic fibrosis
L1.28	Xp21–p22	Duchenne muscular dystrophy
RC8	Xp21–p22	
pERT87	Xp21	Duchenne muscular dystrophy
DX13	Xq28	Hemophilia A
ST14	Xq28	Hemophilia A
G8	4p	Huntington disease
L1.28	Xp21–p22	X-linked retinitis pigmentosa
Glucose-6-phosphate dehydrogenase	Xq28	Adrenoleukodystrophy, fragile X-linked mental retardation
Growth hormone (GH)	17q21–q22	Familial isolated GH deficiency type 1A

(continued)

TABLE 2. Selected Cloned DNAs That Are Potential Tools for Diagnosis of Various Genetic Disorders† (continued)

DNA sequence	Chromosome assignment	Genetic disorder
Hypoxanthine-guanine phosphoribosyl transferase (HGPT)	Xq26–q27	Lesch-Nyhan syndrome
Ornithine transcarbamylase (OTC)	Xp21	OTC deficiency, Duchenne muscular dystrophy
Phenylalanine hydroxylase	12q24	Phenylketonuria
Steroid 21-hydroxylase	6p21	Congenital adrenal hyperplasia
Y-chromosome specific	Yq	Fetal sex determination

†Beaudet AL: Bibliography of cloned human and other selected DNAs. Am J Hum Genet 36:235–251, 1984; McKusick VA: The human gene map. In "Mendelian Inheritance in Man." Baltimore: Johns Hopkins University Press, 1983, pp xxxvii–lxi (also, personal communication).

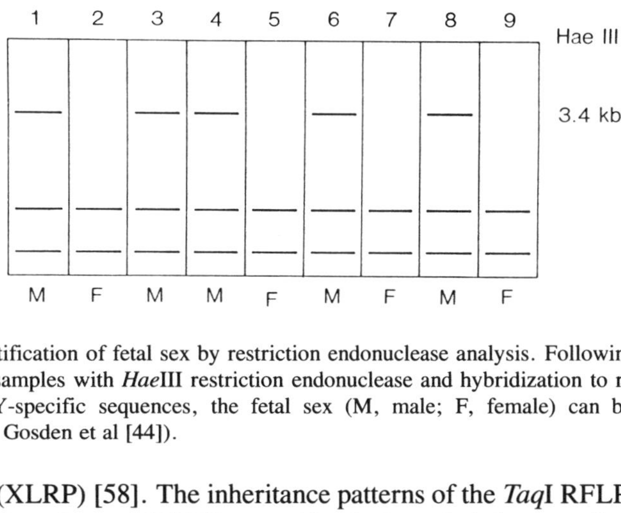

Fig. 12. Identification of fetal sex by restriction endonuclease analysis. Following digestion of fetal DNA samples with *Hae*III restriction endonuclease and hybridization to radiolabeled chromosome Y-specific sequences, the fetal sex (M, male; F, female) can be deduced. (Adapted from Gosden et al [44]).

pigmentosa (XLRP) [58]. The inheritance patterns of the *Taq*I RFLP detected by the L1.28 probe and XLRP were identical, and their map distance was estimated to be about 3 cM. However, it should be pointed out that a limitation in applying this approach is that genetic heterogeneity must be excluded because there is more than one type of XLRP [59]. Since these may be due to genes on different regions of the X chromosome, sufficient family members should be studied to demonstrate that close linkage does in fact exist in each of the different families studied.

Autosome-Specific Probes: Myotonic Dystrophy

Studies of myotonic dystrophy (DM) exemplify the use of specific clones from an autosome that is known to bear the defective locus. DM is a neurologic disease that has an autosomal dominant mode of inheritance and a prevalence of about 1 in 30,000 [60]. The symptoms of facial and distal limb weakness, ptosis, myotonia, cataracts, and testicular atrophy vary in their age of onset and severity. Peculiarly, when the mother is affected, a significant percent of her offspring will be affected with severe congenital DM.

Historically, DM was one of the first autosomal disorders to show genetic linkage. Mohr [61] found that DM was linked to Lutheran and Lewis blood groups. The order of the loci is reported to be Lewis Secretor–C3–DM–(Se-PepD)–Lutheran with an estimated distance of about 6 cM between C3 and DM [62]. Thus, when the autosomal assignment of the linkage group was established, a logical basis for obtaining closely linked markers arose. Techniques of fluorescence-activated chromosome sorting are being used to obtain large numbers of chromosomes 19 to construct genomic libraries enriched for sequences on 19 [37,38,63]. Such libraries should greatly increase the probability of finding clones that are more tightly linked to the DM locus. Currently, the closest known probe is the apolipoprotein CII gene that Shaw et al [64] found to give 4% recombination with the DM locus at the maximum lod score of 7.88.

GENE-SPECIFIC PROBES

This approach involves the use of clones for genes that are suspected on the basis of biochemical or other evidence to be involved in the mutations that lead to certain disorders.

α1-Antitrypsin: α1-Antitrypsin Deficiency

The use of a mutation-specific oligonucleotide to detect the Z allele of α1-AT was discussed above.

Clotting Factors: Hemophilias A and B

Hemophilia occurs in two to ten per 100,000 population [65]. Hemophilia A and B are X-linked recessive disorders resulting from deficiencies of clotting factors VIII:C and IX, respectively. Although both disorders can be treated with appropriate clotting factors, between 5% and 15% of patients will develop antibodies to factor VIII:C or IX after repeated exposures [17]. The use of a closely linked X chromosome-specific probe and direct detection of factor VIII:C mutations were discussed above. A third technique uses DNA sequences that contain exons 17 and 18 of the factor VIII:C gene as a probe [26,66]. Using this clone and the restriction endonuclease *Bcl*I, an RFLP has been identified that is useful in both carrier identification and prenatal diagnosis. Since the exon 17–18 clone is within the factor VIII:C gene, the risk for possible errors because of recombination should be negligible compared to that associated with flanking markers such as DX13 or ST14 [32,33,35,36].

Regarding hemophilia B, clones of factor IX have been used to document deletions of all or portions of the factor IX gene in certain hemophilia B patients with factor IX antibodies [67,68]. Subsequently, an RFLP has been detected using a factor IX clone that allows accurate carrier status determination as well as prenatal diagnosis of hemophilia B in the first trimester [69–71].

Collagen: Osteogenesis Imperfecta (OI) and Marfan Syndrome

Defective expression and alterations of the pro-α1(I) or the pro-α2(I) collagen proteins have been documented in some patients with OI and Marfan syndrome [72,73]. Type I procollagen is a heterotrimer made up of two α1(I) and one α2(I) chains, which are synthesized as precursor polypeptides called pro-α1(I) and pro-α2(I). The locus for the gene for pro-α1(I) maps to chromosome 17 and that for the pro-α2(I) to chromosome 7 [74,75]. In both cases, there is thought to be a single functional locus.

In a family with an autosomal dominant form of OI, Tsipouras et al [76] found that an RFLP detected using a clone for pro-α2(I) cosegregated with the disease phenotype. This established that certain cases of OI are due to mutations within or very near the pro-α2(I) locus. Analysis of the same clone was used to study large families with the Marfan syndrome. The results indicated that in one very large family there was independent segregation of the Marfan syndrome phenotype and the pro-α2(I) collagen locus [77]. In this case, analysis using a "candidate" locus quickly excluded linkage between that locus and the disorder.

Phenylalanine Hydroxylase: Phenylketonuria

Phenylketonuria (PKU) is another example of the use of a gene-specific probe for prenatal diagnosis. PKU is a metabolic disorder in which a deficiency of phenylalanine hydroxylase is inherited in an autosomal recessive mode occurring in about one in 10,000 U.S. whites [78]. Although there is a very effective dietary treatment for PKU, the disorder still constitutes a significant burden for the patients and their families. In studies of families with PKU using a clone for phenylalanine hydroxylase, Woo et al [79] and Woo [80] reported a number of different RFLPs useful in carrier detection and prenatal diagnosis (see Fig. 13). Recently, Lidsky et al [81] reported use of these RFLPs in prenatal diagnosis of two pregnancies at risk. Their findings suggest that the technique should be informative in over 90% of families with a previous affected child.

Steroid 21-Hydroxylase: Congenital Adrenal Hyperplasia

Congenital adrenal hyperplasia (CAH) is a metabolic disorder in which a deficiency of steroid 21-hydroxylation occurs in the adrenal cortex of about 95% of cases [82]. CAH is inherited in an autosomal recessive mode, and it occurs in about one in 5,000 births [83]. The resulting deficiency of cortisol synthesis causes overproduction of precursors proximal to the block including 17-OH progesterone. This, in turn, leads to excessive androgen production that causes masculinization of the external genitalia of affected females in utero. An additional defect in aldosterone biosynthesis occurs in about one-half of the patients, which, if untreated, can result in neonatal death resulting from the inability to retain urinary sodium.

The steroid 21-hydroxylase gene is closely linked to the HLA major histocompatibility complex, and different forms of CAH are associated with

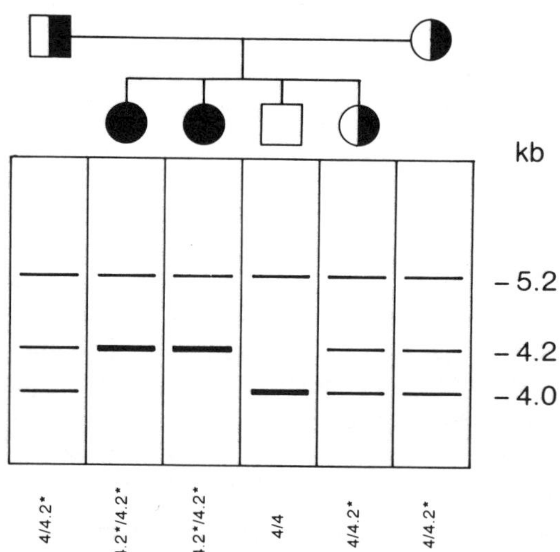

Fig. 13. Autoradiogram patterns of DNA from a family among whom two daughters have phenylketonuria. In this family, the carrier parents are heterozygous for 4/4.2 fragments after digestion with *Hind*III and hybridization to radiolabeled phenylalanine hydroxylase sequences. Since the 4.2* fragment is in coupling with the abnormal allele in both parents, the genotypes of various offspring can be deduced easily. (Adapted from Woo et al [79]).

certain HLA antigen haplotypes [84,85]. Recently, White et al [86] isolated a bovine adrenal cDNA clone encoding a portion of the gene for the cytochrome P-450 specific for steroid 21-hydroxylation. When DNA from controls was digested with *Taq*I and hybridized to this clone, fragments of 3.7 and 3.2 kb were detected. Interestingly, the 3.7 kb fragment was absent in DNA from an individual with CAH resulting from steroid 21-hydroxylase deficiency. Furthermore, the decreased intensity of the 3.7 kb *Taq*I-derived fragment appeared to cosegregate with CAH in a large kindred [86,87]. Although these findings suggest that a gene deletion is responsible for CAH in certain subjects, it remains possible that these different patterns are secondary to an RFLP such as discussed previously under linkage analysis. In either case, it is probable that molecular techniques will clarify the mutation(s) involved and provide tools for detection of CAH in the very near future.

Urea Cycle Disorders

Inborn errors of ureagenesis are associated with hyperammonemic coma in infancy. Very early treatment with appropriate protein restriction as well as the use of medications to conjugate toxic byproducts are effective but difficult [88,89].

Ornithine transcarbamylase (OTC) deficiency. Ornithine transcarbamylase deficiency is X-linked recessive, and affected males often die in the first week after birth. Since OTC is not expressed in fibroblasts, fetal liver biopsy has been used in the past for prenatal diagnosis [90,91]. Old et al [92] used a clone for OTC to exclude the disorder in an at-risk male fetus whose mother was heterozygous for a deletion of Xp21 that included the OTC gene. Deletions and RFLPs detected using the OTC probe should be useful in linkage studies to determine carrier status as well as prenatal diagnosis [93].

Argininosuccinate synthetase: citrullinemia. Argininosuccinate synthetase (AS) deficiency has an autosomal recessive mode of inheritance, and affected infants present with profound elevations of blood ammonia and citrulline. Detection of mutations in the AS gene is made quite difficult by the presence of ten or more copies of AS sequences scattered on different chromosomes [94]. All but one of these, that on chromosome 9, are pseudogenes. Although the presence of multiple copies prevented detection of the DNA mutation in the abnormal AS gene, molecular studies of mRNA from cultured cells demonstrated that in most cases a stable but altered mRNA was detectable [94,95].

Carbamylphosphate synthetase I (CPS): CPS deficiency. CPS catalyzes the first step in the urea cycle, and CPS deficiency has an autosomal recessive mode of inheritance [96]. Fearon et al [97] used a CPS clone to study the CPS genes of affected individuals. Although no CPS gene deletions were detected, an RFLP detected by *Bcl*I was found. Limited family studies suggested that the RFLP segregated with the disorder, but the number of patients studied was insufficient to prove the reliability of this marker in diagnosis of the disorder by linkage analysis.

GENETIC DISORDERS OF UNKNOWN LOCI

In most genetic disorders, the primary gene defect is not yet known, as is the case for adult polycystic kidney disease, cystic fibrosis, and Huntington disease. In all three disorders, DNA techniques have been used to detect the location of the defective gene by linkage analysis.

Huntington Disease

Huntington disease (HD) is an autosomal dominant disorder in which heterozygotes are usually asymptomatic until 30–50 years of age [98]. At that age, those who have the HD gene begin to develop dementia and chorea with progressive deterioration that leads to death within about 15 years of onset of symptoms. The prevalence in the population is one case in 10,000–20,000, and the disease is characterized by the absence of any documented new mutations [99].

To identify the chromosomal location of the HD locus, Gusella et al [99] tested the segregation patterns of multiple clones using samples from very large pedigrees with multiple individuals affected with HD. One clone, G8, showed tight linkage to the HD phenotype in two large families with an original estimate of a map distance of 0 cM. Subsequent study of two additional families with different ethnic and racial backgrounds also showed close linkage between HD and G8 and when combined with the Gusella et al data suggest that the map distance between HD and G8 is about 5 cM [99,100]. Further studies are needed to determine the exact risk for recombination as well as to establish that linkage exists in other large families.

Adult Polycystic Kidney Disease

Adult polycystic kidney disease (APKD) is a common genetic disorder that has an incidence of one in 500–1,000. This disorder has an autosomal dominant mode of inheritance, and heterozygotes develop progressively enlarging cysts in the liver, pancreas, and spleen as well as in the kidneys. The biochemical basis of this disorder is unknown, and the renal involvement leads to irreversible renal failure and death at a mean age of 51 years unless dialysis or transplantation is done [101]. Exclusion of heterozygosity by renal ultrasound is reported to have a high rate of false negatives; 34% of patients at risk have been shown to develop the disease despite negative renal ultrasound examinations in the second decade [102].

Recently, Reeders et al [103] reported their linkage studies of APKD in nine different families using a highly polymorphic area that is approximately 8 kb from the α-globin gene. This clone (3′ HVR) demonstrated seven recombinants in 127 meioses (Θ = .05 at a lod score of 25.85). These findings suggest DNA analysis may soon provide clues to the basic cause of APKD as well as allow prenatal detection of APKD in selected families.

Cystic Fibrosis

In the last 3 months of 1985, five reports were published that localized the CF locus and provided RFLPs that could be used in prenatal detection of this disorder. First, Eiberg et al [104] reported that the serum arylesterase paraoxonase (PON) is linked to the CF locus with a lod score of 2.40 at Θ = 0.04. Next, Tsui et al [105] reported a DNA marker (DOCR1-917) that detected RFLPs following digestion with *Hind*II. Mapping studies indicated that the order was DOCR1-917:PON-CF with map distances of 15 and 10 cM from the CF locus, respectively, and these genes were assigned to chromosome 7 [106]. Since DOCR1-917 and PON were $\geqslant$10 cM from the CF locus, considerable error would result from their use as linkage markers in prenatal diagnosis. This limitation was at least partially overcome by reports of two additional DNA markers, the *met* oncogene locus and an anonymous clone (pJ3.11), which were found to be tightly linked to CF (Θ = 0 at lod 8.65 for *met* and Θ = 0 at lod 5.24 for pJ3.11) [107,108]. Although determination of the exact error rate associated with linkage diagnosis of CF using these latter two markers must await further studies, it is very likely that these probes will be clinically useful. As in other linkage analysis applications, DNA samples will be needed from a previous affected child as well as both parents to establish the linkage phase between the informative marker(s) and the CF alleles for a given family at risk for CF.

SUMMARY

There are five major points to be brought out here.

1. Normal variation in DNA is extensive and ancient. This refers to the finding of RFLPs that are inherited along with the mutations that have arisen on those chromosomes in a consistent manner. This variation is useful in identifying different mutant alleles through the use of haplotype analysis.

2. Knowledge of normal variation is necessary in identifying mutant genes. The patterns seen in nonaffected controls must be compared with those seen in affected individuals to determine if the alterations seen in the affecteds represent polymorphisms or actually represent the result of DNA substitutions, insertions, or deletions that are affecting the gene under study.

3. Multiple methods of detecting mutant genes exist. These include restriction analysis, cloning, and sequencing and the use of oligonucleotide probes.

4. Molecular heterogeneity is the rule rather than the exception. Molecular heterogeneity refers to the demonstrated observations that many affecteds will be genetic compounds rather than true homozygotes. This results from the fact that within the same population there may exist many different alleles, which in the homozygous state or in combination with different mutant alleles can cause a disease state.

5. Methods of DNA analysis are applicable to all inherited diseases. The prerequisite for the study of any given genetic disorder by DNA analysis is either a copy of the gene that is involved or a DNA fragment that is close to the gene. Given either of these tools, linkage analysis can be done preliminarily to demonstrate that cosegregation of the abnormal phenotype with the DNA sequences being studied and the gene can be cloned and sequenced.

FUTURE DEVELOPMENTS

Future applications of molecular genetic techniques to gene diagnosis will result from the development of 1) additional disease-specific probes, 2) construction of linkage maps with DNA markers for human chromosomes, and 3) technical improvements.

Additional Disease-Specific Probes

The number of diseases that can be detected is a function of the number of available probes. The prerequisite for a probe to be useful in the detection of a given genetic disorder is that it is a copy of the gene that is changed by the mutation or that it is sufficiently close to serve as a reliable marker in linkage studies.

Construction of a Human Linkage Map

The human genome contains about 3,300 cM, and 100 evenly spaced DNA marker loci would provide a marker every 33 cM [109]. This, in turn, would mean that any defective locus would be no more than about 16 cM from a marker locus. Although such markers have not been found and mapped for the great majority of human chromosomes, selected maps are available for regions of certain chromosomes. Two such regions span significant portions of 11p and the X chromosomes [110,111]. If such maps and probes were available for all chromosomes, determining the location of any disease locus would be much easier.

Technical Improvements

New DNA markers. The identification and mapping of new DNA markers is occurring at an increasing rate. This is being achieved by isolation of probes from purified chromosomes or chromosome fragments and by intensive use of these clones in linkage studies [112,113]. In addition, chromosomal localization of clones can now be done reliably by in situ hybridization [114]. The combination of these factors ensures that the number of probes applicable to prenatal diagnosis will increase steadily.

New methods of labeling probes. Currently, most DNA probes are radiolabeled with ^{32}P. Whereas this method is convenient for researchers, it is awkward for diagnostic laboratories where quality control and long reagent shelf-life are needed. One alternative already developed is the use of biotinylated nucleotides in standard nick translations. Subsequent detection of the signal is achieved by enzymatic detection [115]. It is hoped that this or other alternatives will provide "cold" probes of very high sensitivity.

Denaturing gradient electrophoresis. Agarose or polyacrylamide gels are currently used to separate DNA fragments, but these systems do not give altered mobilities for molecules with single base substitutions. Using denaturing gradient gels, Lerman et al [116] have shown that mutants with one or two base substitutions can give mobilities quite different from those of normal alleles. This methodology has the potential to identify mutations that do not alter the size of the DNA fragment or a restriction site. Since over 94% of base substitutions are thought not to occur within recognition sites, their detection by altered mobility would greatly enhance gene diagnoses.

REFERENCES

1. Paulson JR, Laemmli UK: The structure of histone-depleted metaphase chromosomes. Cell 12:817–828, 1977.
2. Watson JD, Crick FHC: Molecular structure of nucleic acids. Nature 171:737–738, 1953.
3. Lewin B: What is a gene? A biochemical view. In "Genes." New York: John Wiley & Sons, 1983, pp 21–48.
4. Tilghman SM, Tiemeier DC, Seidman JG et al: Intervening sequence of DNA identified in the structural portion of a mouse β-globin gene. Proc Natl Acad Sci USA 75:725–729, 1978.

5. Maniatis T, Fritsch EF, Lauer J et al: The molecular genetics of human hemoglobins. Annu Rev Genet 14:145–178, 1980.
6. Deisseroth A, Nienhuis A, Turner P et al: Localization of the human α-globin structural gene to chromosome 16 in somatic cell hybrids by molecular hybridization assay. Cell 12:205–218, 1977.
7. Deisseroth A, Nienhuis A, Lawrence J et al: Chromosomal localization of human β-globin gene on human chromosome 11 in somatic cell hybrids. Proc Natl Acad Sci USA 75:1456–1460, 1978.
8. Barsh G, Seeburg P, Gelinas R: Structure and expression of the human growth hormone-like gene cluster. Nucleic Acids Res 11:3939–3958, 1983.
9. Arber W: Promotion and limitation of genetic exchange. Science 205:361–365, 1979.
10. Smith HO: Nucleotide sequence specificity of restriction endonucleases. Science 205:455–462, 1979.
11. Nathans D: Restriction endonucleases, simian virus 40 and the new genetics. Science 206:903–909, 1979.
12. Malcolm ADB: The use of restriction enzymes in genetic engineering. Genet Eng 2:129–173, 1981.
13. Phillips JA III, Hjelle BL, Seeburg PH et al: Molecular basis for familial isolated growth hormone deficiency. Proc Natl Acad Sci USA 78:6372–6375, 1981.
14. Southern E: Detection of specific sequences among DNA fragments separated by gel electrophoresis. J Mol Biol 98:503–517, 1975.
15. Rivarola MA, Phillips JA III, Migeon CJ et al: Phenotypic heterogeneity in familial isolated growth hormone deficiency (IGHD) type 1A. J Clin Endocrinol Metab 59:34–40, 1984.
16. Dobs AS, Phillips JA III: Diagnosis of human endocrine disorders using recombinant DNA techniques. In Ney RL (ed): "Clinics in Endocrinology and Metabolism." London: WB Saunders, 1985, pp 273–294.
17. Shapiro SS: Genetic predisposition to inhibitor formation. In Hoyer LW (ed): "Factor VIII Inhibitors." New York: Alan R. Liss, Inc, 1984, pp 45–55.
18. Dozy AM, Forman EN, Abuelo DN et al: Prenatal diagnosis of homozygous α-thalassemia. JAMA 241:1610–1612, 1979.
19. Pressley L, Higgs DR, Clegg JB et al: Gene deletions in α-thalassemia prove that the 5′ ζ locus is functional. Proc Natl Acad Sci USA 77:3586–3589, 1980.
20. Antonarakis SE, Phillips JA III, Kazazian HH Jr: Genetic diseases: Diagnosis by restriction endonuclease analysis. J Pediatr 100:845–856, 1982.
21. Wilson JT, Milner PF, Summer ME et al: Use of restriction endonucleases for mapping the allele for β^s-globin. Proc Natl Acad Sci USA 79:3628–3631, 1982.
22. Chang JC, Kan YW: A sensitive new prenatal test for sickle-cell anemia. N Engl J Med 307:30–32, 1982.
23. Orkin SH, Little PFR, Kazazian HH Jr et al: Improved detection of the sickle mutation by DNA analysis: Application to prenatal diagnosis. N Engl J Med 307:32–36, 1982.
24. Phillips JA III, Scott AF, Kazazian HH Jr et al: Prenatal diagnosis of hemoglobinopathies by restriction endonuclease analysis: Pregnancies at risk for sickle cell anemia and SO^{Arab} disease. Johns Hopkins Med J 145:57–60,1979.
25. Gitschier J, Wood WI, Tuddenham EGD et al: Dectection and sequence of mutations in the factor VIII gene of haemophiliacs. Nature 315:427–430, 1985.
26. Gitschier J, Drayna D, Tuddenham EGD et al: Genetic mapping and diagnosis of haemophilia A achieved through a *Bcl*I polymorphism in the factor VIII gene. Nature 314:738–740, 1985.

27. Jeffreys AJ: DNA sequence variants in the $^{G}\gamma$, $^{A}\gamma$, δ and β-globin genes of man. Cell 18:1–10, 1979.
28. Ewens WJ, Spielman RS, Harris H: Estimation of genetic variation at the DNA level from restriction endonuclease data. Proc Natl Acad Sci USA 78:3748–3750, 1981.
29. Orkin SH, Kazazian HH Jr, Antonarakis SE et al: Linkage of β-thalassemia mutations and β-globin gene polymorphisms with DNA polymorphisms in the human β-globin gene cluster. Nature 296:627–631, 1982.
30. Orkin SH, Antonarakis SE, Kazazian HH Jr: Polymorphism and molecular pathology of the human beta-globin gene. In Brown EB (ed): "Progress in Hematology XIII." New York: Grune & Stratton, 1983, pp 49–73.
31. Boehm CD, Antonarakis SE, Phillips JA III et al: Prenatal diagnosis using DNA polymorphisms. Report in 95 pregnancies at risk for sickle-cell disease or β-thalassemia. N Engl J Med 308:1054–1058, 1983.
32. Harper K, Pembrey ME, Davvies KE et al: A clinically useful DNA probe closely linked to haemophilia A. Lancet 2:6–8, 1984.
33. Tonnessen T, Sondergaard F, Mikkelsen M et al: X-chromosome-specific probe DX13 for carrier detection and first trimester prenatal diagnosis in haemophilia A. Lancet 2:1269–1270, 1984.
34. Kurnit DM, Hoehn H: Prenatal diagnosis of human genome variation. Annu Rev Genet 13:235–258, 1979.
35. Janco RL, Phillips JA III, Orlando P et al: Carrier testing strategy in haemophilia A. Lancet 1:148–149, 1986.
36. Oberle I, Camerino G, Heilig R: Genetic screening for hemophilia A (classic hemophilia) with a polymorphic DNA probe. N Engl J Med 312:682–686, 1985.
37. Steel CM: DNA in medicine: The tools (part I). Lancet 2:908–911, 1984.
38. Steel CM: DNA in medicine: The tools (part II). Lancet 2:966–968, 1984.
39. Sanger F, Nicklen S, Coulson AR: DNA sequencing with chain-terminating inhibitors. Proc Natl Acad Sci USA 74:5463–5467, 1977.
40. Messing JR, Crea R, Seeburg PH: A system for shotgun DNA sequencing. Nucleic Acids Res 9:309–321, 1981.
41. Kidd VJ, Wallasce RB, Itakura K et al: α1-Antitrypsin deficiency detection by direct analysis of the mutation in the gene. Nature 304:230–234, 1983.
42. Laurell C-B, Sveger T: Mass screening for newborn Swedish infants for α1-antitrypsin deficiency. Am J Hum Genet 27:213–217, 1975.
43. Orkin SH, Markham AF, Kazazian HH Jr: Direct detection of the common Mediterranean β-thalassemia gene with synthetic DNA probes: An alternative approach for prenatal diagnosis. J Clin Invest 71:775–779, 1983.
44. Gosden JR, Mitchell AR, Gosden CM: Direct vision chorion biopsy and chromosome-specific DNA probes for determination of fetal sex in first-trimester prenatal diagnosis. Lancet 2:1416–1419, 1982.
45. Disteche C, Luthy D, Haslam DB et al: Prenatal identification of a deleted Y chromosome by cytogenetics and a Y-specific repetitive DNA probe. Hum Genet 67:222–224, 1984.
46. Gibbs DA, Crawfurd MD'A, Headhouse-Benson CM et al: First trimester diagnosis of Lesch-Nyhan syndrome. Lancet 2:1180–1183, 1984.
47. Antonarakis SE, Carpenter RJ Jr, Hoyer LW: Prenatal diagnosis of haemophilia A by factor VIII gene analysis. Lancet 1:1407–1409, 1985.
48. Gosden JR, Christie S, Gosden CM et al: Rapid fetal sex determination in first trimester prenatal diagnosis by dot hybridization of DNA probes. Lancet 1:540–541, 1984.

49. Lau Y-F, Dozy AM, Huang JC et al: A rapid screening test for antenatal sex determination. Lancet 1:14–16, 1984.
50. Emery AEH: The muscular dystrophies. In Emery AEH, Rimoin DL (eds): "Principles and Practice of Medical Genetics." Edinburgh: Churchill Livingstone, 1983, pp 396–402.
51. Lindenbaum RH, Clarke G, Patel C et al: Muscular dystrophy in an X;1 translocation female suggests that Duchenne locus is on X chromosome short arm. J Med Genet 16:389–392, 1979.
52. Zatz M, Vianna-Morgante AM, Campos D et al: Translocation (X;6) in a female with Duchenne muscular dystrophy: Implications for the localization of the DMD locus. J Med Genet 18:442–447, 1981.
53. Murray JM, Davies KE, Harper PS et al: Linkage relationship of a cloned DNA sequence on the short arm of the X chromosome to Duchenne muscular dystrophy. Nature 300:69–71, 1982.
54. Davies KE, Pearson PL, Harper PS et al: Linkage analysis of two cloned DNA sequences flanking the Duchenne muscular dystrophy locus on the short arm of the human X chromosome. Nucleic Acids Res 11:2303–2311, 1983.
55. Kunkel LM, Monaco AP, Middleworth W et al: Specific cloning of DNA fragments absent from the DNA of a male patient with an X chromosome deletion. Proc Natl Acad Sci USA 82:4778–4782, 1985.
56. Kohne DE, Levinson SA, Byers MJ: Room temperature method for increasing the rate of DNA reassociation by many thousandfold: The phenol emulsion, reassociation technique. Biochemistry 16:5329–5341, 1977.
57. Monaco AP, Bertelson CJ, Middlewsworth W et al: Detection of deletions spanning the Duchenne muscular dystrophy locus using a tightly linked DNA segment. Nature 316:842–845, 1985.
58. Bhattacharya SS, Wright AF, Clayton JF et al: Close genetic linkage between X-linked retinitis pigmentosa and a restriction fragment length polymorphism identified by recombinant DNA probe L1.28. Nature 309:253–255, 1984.
59. Heckenlively J: The hereditary retinal and choroidal degenerations. In Emery AEH, Rimoin DL (eds): "Principles and Practice of Medical Genetics." Edinburgh: Churchill Livingstone, 1983, pp 522–528.
60. Harper PS: Myotonic dystrohy and related disorders. In Emery AEH, Rimoin DL (eds): "Principles and Practice or Medical Genetics." Edinburgh: Churchill Livingstone, 1983, pp 426–433.
61. Mohr J: "A Linkage Study in Man." Copenhagen: Munksgard, 1954, pp 1–119.
62. Eiberg H, Mohr JA, Nielsen LS, Simonsen N: Genetics and linkage relationships of C3 polymorphism: Discovery of C3-Se linkage and assignment of LES-C3-DM-Se-PEPD-Lu synteny to chromosome 19. Clin Genet 24:159–170, 1983.
63. Latt SA, Kunkel LM, Tantravahi U et al: Construction, analysis, and utilization of recombinant phage libraries obtained using fluorescence activated flow sorting in recombinant phage libraries. In Messer A, Porter IH (eds): "Recombinant DNA and Medical Genetics." New York: Academic Press, 1983, pp 35–47.
64. Shaw DJ, Meredith AL, Sarfarazi M et al: The apolipoprotein II gene: Subchromosomal localization and linkage to the myotonic dystrophy locus. Hum Genet 70:271–273, 1985.
65. Biggs R: Defects in coagulation. In Emery AEH, Rimoin DL (eds): "Principles and Practice of Medical Genetics." Edinburgh: Churchill Livingstone, 1983, pp 1068–1072.
66. Antonarakis SAE, Waber PG, Kittur SD et al: Hemophilia A: Detection of molecular defects and of carriers by DNA analysis. N Engl J Med 313:842–848, 1985.

67. Giannelli F, Choo KH, Rees DJG et al: Gene deletions in patients with hemophilia B and anti-factor IX antibodies. Nature 303:181–182, 1983.
68. Chen S-H, Yoshitake S, Chance PF et al: An intragenic deletion of the factor IX gene in a family with hemophilia B. J Clin Invest 76:2161–2164, 1985.
69. Tonnesen T, Sondergaard F, Guttler F et al: Exclusion of haemophilia B in male fetus by chorionic villus biopsy. Lancet 2:932, 1984.
70. Giannelli F, Choo KH, Winship PR et al: Characterization and use of an intragenic polymorphic marker for detection of carriers of haemophilia B (factor IX deficiency). Lancet 1:239–241, 1984.
71. Grunebaum L, Cazenave J-P, Camerino G et al: Carrier detection of hemophilia B using a restriction site polymorphism associated with the coagulation factor IX gene. J Clin Invest 73:1491–1495, 1984.
72. Williams CJ, Prockop DJ: Synthesis and processing of a type I procollagen containing shortened proα(I) chains by fibroblasts from a patient with osteogenesis imperfecta. J Biol Chem 258:5915–5921, 1983.
73. Henke E, Leader M, Tajima S et al: A 38 base pair insertion in the proα2(I) collagen gene of a patient with Marfan syndrome. J Cell Biochem 27:169–174, 1985.
74. Huerre C, Junien C, Weil D et al: Human type I procollagen genes are located on different chromosomes. Proc Natl Acad Sci USA 79:6627–6630, 1982.
75. Junien C, Weil D, Myers JC et al: Assignment of the human proα2(I) collagen structural gene (COLIA2) to chromosome 7 by molecular hybridization. Am J Hum Genet 34:381–387, 1982.
76. Tsipouras P, Myers JC, Ramirez AF et al: Restriction fragment length polymorphism associated with the proα2(I) gene of human type I procollagen: Application to a family with an autosomal dominant form of osteogenesis imperfecta. J Clin Invest 72:1262–1267, 1983.
77. Tsipouras P, Francomano C, Pyeritz RE et al: Genetic linkage studies in the Marfan syndrome. Pediatr Res 18:226, 1984.
78. Scott CR: Disorders of amino acid metabolism. In Emery AEH, Rimoin DL (eds): "Principles and Practice of Medical Genetics." Edinburgh: Churchill Livingstone, 1983, pp 1241–1244.
79. Woo SLC, Lidsky AS, Guttler F et al: Cloned human phenylalanine hydroxylase gene allows prenatal diagnosis and carrier detection of classical phenylketonuria. Nature 306:151–155, 1983.
80. Woo SLC: Prenatal diagnosis and carrier detection of classic phenylketonuria by gene analysis. Pediatrics 74:412–423, 1984.
81. Lidsky AS, Guttler F, Woo SLC: Prenatal diagnosis of classic phenylketonuria by DNA analysis. Lancet 2:549–551, 1985.
82. Kuhnle U, Chow D, Rapaport R et al: The 21-hydroxylase activity in the glomerulosa and fasciculata of the adrenal cortex in congenital adrenal hyperplasia. J Clin Endocrinol Metab 52:534–544, 1981.
83. New MI, Dupont B, Grumbach MM et al: Congenital adrenal hypoplasia and related conditions. In Standbury JB, Wyngaarden JB, Fredrickson DS, Goldstein JL, Brown MS (eds): "The Metabolic Basis of Inherited Disease." NewYork: McGraw-Hill, 1983, pp 973–1000.
84. Dupont B, Oberfield SE, Smithwick EM et al: Close genetic linkage between HLA and congenital adrenal hyperplasia (21-hydroxylase deficiency). Lancet 2:1309–1312, 1977.

85. Carroll MC, Campbell RD, Porter RR: Mapping of steroid 21-hydroxylase genes adjacent to complement component C4 genes in HLA, the major histocompatibility complex in man. Proc Natl Acad Sci USA 82:521–525, 1985.
86. White PC, New MI, Dupont B: HLA-linked congenital adrenal hyperplasia results from a defective gene encoding a cytochrome P-450 specific for steroid 21-hydroxylation. Proc Natl Acad Sci USA 81:7505–7509, 1984.
87. White PC, Grossberger D, Onufer BJ et al: Two genes encoding steroid 21-hydroxylase are located near the genes encoding the fourth component of complement in man. Proc Natl Acad Sci USA 82:1089–1093, 1985.
88. Batshaw ML, Brusilow S, Waber L et al: Treatment of inborn errors of urea synthesis. N Engl J Med 306:1387–1392, 1982.
89. Msall M, Batshaw ML, Suss R et al: Neurologic outcome in children with inborn errors of urea synthesis. N Engl J Med 310:1500–1505, 1984.
90. Rodeck CH, Patrick AD, Pembrey ME et al: Fetal liver biopsy for prenatal dignosis of ornithine carbamyl transferase deficiency. Lancet 2:297–300, 1982.
91. Holzgreve W, Golbus AMS: Prenatal diagnosis of ornithine transcarbamylase deficiency utilizing fetal liver biospy. Am J Hum Genet 36:320–328, 1984.
92. Old JM, Purvis-Smith S, Wilcken AB: Prenatal exclusion of ornithine transcarbamylase deficiency by direct gene analysis. Lancet 1:73–75, 1985.
93. Rozen R, Fox J, Fenton WA: Gene deletion and restriction fragment length polymorphisms at the human ornithine transcarbamylase locus. Nature 313:815–817, 1985.
94. Su T-S, Bock H-GO, Beaudet AL et al: Molecular analysis of argininosuccinate synthetase deficiency in human fibroblasts. J Clin Invest 70:1334–1339, 1982.
95. Su T-S, Beaudet AL, O'Brien WE: Abnormal mRNA for argininosuccinate synthetase in citrullinaemia. Nature 301:533–534, 1983.
96. Geleuhrter TD, Snodgrass PJ: Lethal neonatal deficiency of carbamyl phosphate synthetase. N Engl J Med 290:430–433, 1974.
97. Fearon ER, Mallonee RL, Phillips JA III et al: Genetic analysis of carbamyl phosphate synthetase I deficiency. Hum Genet 70:207–210, 1985.
98. Tyler A, Harper PS: Attitudes of subjects at risk and their relatives toward genetic counselling in Huntington's chorea. J Med Genet 20:179–188, 1983.
99. Gusella JF, Wexler NS, Conneally PM et al: A polymorphic DNA marker genetically linked to Huntington's disease. Nature 306:234–38, 1983.
100. Folstein SE, Phillips JA III, Meyers DA et al: Huntington's disease: Two families with differing clinical features show linkage to the G8 probe. Science 209:776–779, 1985.
101. Dalgaard OZ: Bilateral polycystic disease of the kidneys. A follow-up of 284 patients and their families. Acta Med Scand 328 [Suppl]:1–233, 1957.
102. Bear JC, McManamon P, Morgan J et al: Age at clinical onset and at ultrasonographic detection of adult polycystic kidney disease: Data for genetic counseling. Am J Med Genet 18:45–53, 1984.
103. Reeders ST, Breuning MH, Davies KE et al: A highly polymorphic DNA marker linked to adult polycystic kidney disease on chromosome 16. Nature 317:542–544, 1985.
104. Eiberg H, Mohr J, Schmiegelow LS et al: Linkage relationships of paraoxonase (PON) with other markers: Indication of PON-cystic fibrosis synteny. Clin Genet 28:265–271, 1985.
105. Tsui L-C, Buchwald M, Barker D et al: Cystic fibrosis locus defined by a genetically linked polymorophic DNA marker. Science 230:1054–1057, 1985.
106. Knowlton RG, Cohen-Hauguenauer O, Van Cong N et al: A polymorphic DNA marker linked to cystic fibrosis is located on chromosome 7. Nature 318:380–382, 1985.

107. White R, Woodward S, Leppert M et al: A closely linked genetic marker for cystic fibrosis. Nature 318:382–384, 1985.
108. Wainwright BJ, Scambler PJ, Schmidtke J et al: Localization of cystic fibrosis locus to human chromosome 7cen–q22. Nature 318:384–385, 1985.
109. Botstein D, White RL, Skolnick M et al: Construction of a genetic linkage map in man using restriction fragment length polymorphisms. Am J Hum Genet 32:314–331, 1980.
110. Drayna D, Davies K, Hartley D et al: Genetic mapping of the human X chromosome by using restriction fragment length polymorphisms. Proc Natl Acad Sci USA 81:2836–2839, 1984.
111. White R, Leppert M, Bishop DT et al: Construction of linkage maps with DNA markers for human chromosomes. Nature 313:101–105, 1985.
112. Krumlauf R, Jeqanpierre M, Young BD: Construction and characterization of genomic libraries from specific human chromosomes. Proc Natl Acad Sci USA 79:2971–2975, 1982.
113. Rohme D, Fox H, Herrmann B et al: Molecular clones of the mouse t complex from microdissected metaphase chromosomes. Cell 36:783–788, 1984.
114. Morton CC, Kirsch IR, Nance WE et al: Orientation of loci within the human major histocompatibility complex by chromosomal in situ hybridization. Proc Natl Acad Sci USA 81:2816–2820, 1984.
115. Leary JJ, Brigati DJ, Ward D: Rapid and sensitive colorimetric method for visualizing biotin-labeled DNA probes hybridized to DNA or RNA immobilized on nitrocellulose: Bio blots. Proc Natl Acad Sci USA 80:4045–4049, 1983.
116. Lerman LS, Fischer G, Lumelsky N: Inferring variations in genotype from properties of DNA. In Messer A, Porter IH (eds): "Recombinant DNA and Medical Genetics." New York: Academic Press, 1983, pp 157–186.

Prospects for Gene Replacement Therapy

Stephen M.W. Chang, PhD, and C. Thomas Caskey, MD

Howard Hughes Medical Institute/Institute for Molecular Genetics, Baylor College of Medicine, Houston, TX 77030

INTRODUCTION

Recent advances in molecular biology have raised the prospect of treating inherited diseases through gene replacement therapy. We are on the verge of managing inborn errors of metabolism at the level of the genetic defect rather than at the symptomatic or biochemical level. Current treatments of human heritable disease are directed at correction of symptoms or secondary biochemical metabolic defects. Medicine has made significant advances in the pharmacologic and surgical treatment of many acquired diseases, yet therapy and management of most genetic disease are inadequate and ineffective. Because gene replacement therapy offers a new approach toward the treatment of inborn errors of metabolism it deserves intensive research efforts. Gene replacement therapy is directed at the mutant gene instead of peripheral or secondary events in the pathogenesis of disease. Impetus for the development of this therapeutic approach is a direct result of the developments in molecular biology allowing the cloning of disease-related genes. Presently,

This work was supported by the Howard Hughes Medical Institute, NIH grant AM31428 to C.T.C., and NIH fellowship GM10149 to S.M.W.C.

Birth Defects: Original Article Series, Volume 23, Number 3, pages 297–321

over 400 human genes, many disease-related, have been isolated and characterized. Utilizing molecular biology techniques, a number of these genes have been introduced and expressed in cultured cells and mice. These exciting developments are the factual basis on which the forecast of gene replacement therapy for man is made.

HISTORICAL PERSPECTIVE: EVOLUTION OF TECHNOLOGY

Wigler et al [1977] were the first to transfer a cloned gene into animal cells in tissue culture. By modifying the original procedure of Graham and van der Eb [1973] the herpes simplex thymidine kinase (tk) gene was successfully transfected into thymidine kinase-deficient (tk^-) mouse cells. This genetic transfer technique utilized recipient monolayers of tk^- cells in culture onto which a suspension of viral tk DNA complexed to calcium phosphate ($CaPO_4$) was overlayed. The tk^- cells (HAT^s) could not grow in selective HAT medium, and when the HSV tk gene was introduced by this method the genetic defect (tk^-) was overcome, thereby allowing recovery of cells harboring the thymidine kinase-positive phenotype (tk^+, HAT^r) [Littlefield, 1964]. The frequency of HAT^r, tk^+ transfected cells was 10^{-3} to 10^{-4}. Furthermore, Wigler et al [1979] showed that other added DNA could be cotransfected as large concatemers, which integrated into chromosomes of the recipient cell. If the added DNA was an active gene, then it frequently was expressed. This observation prompted development of cloned genes with dominant selectable properties, such as the mouse dihydrofolate reductase gene ($methotrexate^r$), the neomycin-resistance gene ($G418^r$), the hygromycin B gene (hygromycin B^r), and the *Escherichia coli* gpt gene (mycophenolic $acid^r$) [Simonsen and Levinson, 1983; Southern and Berg, 1982; Santerre et al, 1984; Mulligan and Berg, 1980]. These techniques now provide a means of selection for cells that can acquire active genes without selectable properties (ie, the selectable gpt cotransfected with human α-globin).

The parallel development of microinjection as a means of introducing cloned genes into cells has been tremendously successful. Capecchi [1980] demonstrated by microinjection of the herpes simplex virus (HSV) tk gene into the nucleus of mouse LTK^- cells that the cells can be rescued to the tk^+ phenotype in HAT medium at an efficiency of 20% or greater. Microinjection gene transfer efficiency was higher by 10^2 to 10^3-fold than calcium phosphate DNA-mediated transfer. Conformation of the injected DNA was found to be important for high-frequency gene transfer [Folger et al, 1982]. Capecchi [1980] showed that linear forms of tk plasmid DNA integrated into chromosomes in head-to-tail tandem array more efficiently than supercoiled plasmids. By controlling the amount of DNA injected, cells with single copies of foreign genes could be isolated.

The study of viruses has contributed significantly to the development of methods for introducing foreign genetic material into cells. The biology of both DNA and RNA viruses has been extensively studied for more than five decades. Viruses are naturally occurring "vectors" that introduce their genetic material into mammalian cells and then utilize either their own and/or the host cell's genes for propagation and chromosomal integration. The modification of both DNA and RNA viruses to include additional cloned genes has permitted highly efficient transfer and expression of the cloned gene into a variety of cells [Gluzman, 1982].

The viral vector initially used for this purpose was simian virus 40 (SV40). Methods were developed for the growth of defective viral genomes by complementation with a conditionally lethal, temperature-sensitive helper virus. When a cell is coinfected with a defective SV40 recombinant in which the late region has been replaced by foreign DNA and a helper tsA mutant that is defective for early function at 41°C, complementation can occur at the restrictive temperature, giving rise to virus production. The recombinant provides the early functions in *trans*, whereas the tsA helper virus provides late functions in *trans*, resulting in virus stocks containing both helper and the defective recombinant [Tegtmeyer, 1972]. Mulligan et al [1979] have utilized this strategy to replace the late-coding gene VPI of SV40 with rabbit β-globin cDNA. This construction, SVGT5, has been grown in permissive monkey cells by complementation with a *tsA* helper virus. The monkey cells synthesize rabbit β-globin as ascertained by various biochemical parameters. This initial development led to the ongoing development of other DNA viral vectors such as adenovirus, herpes simplex, polyoma, bovine papilloma, and vaccina virus vectors [Gluzman, 1982]. The establishment of these gene-delivery systems has broadened the expansion of the host range and tissue specificity of DNA viral vectors.

Retroviruses presently appear to be the preferred vectors for delivering exogenous genes into cells in vitro as well as in vivo. There are several advantages of these viruses as gene-delivery systems. Cells are infected and express the integrated viral gene at high efficiency (100%). Many cells can be treated simultaneously. The DNA integrates as a single copy at a single random site. The infection and maintenance of a retrovirus (RNA) does not appear to harm cells, whereas DNA viral vectors replicate within cells, transform, and frequently kill their host cell. Finally, the host ranges of these retroviruses in general are broader than DNA viruses [Weiss et al, 1984].

The development of this viral delivery method has been fostered by both R. Mulligan at MIT [1984] and I. Verma at the Salk Institute [1983] and their co-workers. They utilize the retrovirus as a viral delivery system by replacing the viral genes with the exogenous cloned gene by recombinant DNA techniques, resulting in a defective virus. This recombinant vector is used to transfect tissue culture cells permissive to viral infection by $CaPO_4$ DNA-mediated gene transfer. Twenty-four to forty-eight hours later, the cells are infected with helper virus, which provides the genes necessary to produce the viral proteins. The cells bud off infectious viral particles that contain both the defective recombinant virus and helper virus. This mixed viral stock containing both helper and defective virus is then capable of tranferring cloned genes into defective cells.

Mulligan and co-workers have sought to develop a system in which the recombinant defective virus could be produced, but not the wild-type helper virus. The regulatory sequence ψ must be present in viral RNA for viral particle packaging. Mann et al [1983] constructed an ecotropic recombinant virus missing this critical sequence. This ψ^- helper virus, when introduced into cells, is able to produce all the viral proteins necessary for making an intact viral particle but lacks the property of viral packaging and viral production. Recombinant retroviruses containing the recombinant gene of interest possess the complete ψ sequence and therefore could be packaged as a virion. The defective viral particle can infect only once, delivering the recombinant gene to the nucleus without capacity to replicate. Mann et al [1983] have established a cell line, Ψ_2, in which Ψ^- helper virus is integrated in proviral form. When this cell line is transfected with the defective but ψ^+ retroviral recombinants, the cells produce defective viral particles containing the recombinant retroviral gene within 48 hr. However, this ecotropic recombinant retrovirus is capable of infecting only mouse and rat cells. Utilizing the same basic principles utilized for ψ_2, Cone and Mulligan [1984] have developed a helper-free cell line ψ_{am} that allows production of helper-free amphotropic retroviruses capable of infecting a broad range of cells, including human cells. Similar amphotropic cell lines have been developed by Sorge et al [1984] and Miller et al [1985].

The combined technologic advances of DNA-mediated gene transfer, microinjection, and retroviral vectors have been utilized for somatic gene transfer in bone marrow cells of mice. The first demonstration of gene transfer via retroviruses into hematopoietic progenitor cells was by Joyner, Bernstein, and co-workers in 1983. They utilized a Moloney retrovirus vector which contained the neomycin resistant (*neo*) gene. The *neo* gene conferred dominant selectable resistance on the antibiotic G418 in mammalian cells. Moloney leukemia virus (MLV) encoding for neo^r was generated by cotransfection into NIH/3T3 cells of a plasmid clone of the replication-competent

MLV and a plasmid containing the Moloney 5′ long terminal repeat (LTR) and neo^r gene. During the cotransfection, recombination occurred between the two input molecules producing a replication-competent retroviral containing the neo^r gene and helper virus [Joyner et al, 1983]. They utilized this neo vector virus to infect mouse bone marrow cells in vitro. The cells were maintained in culture and neo^r colonies isolated. Joyner and Bernstein [1983] demonstrated that the vast majority of neo^r colonies were the granulocyte macrophage progenitors CFU-GM.

Williams and co-workers [1984] have further extended Bernstein's initial experiments to in vivo gene transfer of hematopoietic cells. Bone marrow cells were infected in vitro utilizing a helper-free Moloney sarcoma retrovirus (MSV) that encodes both the neo^r and dihydrofolate reductase (DHFR) genes. Marrow cells and MSV-DHFR NEO producing cells were cocultivated. The marrow cells were then isolated and injected IV into lethally irradiated mice. Analysis of the regenerated spleens of these mice demonstrated that the mice harbored the MSV-DHFR NEO proviral DNA in their hematopoietic cells. Subsequent analysis and experimentation have demonstrated that the initial colonies containing MSV-DHFR NEO were able to produce similar spleen colonies in a secondary group of lethally irradiated mice. Southern blot analysis demonstrated that the restriction pattern of the colonies showed only a small number of proviral integration sites, suggesting that a few infected stem cells were proliferating to repopulate the irradiated spleen.

The development of embryo gene transfer allows the generation of transgenic animals. Gordon and Ruddle [1981] have demonstrated that if plasmid DNA is microinjected into one of the pronuclei of a recently fertilized mouse egg, the egg can develop into a mouse carrying the plasmid DNA in every cell, including germ cells. Utilizing this technique Hammer, Palmiter, and Brinster [1984] microinjected the rat growth hormone gene under the control of the mouse metallothionein promoter into the embryos of growth hormone-deficient mutant mice called *little*. These transgenic dwarf mice grow rapidly; however, since the hormone gene is not properly regulated, gigantism is produced. This was the first demonstration of a genetic cure of a heritable defect in a mammal.

Table 1 summarizes the various techniques utilized for gene transfer into cells or animals. Each has advantages and disadvantages. Successful gene replacement therapy requires that three criteria be met: high efficiency of integration, functional expression, and relative safety. The helper-free retroviral systems seem to meet these criteria and surpass the other systems. The efficiency of transfer approaches 100%. In addition, it is essentially nonin-

TABLE 1. Summary of Various Techniques for Gene Transfer Into Cultured Cells or Animals

Technique	Cell recipients	Relative efficiency	Presence of infections viral elements
Calcium phosphate	Cultured cells	10^{-3}–10^{-5}	−
Microinjection	Cultured cells Embryos	1–10^{-2}	−
SV40	Cultured cells	1–10^{-1}	+
Retrovirus with helper	Cultured cells Bone marrow Embryos	1	+
Retrovirus utilizing helper-free system	Cultured cells Bone marrow Embryos	1	±

fectious, making it relatively safe. For these reasons, gene replacement therapy is presently focused on the helper-free retroviral system.

These advances have led us to a point at which gene therapy for humans seems feasible reasonably soon. The ethics of gene therapy are being widely debated. Do we need such therapy, or is it best left alone? Are there potential alternatives?

THERE IS A NEED FOR GENE THERAPY

There are over 3,000 different human genetic diseases resulting from single gene defects [Friedmann, 1983]. Man has an estimated 50,000–400,000 genes. Some new mutations are expressed in the heterozygote (eg, achondroplasia) or the hemizygous male (eg, Duchenne muscular dystrophy) and cannot be predicted or prevented by any available means. Population carrier detection methods for recessive traits are presently limited to a small number of diseases (eg, sickle cell anemia, the thalassemias, Tay-Sachs disease, and α1-antitrypsin deficiency). Carrier detection in selected families is available for a number of less common diseases of X-linked and autosomal recessive types (eg, hemophilia, Lesch-Nyhan syndrome, Hurler syndrome, Sandhoff disease, etc). However, initiation of carrier studies for these diseases follows identification of an affected family member. Even with development of new recombinant DNA methods for carrier detection (Huntington disease, phenylketonuria, ornithine transcarbamylase deficiency, cystic fibrosis), it appears unlikely that a modern medical community can apply carrier detection and genetic counseling to the objective of avoiding recessive heritable diseases. Not only is it not technically feasible but public compliance with carrier screening is less than desired.

Prenatal diagnosis (PND) utilizing amniocentesis and chorionic villus biopsy is a very effective means of avoiding serious heritable disease. Over 400 different mendelian diseases can now be detected by these methods [*Antenatal Diagnosis*, 1979]. Over 30,000 prenatal diagnosis studies were performed in the United States in 1984. Since PND can be applied only to pregnancies with an established specific genetic risk, it also lacks the ability to prevent occurrence of all or most recessive diseases. It is unlikely that this status will change dramatically.

Present therapy is very limited for management of genetic diseases. Dietary management for a genetic disease has been employed most successfully for PKU [Tourian and Sidbury, 1983]. Classic PKU is a genetic disorder in which the enzyme phenylalanine hydroxylase is absent or deficient. This deficiency results in abnormal metabolism of phenylalanine, resulting in elevated serum levels of phenylalanine and its metabolites. Untreated individuals develop severe mental retardation. However, early application of dietary restriction of phenylalanine lowers serum metabolites and avoids the complication of mental retardation. Other inborn errors of metabolism, such as the urea cycle defects, citrullinemia (argininosuccinate synthetase deficiency) and ornithine transcarbamylase deficiency, have proved resistant to equivalent dietary control despite our knowledge of their biochemical defect and rational dietary management. Recent reviews of the dietary management of inborn errors of metabolism indicate an overall marginal success for the approach, with an abundance of treatment failures leading to multiple hospitalizations, retardation, and death [Stanbury et al, 1983].

Peptide and protein replacement for heritable defects have been successful for the endocrine disorders of heritable growth hormone deficiency, juvenile onset diabetes mellitus, and clotting disorders such as hemophilias A and B. It is of special interest that recombinant DNA techniques have succeeded in the large-scale production of human growth hormone, insulin, and factor VIII [Goeddel et al, 1979; Sures et al, 1980; Wood et al, 1984]. The first two have received FDA approval and are in current clinical use. Enzyme replacement therapy has been attempted for several lysosomal storage diseases (Gaucher, Pompe, and Fabry diseases) with limited success [Brady et al, 1974; Williams and Murray, 1980; Desnick et al, 1980]. It is unclear if this treatment modality will improve with enzymes produced by recombinant DNA methods. All efforts thus far have utilized tissue sources for the purified enzymes.

Organ transplantation has an increasing role in experimental efforts to treat heritable diseases. This is due to the improved surgical techniques of organ transplantation, increased availability of transplant organs, and improved immunologic procedures for tissue matching and posttransplantation therapy (eg, cyclosporin) of rejection reactions. The experience is greatest with bone marrow transplantation [O'Reilly et al, 1984]. Successful treatment of the following heritable disorders is being reported: 1) adenosine deaminase deficiency associated with severe combined immune deficiency [O'Reilly et al, 1979], 2) X-linked severe combined immune deficiency [O'Reilly et al, 1979], 3) sickle cell anemia [Johnson et al, 1984], 4) several lysosomal storage diseases including the Maroteaux-Lamy syndrome and Gaucher disease [Rappeport and Ginns, 1984; Krivit et al, 1984], 5) thalassemia, and 6) osteopetrosis [Coccia et al, 1980]. The most extensively used organ is the kidney. Successfully treated heritable diseases include polycystic renal disease and Fabry disease but not cystinosis or oxalosis. In recent years, Starzl has succeeded in liver transplantation in children [Bilheimer et al, 1984]. Definite success has been achieved for α1-antitrypsin deficiency [Putnam et al, 1977]. The urea cycle enzyme deficiencies as well as disorders of glycogen catabolism appear to be ideal candidates for liver transplantation. Each of these transplantation approaches is made difficult by the need for HLA matching and organ availability. Failure because of graft-versus-host disease has limited the successful use of bone marrow transplantation for many genetic diseases.

In summary, presently genetic screening, counseling, prenatal diagnosis, dietary management, and tissue or organ transplantation are inadequate for the successful treatment of inborn errors of metabolism. It is reasonable to proceed with the development of new treatment modalities directed at gene replacement.

GERM LINE THERAPY VS SOMATIC THERAPY

The ethical considerations of gene therapy in humans are often debated and highly controversial. Attempts to correct genetic defects in embryos by addition of new genetic material to the nucleus has societal and religious implications, since it would alter the germ cell lines of the treated embryo.

Embryo gene therapy suffers from deficiencies other than societal and religious objections. Since most inborn errors of metabolism are inherited in a recessive manner, only one in four pregnancies is at risk for disease. If gene therapy efforts were attempted on all embryos of a couple at risk, three of four would be given exogenous genetic material unnecessarily. Of those deficient embryos treated appropriately, we would expect a success rate for gene expression of 5% to 25%. By present techniques, exogenous genes introduced by microinjection or defective retroviruses integrate illegitimately. Thus correction of an X-linked recessive disease by an autosomal or otherwise unlinked integrated transgene would lead to segregation of the gene defect from the "correcting transgene" in progeny. Chance integration of the transgene into an essential gene by illegitimate recombination could engender new genetic defects. This concern has been realized in mouse studies, where interruption of a collagen gene by a retrovirus led to an autosomal recessive embryonic lethal genetic defect [Jaenisch et al, 1983]. In light of the problems associated with embryo therapy, somatic cell gene therapy is the approach of choice.

What are the considerations in developing a gene delivery system to a human being? The major ones are 1) delivery, 2) expression, and 3) safety [Anderson, 1984]. Delivery encompasses the method for introducing the foreign genetic material into the appropriate cell. Expression relates to the appropriate regulation of the transgene in its proper cell. Safety of the patient and society are of paramount consideration. These issues will be more fully discussed below.

CANDIDATES FOR GENE REPLACEMENT THERAPY

The selection of a disease for gene replacement therapy should encompass a number of essential features. Since all efforts at gene therapy thus far have been directed toward the addition of normal genes to cells, recessively inherited diseases are the logical candidates. The present techniques are not yet amenable for the treatment of dominant disorders. We are obliged to work with inborn errors of metabolism for which the gene defect is known and strategies are available for molecular cloning. A less desirable candidate would be inborn errors of metabolism for which the enzyme is multimeric for different subunits (heteromeric). Several recent publications have summarized the rapidly growing list of cloned disease-related genes [Schmidtke and Cooper, 1983; Beaudet, 1985]. Given the improved methods of molecular cloning, it appears that the isolation of the desired gene will be a minor obstacle compared to the biology and pathophysiology of the disease entity.

Inborn errors of metabolism with a high morbidity and/or mortality are likely candidates for initial trials given the uncertainties of the new treatment. Such inborn errors of metabolism number more than 400. Many are associated with mental retardation and early death. We have selected several exemplary diseases that are likely candidates according to the above considerations.

Since the clinical spectrum of severity for sickle cell anemia and β-thalassemias differs from patient to patient, deliberate selection of patients such as was done for the recent 5-azacytidine drug trials would be indicated. These diseases are ideal from the vantage point that a single tissue, bone marrow, is required for gene transfer [Winslow and Anderson, 1983; Kan, 1983]. It is also encouraging that bone marrow replacement has been successful in treating both sickle cell disease and β-thalassemia. The human globin cDNA and genomic sequences were among the earliest genes cloned and characterized. Furthermore, cloned β-globin sequences were among the first genes to be transferred and expressed in mammalian cells using SV40 viral vectors.

Gene replacement therapy for both β-thalassemia and sickling hemoglobinopathies appears to be difficult because the disease genes direct the synthesis of one of the two subunits of the hemoglobin tetramer. Synthesis of these subunits is known to be tightly regulated in a coordinate manner. The pathophysiology of hemoglobin aggregation and red cell deformation is known to be related to the unequal synthesis of the subunits. The regulatory elements and factors controlling this well balanced synthesis are presently poorly understood. Charhay, Maniatis, and co-workers [1984] have clearly demonstrated the presence of both *cis* and *trans* regulatory sequences and factors. Considerable research effort is now directed toward understanding globin gene regulation prior to gene replacement therapy attempts. Thus, although the hemoglobinopathies remain ideal candidates for gene replacement therapy attempts, regulation of the β-globin gene remains difficult and will limit early clinical trials. It should be pointed out that recent transgenic mice studies have demonstrated tissue-specific regulation of β-globin [Chada et al, 1985]. Furthermore, β-globin transfer into erythroleukemia cells results in appropriate regulation of the transgene [Wright et al, 1984].

Heritable severe combined immune deficiencies (SCIDs) have been treated successfully by bone marrow transplantation [O'Reilly et al, 1979, 1984]. Presently, the enzymic defect is known for only two of these genetically heterogeneous disorders. Adenosine deaminase (ADA) and purine nucleoside phosphorylase (PNP) deficiencies both result in T- and B-cell dysfunction with SCID. Both are recessively inherited and account for ≃10–20% of all SCID, with ADA deficiency being five- to tenfold more common than PNP deficiency. ADA deficiency is a life-threatening SCID syndrome with death frequently occurring within the first year despite therapeutic efforts by bone marrow transplantation and blood transfusion. The level of ADA is ≃ 100-fold elevated in normal T cells. Deficient patients have markedly elevated tissue and serum levels of deoxyadenosine. Although the mechanism of immune toxicity has not been precisely defined, all evidence suggests that the elevation of deoxyadenosine is associated with T-cell dysfunction. Similar evidence exists for deoxyguanine in PNP deficiency. Although gene therapy efforts for these disorders are directed toward restoration of enzymatic activity of immune cells, the possibility exists that enzyme restoration to other marrow cells will lower deoxynucleotide levels and reduce immune cell toxicity. These diseases appear to be very suitable for gene therapy based on the clinical and pathophysiologic characteristics.

The ADA enzyme is composed of a single polypeptide of 36,000–38,000 mw encoded by a gene that maps to 20q13.2→qter. The complete ADA cDNAs of mouse and man are available and have been successfully placed into mammalian expressing vectors [Yeung et al, 1983; Orkin et al, 1983]. PNP is composed of a single subunit of 30,000–32,000 mw. A full-length cDNA for human PNP has been cloned [Goddard et al, 1983].

A large number of investigative groups have devoted effort particularly to ADA. It is likely that their combined efforts will contribute significantly to early clinical trials for ADA deficiency. The rarity of PNP deficiency is likely to retard its clinical trial of gene therapy.

The Lesch-Nyhan (L-N) syndrome has been proposed as an inborn error of metabolism particularly suitable for gene replacement therapy [Kelley and Wyngaarden, 1983]. This X-linked inherited disease is associated with deficiency of hypoxanthine-guanine phosphoribosyltransferase (HPRT), an enzyme found in all tissues. There is no question regarding the morbidity of the L-N syndrome. The disease is associated with mental retardation, self-mutilating behavior, choreoathetosis, and hyperuricemia. Hyperuricemia, if left unattended, leads to early renal failure. Allopurinol, a xanthine oxidase inhibitor, corrects the hyperuricemia without affecting the associated neurologic defects.

The pathobiology accounting for the CNS dysfunction in L-N syndrome is presently unresolved. The level of HPRT is high in the basal ganglia of the CNS. Unlike ADA and PNP, there is no evidence that a soluble metabolite is toxic to the CNS. Since the basal ganglia have a low level of de novo purine biosynthesis, the high level of HPRT may be required for maintenance of nucleotides. Depressed dopamine pathway enzymes in the CNS of L-N patients as well as induction of self-mutilating behavior in mice by dopaminergic antagonist suggest an association of HPRT and dopamine depression. If this is correct, the restoration of HPRT to the CNS of L-N patients early in their development may be requisite for successful gene therapy.

The HPRT enzyme is a tetramer composed of identical polypeptides of 25,500 mw encoded by a gene that maps to Xq26. Full-length HPRT cDNA clones of mouse and man have been isolated and characterized [Brennand et al, 1982; Jolly et al, 1983]. The genomic organization and transcriptional regulatory elements have been determined [Melton et al, 1984; Patel et al, 1986]. Mammalian expressing vectors are now available [Miller et al, 1984a; Brennand et al, 1983].

The development of expressing vectors toward the objective of gene replacement therapy has been greatly facilitated by the availability of a selective medium (HAT) permitting accurate measurement of genetic correction of $HPRT^-$ (HAT^s) to $HPRT^+$ (HAT^r). This feature will permit accurate quantitation in correction of cells in peripheral blood specimens of L-N patients. It remains to be determined if CNS improvement for L-N disease will be observed if bone marrow is the only tissue corrected. Correction in CNS tissue at early stages of brain development may be needed.

Heritable hyperammonemia in newborns occurs secondary to urea cycle enzyme deficiencies. Citrullinemia (argininosuccinate synthetase [ASA] deficiency) and ornithine transcarbamylase (OTC) deficiency are two of the most severe and common inborn errors in the pathway [Walser, 1983]. Dietary restrictions of protein, supplementation of ketoacids, and addition of benzoic acid ameliorate hyperammonemia under normal metabolic conditions. During catabolic phases (illness and infection) hyperammonemia is less well controlled, however, and leads to frequent hospitalizations and death. There are no reported attempts at liver transplantation for these disorders. Mental retardation commonly accompanies these disorders. The complete human cDNA and genomic structure for ASA synthetase have been determined [Su et al, 1981; Freytag et al, 1984]. The complete human OTC cDNA has been characterized [Horwich et al, 1984]. Both cDNAs have been used successfully for minigene constructions that express in mammalian cells [Veres et al, 1986] (also, G. Veres, personal communication).

The transfer of genes into hepatic tissue, the site of urea synthesis, seems ideal. This would avoid the potential problems of tissue-specific regulatory factors and the necessity for intercellular movement of pathway metabolites. It is uncertain if expression of the transgene in hematopoietic cells would permit adequate urea synthesis.

α1-Antitrypsin deficiency is an autosomal recessively inherited disease with newborn and adult medical complications [Gadek and Crystal, 1983]. This relatively common disorder is associated with severe cholestatic jaundice and cirrhosis in 30% of newborns with certain mutant alleles. Adult survivors invariably have early-onset emphysema and shortened life-span. Thus morbidity and mortality are significant. A number of infants and children with liver cirrhosis have been successfully treated by liver transplantation. It is too early to determine if these patients will avoid emphysema given their normal serum levels of α1-antitrypsin.

The pathophysiology of α1-antitrypsin (AAT) is not completely understood. It is clear however that patients with the S and Z mutant alleles accumulate α1-antitrypsin in hepatic cells and have low circulating serum levels. It appears that AAT hepatic accumulation is associated with cellular dysfunction. AAT is an acute phase reactant thought to be involved as the major serum antiprotease in regulating inflammatory responses. Its anti-elastase function may protect pulmonary parenchyma from damage during inflammation. The full-length cDNA and genomic sequences of AAT have been isolated and characterized [Long et al, 1984].

One could anticipate that gene replacement therapy for AAT would require expression in the liver, since the protein requires extensive posttranslational modification before its secretion. It is unlikely that gene transfer will ameliorate the hepatic damage occurring in 30% of patients. For those escaping hepatic damage, adequate serum leads of AAT may protect from emphysema.

CURRENT STATE

Most treatment of genetic diseases has focused on the use of retroviral vectors and their introduction into bone marrow cells. Willis, Friedmann, and their colleagues [1984] have used a pseudotyped retroviral vector containing human HPRT and introduced it in vitro into defective human lymphoblast cells that were HPRT-deficient. These studies resulted in the characterization of 17 $HPRT^+$ lymphoblast clones in which the defect was corrected and in which enzyme levels range from 4% to 23%. Subsequent follow-up work has shown that, although the virus is inserted stably under selection, there seems to be a propensity for instability when taken out of selection [Jolly et al, 1986]. Additional studies have shown that this HPRT vector is capable of being introduced into mouse bone marrow cells, and, when injected into lethally irradiated mice, it is capable of repopulating and expressing the human HPRT [Miller et al, 1984a]. Gruber et al [1985] have also shown the ability to infect human bone marrow with pseudotyped HPRT retroviruses.

Williams, Mulligan, and their colleagues [1984] have successfully transferred the neo^r gene into marrow cells using their MSV-DHFR NEO vector. They were able to demonstrate that the transferred hematopoietic cells repopulate the bone marrow of irradiated mice. Furthermore, Lemischka et al [1986] were able to demonstrate that the stem cells that repopulated the spleen were of several different classes and that very few stem cell clones accounted for the majority of mature hematopoietic cells.

Joyner and Bernstein [1983] and Dick et al [1985] initially showed that the neo^r gene was transferable to about 0.3% of all granulocytic cells by the use of MoLV retroviral vectors. Subsequently, they improved their techniques of transfer through the use of cocultivation of virus-producing cells with 5-flurouracil-treated bone marrow and have demonstrated the transfer of neo^r to many marrow cell types including the stem cells, CFU-S.

Dick et al [1985] made some dramatic changes in growth and retroviral infection protocols such that the efficiency of infecting bone marrow cells approached 100%. This high efficiency would be attributed to the use of interleukin-3 for bone marrow growth and the technique of preselecting progenitor cells by selection before bone marrow reconstitution was attempted. These protocols have been utilized also by Keller et al [1985] and Eglitis et al [1985] with similar results. Hock and Miller [1986] have initiated the first in vitro human experiments by introducing the neo^R and $dHFR^r$ gene into human bone marrow cultures. Furthermore, Kwok et al [1986] have begun to utilize canine animal models and have shown the transfer of neo^R and dHFR into canine hemopoietic progenitor cells.

Metabolic genes corresponding to common genetic defects have also been inserted into retroviral vectors. Ledley et al [1986] have inserted the cDNA encoding human phenylalanine hydroxylase into a retroviral vector and were able to demonstrate phenylalanine hydroxylase in infected hepatoma cells.

Choudary et al [1986] have generated a recombinant retrovirus carrying the cDNA encoding human glucocerebrosidase. Their study showed that they could infect type 2 Gaucher patient cells and generate cross-reacting material.

Recent technical advances in both viral-packaging cell lines and suicide vectors will probably accelerate the attempts at somatic gene therapy. Miller and Buttimore [1986] have recently developed an amphotropic packaging line PA-317 that is capable of packaging viral vectors at a high frequency that is clearly helper-free under the most stringent of helper-virus assay conditions. Another significant technical advance has been the development of viral vectors that, upon delivery of the gene, self-inactivate the viral LTR [Yu et al, 1986].

The most significant advance in somatic gene therapy has been the intensive effort to introduce ADA into hematopoietic cells. The various laboratories of Verma, Martin, Anderson, Gilboa, Mulligan, Orkin, and Caskey have been concentrating on insertion of ADA into viral vectors. R. Friedman [1985] as well as Valerio et al [1985] showed that human ADA could express in mouse fibroblasts. Williams et al [1986] confirmed those initial observations and extended the experiments to actual in vivo infection of the ADA virus in reconstituted mouse CFU-S-derived colonies. However, no protein expression was demonstrable, although a low level of ADA RNA could be detected. Belmont et al [1986] have shown, utilizing an in vitro system, that murine hematopoietic progenitor cells infected with a retroviral vector containing human ADA are capable of expression. Furthermore, in vivo expression experiments (Belmont, Chang, and Caskey, unpublished data) show a very low level of ADA expression in mouse spleens.

Our laboratory has pursued gene therapy by developing plasmid-borne vectors in which we could begin to investigate the regulation of HPRT and we have subsequently introduced it into retroviral vectors. Initially, human HPRT minigenes were constructed using LTR sequences from spleen focus-forming virus as promoters [Brennand et al, 1983]. Subsequently, we demonstrated that minigenes containing the mouse metallothionein (MT-I) and both the rodent and human HPRT promoter sequences are able to correct $HPRT^-$ fibroblasts to $HPRT^+$ (Fig. 1). these recombinants, when microinjected into the nucleus of $HPRT^-$ rodent cells, transiently express HPRT as demonstrated by cellular uptake of [^{3}H]hypoxanthine. Integrants have been isolated that stably maintain single or multiple copies of these minigenes. Each of these minigene sequences has been cloned into a viral vector, pZIP NEO(X), that contains the gene for neomycin resistance (Fig. 2) [Cepko et al, 1984]. The viral recombinant SVX HPRT B, has then been transferred via $CaPO_4$ DNA-mediated gene transfer into ψ_2 cells. Neomycin-resistant cells have been selected and defective ecotropic virus producing colonies isolated (Fig. 3). Subseqently, these defective viral stocks were utilized to infect $HPRT^-$ rodent cells and effectively demonstrate their genetic correction to

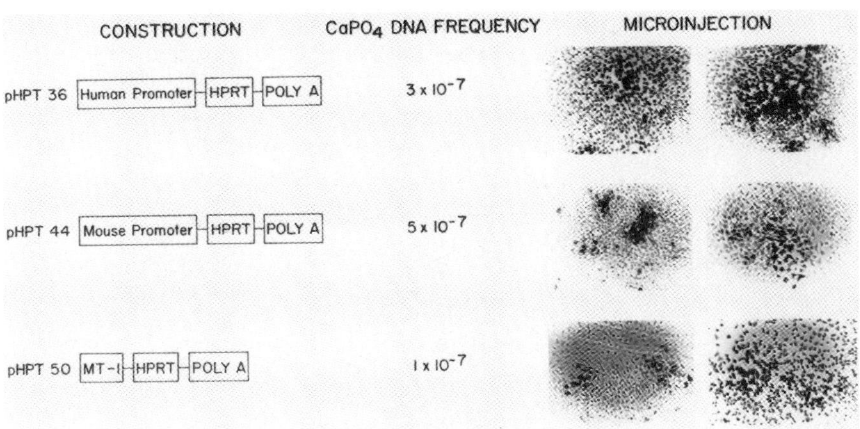

Fig. 1. Minigenes containing either the metallothionein (pHPT 50) or natural human (pHPT 36) or mouse HPRT (pHPT 44) promoter controlling the expression of the human HPRT cDNA. The figure shows the minigene construction, relative transformation frequency to stable HAT^r, and ability to take up [^{3}H]hypoxanthine.

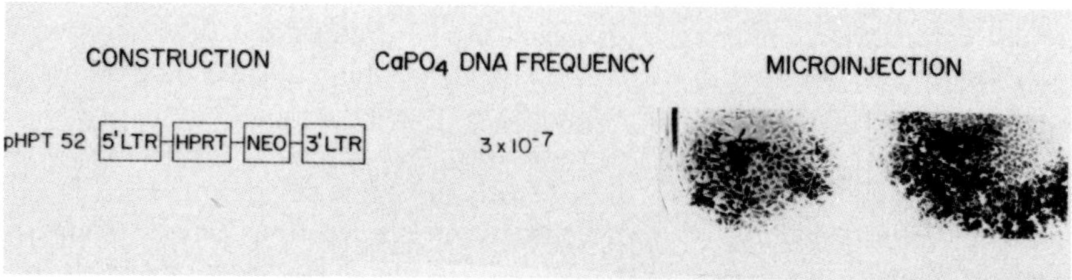

Fig. 2. Retroviral vector ZIP NEO (X) containing the human HPRT cDNA (pHPT 52). The figure shows the construction, relative transformation frequency to stable HAT^r, and ability to take up [^{3}H] hypoxanthine. This virus has been designated SVX HPRT B.

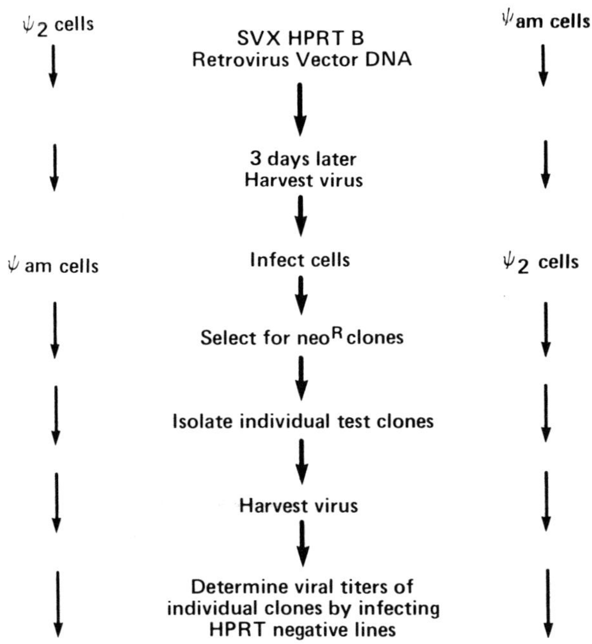

Fig. 3. Details the protocol utilized in making either ecotropic or amphotropic retroviruses.

$HPRT^+$. We have further generated replication-defective amphotropic viruses by infecting ψ_{am} cells with those ecotropic viruses. Results from several ψ_2 and ψ_{am} cell lines producing HPRT ecotropic and amphotropic viruses are shown in Table 2. Enzymatic expression of human HPRT in $HPRT^-$ cells infected with defective viruses is shown in Table 3. In experiments characterizing the HPRT retroviral vector, we noted a higher titer of G418 production from ψ_2 and ψ_{am} cells when the cells received the HPRT retroviral construct through infection rather than transfection. For example, in ψ_2 cells of 24 transfected clones, 17 had titers between 10^2 and 10^3 G418 resistance CFU/ml and five had titers between 10^3 and 10^4, with no titers greater than 10^4. This greatly contrasts with the infected clones, where 76 of 103 clones assayed had titers between 10^3 and 10^4, eight clones were measured above 10^4, and no clones were found to have titers below 10^2. Since high viral titers are important in bone marrow infections, we have adopted the use of viral infection as a routine method of creating virus-producing cell lines. Even utilizing viral infection, the highest titer virus stock was 5×10^4, and we sought to increase this titer. We reasoned that the amount of vector RNA available to be packaged by the cell would be one of the limiting factors in producing high titers and that position of integration of the vector DNA in the genome of the viral-producing cell would alter the packageable vector RNA. Knowing that the ecotropically packaged virus from ψ_2 cells can infect amphotropic-producing ψ_{am} (or in the equivalent PA-12) cells and that PA-12 will infect ψ_2 cells, we cocultivated ψ_2 producers of HPRT with PA-12 cells. The ecotropic titers rose dramatically, with the viral titer measuring at best 8

TABLE 2. Virus Production From SVX HPRT B-Infected ψ_2 and ψ_{am} Cells

Helper clones producing SVX HPRT B	Viral titer	
	HAT^r	$G418^r$
$\psi 2^\dagger$		
1	3.3×10^4	3.2×10^4
2	9.7×10^3	1.1×10^4
3	6.1×10^3	6.4×10^3
4	8.4×10^3	7.6×10^3
5	4.6×10^3	5.2×10^3
$\psi_{am}^\ddagger$		
6	3.2×10^3	3.0×10^3
7	4.9×10^3	5.0×10^3
8	1.3×10^3	1.1×10^3
9	1.8×10^3	2.1×10^3

$^\dagger\psi_2$ Titers were performed on NIH 3T3 HPT^- cells.
$^\ddagger\psi_{am}$ Titers were performed in Hela HPT^- cells.

TABLE 3. Enzymatic Activity of Hypoxanthine Phosphoribosyltransferase and Adenine Phosphoribosyltransferase in Cell Lines Infected With Defective HPRT Retrovirus

Cell line	HPRT activity (pmol/min/mg protein)	% Norm	APRT activity (pmol/min/mg protein)
NIH 3T3 wild-type	7,876 ± 182	100	9,138 ± 2,160
NIH 3T3 HPRT	0.1	0.1	8,892 ± 2,282
A	1,648 ± 161	21	8,308 ± 485
B	3,308 ± 236	42	8,285 ± 562
C	3,702 ± 321	47	9,015 ± 788
D	3,544 ± 266	45	9,455 ± 1,078
E1	1,722 ± 50	22	8,365 ± 1,361
E2	2,836 ± 422	36	7,265 ± 1,265
E3	1,969 ± 240	25	8,488 ± 908
E4	709 ± 40	9	9,563 ± 625
F	4,411 ± 241	52	8,770 ± 676
G	4,411 ± 241	56	9,065 ± 397
H	1,102 ± 188	14	8,675 ± 722
I	315 ± 65	4	7,738 ± 644
J	598 ± 96	8	8,316 ± 672
Normal human lymphoblasts	6,245 ± 377	100	8,442 ± 1,106
Human HPRT lymphoblasts	0.1	0.1	8,761 ± 809
L1	1,802 ± 431	29	9,264 ± 644
L2	1,465 ± 277	23	7,651 ± 432

TABLE 4. Helper-Free Ecotropic Producer Line Cocultivation With Amphotropic Helper Line PA-12

Line	Initial titer	Titers after cocultivation with PA-12	
		Ecotropic†	Amphotropic‡
W	1×10^3	1×10^4	ND
X	2×10^4	5×10^4	$>10^4$
Y	3×10^4	2×10^5	6×10^4
Z	5×10^4	8×10^5	1×10^4

†Ecotropic titers were determined on rat HPT⁻ cells.
‡Amphotropic titers were determined on Hela HPT⁻ cells.

$\times 10^5$ (Table 4). Titers of close to 10^7 have been achieved by the addition of Moloney murine leukemia virus by superinfection of ψ_2 producer lines (Table 4). Thus, when presence of competent helper virus can be tolerated, these types of viral stocks are preferred.

The construction of the HPRT/neo-expressing retrovirus facilitates the investigation of retroviral stability after infection of cells in culture, since HPRT can be selected for its presence or absence. $HPRT^-$ NIH 3T3 cells were rescued from HAT selection by infection with the HPRT virus. After expression, clonal isolates expressing the introduced HPRT and neo gene were passed in the absence of selective pressure. Clones were plated into four different media: no selection, 6-thioguanine (selecting for the absence of HPRT expression), HAT, and G418. Three lines appear to be very stable with few or no 6TG-resistant colonies. The remaining three lines show a highly unstable phenotype. We have demonstrated two mechanisms by which loss of HPRT activity can occur. The first is through loss or rearrangement of the provirus. The second mechanism has been found in a 6-TG-resistant derivative of the E4 line. This clone (E4 $HPRT^-$) showed high plating efficiency in 6-TG and no plating in HAT or G418, yet the proviral sequence remained as shown by Southern analysis and by the fact that rare $HPRT^+$ clones could be isolated. In addition, these clones showed high plating efficiency in G418, demonstrating a cis-acting regulatory effect on both the HPRT and neo genes.

Although somatic gene replacement therapy for L-N patients will likely involve tissues other than just bone marrow, we wished to determine whether detectable levels of human HPRT could be found in mouse bone marrow cells infected in vitro. Figure 4 shows an isoelectric focusing gel of HPRT-infected mouse bone marrow cells. This shows that human HPRT is detectable in mouse bone marrow cells and that the HPRT vector is functional in mouse hematopoietic cells.

These types of observations are being used in the design and construction of additional retroviral vectors. Additionally, regulation of HPRT or for that matter any gene in retroviral vectors should be possible by genomic control signals. Recently, Miller et al [1984b] succeeded in placing a rat growth hormone gene in a retroviral vector. This growth hormone gene expressed from its own promoter and regulatory sequence. These data provide encouragement that vectors can be built that contain all the genomic regulatory signals necessary to produce controlled expression in targeted cells.

PRECAUTIONS

Although retroviruses have all the advantages of a gene transfer system, their most serious drawback is that they recombine at a high rate with other retroviruses via homologous recombination. The properties of these recombinant viruses are basically unknown, yet they have a striking similarity to known oncogenic retroviruses as well as the AIDS-associated HIV retrovirus. The following prerequisites should be satisfied before we use gene transfer techniques on patients. 1) The retrovirus vectors should be essentially free of

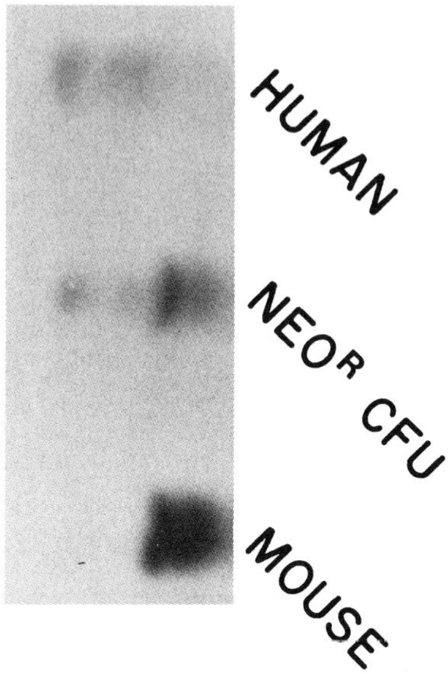

Fig. 4. Isoelectric focusing of neo^R mouse bone marrow CFU. The lane marked mouse is an extract from a pool of picked CFUs from ψ_2 cocultivation. The NEO^R lane is a pool of picked NEO^R CFUs from cocultivation with SVX HPRT B producting ψ_2. The human lane is a human lymphoblast HPRT extract.

competent virus. 2) Long-term studies in vivo with mice would help ascertain the target tissue infections and the possible pathologic manifestions. 3) In vivo studies with primates could be used to mimic the protocols considered for humans to ascertain the potential complications. 4) Successful in vitro studies should be satisfied with human bone marrow or other target cells to ascertain the feasibility of gene transfer.

It is highly unlikely that the illegitimate recombination property of the retroviral vector system can be avoided. Somatic mutations are not considered to be a health hazard, since disadvantaged mutant cells simply would be lost. Mutations that give proliferative advantage (oncogene activation) are the greatest hazard. Use of defective viral vectors with altered LTR sequences should minimize this risk. The severity of certain heritable diseases and the lack of therapeutic alternatives will weigh heavily in our decision for an initial trial of gene replacement therapy.

REFERENCES

Anderson WF: Prospects for human gene therapy. Science 226:401–409, 1984.

Antenatal Diagnosis: "Report of a Consensus Development Conference, NICHD." NIH Publication No. 79-1973. Washington, DC: U.S. Government Printing Office, 1979.

Beaudet AL: Bibliography of cloned human and other selected DNAs. Am J Hum Genet 37:386, 1985.

Belmont JW, Henkel-Tigges J, Chang SMW, Wager-Smith K, Kellems RE, Dick JE, Magli MC, Phillips RA, Bernstein A, Caskey CT: Expression of human adenosine deaminase in murine haematopoietic progenitor cells following retroviral transfer. Nature 322:385–387, 1986.

Bilheimer DW, Goldstein JL, Grundy SM, Starzl TE, Brown MS: Liver transplantation to provide low density-lipoprotein receptors and lower plasma cholesterol in a child with homozygous familial hypercholesterolemia. N Engl J Med 311:1658–1664, 1984.

Brady RO, Pentchev PG, Gal AE, Hibbert SR, Dekaban AS: Replacement therapy for inherited enzyme deficiency. Use of purified glucocerebrosidase in Gaucher's disease. N Engl J Med 291:989–993, 1974.

Brennand J, Chinault AC, Konecki D, Melton D, Caskey CT: Cloned cDNA sequences of the hypoxanthine-guanine phosphoribosyltransferase gene from a mouse neuroblastoma cell line found to have amplified genomic sequences. Proc Natl Acad Sci USA 79:1950–1954, 1982.

Brennand J, Konecki D, Caskey CT: Expression of human and Chinese hamster hypoxanthine-guanine phosphoribosyltransferase cDNA recombinants in cultured Lesch-Nyhan and Chinese hamster fibroblasts. J Biol Chem 258:9593–9596, 1983.

Capecchi MR: High efficiency transformation by direct microinjection of DNA into cultured mammalian cells. Cell 22:479–488, 1980.

Cepko C, Roberts B, Mulligan R: Construction and applications of a highly transmissible murine retrovirus shuttle vector. Cell 37:1053–1062, 1984.

Chada K, Magram J, Raphael K, Radice G, Lacy E, Costantini F: A foreign β-globin gene expressed specifically in erythroid cells of transgenic mice. Nature 314:377–380, 1985.

Charhay P, Treisman R, Mellon P, Chao M, Axel R, Maniatis T: Differences in human α- and β-globin gene expression in mouse erythroleukemia cells. The role of intragenic sequences. Cell 38:251–263, 1984.

Choudary P, Barranger J, Tsvji S, Mayor J, LaMarca M, Cepko C, Mulligan R, Ginns E: Retrovirus-mediated transfer of the human glucocerebrosidase gene to Gaucher fibroblasts. Mol Biol Med 3:293–299, 1986.

Coccia PF, Krivit W, Cervenka J et al: Successful bone marrow transplantation for infantile malignant osteopetrosis. N Engl J Med 302:701–708, 1980.

Cone R, Mulligan RC: High efficiency gene transfer into mammalian cells: Generation of helper-free recombinant retrovirus with broad mammalian host range. Proc Natl Acad Sci USA 81:6344–6353, 1984.

Desnick RJ, Dean KJ, Grabowski G, Bishop DF, Sweeley CC: Enzyme therapy XVII: Metabolic and immunologic evaluation of α-galactosidase. A replacement in Fabry disease. In Desnick RJ (ed): "Enzyme Therapy in Genetic Diseases: 2." New York: Alan R. Liss, Inc., for The National Foundation—March of Dimes. BD:OAS XVI(1):393–413, 1980.

Dick J, Mogli M, Huszer D, Phillips R, Bernstein A: Introduction of a selectable gene into primitive stem cells capable of long-term reconstitution of the hematopoietic system w/w^{r} mice. Cell 42:71–79, 1985.

Eglitis M, Kantoff P, Gilboa E, Anderson W: Gene expression in mice after high efficiency retroviral mediated gene transfer. Science 230:1395–1398, 1985.

Folger KR, Wong EA, Wahl G, Capecchi MR: Patterns of integration of DNA microinjected into cultured mammalian cells: Evidence for homologous recombination between injected plasmid DNA molecules. Mol Cell Biol 2:1372–1387, 1982.

Freytag S, Beaudet A, Bock H, O'Brien W: Molecular structure of the human argininosuccinate synthetase gene: Occurrence of alternative mRNA splicing. Mol Cell Biol 4:1978–1984, 1984.

Friedman R: Expression of human adenosine deaminase using a transmissible murine retrovirus vector system. Proc Natl Acad Sci USA 82:703–607, 1985.

Friedmann T (ed): "Gene Therapy: Fact and Fiction in Biology's New Approaches to Disease." Cold Spring Harbor, NY: Cold Spring Harbor Laboratory, 1983.

Gadek J, Crystal R: α-l-Antitrypsin deficiency. In Stanbury JB, Wyngaarden JB, Fredrickson DS, Goldstein JL, Brown M (eds): "Metabolic Basis of Inherited Disease," 5th Ed. New York: McGraw-Hill, 1983, pp 1450–1467.

Gluzman Y (ed): "Eukaryotic Viral Vectors." Cold Spring Harbor, NY: Cold Spring Harbor Laboratory, 1982.

Goddard J, Caput D, Williams S, Martin D: Cloning of human purine-nucleoside phosphorylase cDNA sequences by complementation in *Escherichia coli*. Proc Natl Acad Sci USA 80:4281–4285, 1983.

Goeddel D, Heyneker H, Hozumi T, Arentzen R, Itukura K et al: Direct expression in *Escherichia coli* of a DNA sequence coding for human growth hormone. Nature 281:544–548, 1979

Gordon JW, Ruddle FH: Integration and stable germ line transmission of genes injected into mouse pronuclei. Science 214:1244–1246, 1981.

Graham FL, van der Eb AJ: A new technique for the assay of infectivity of human adenovirus 5 DNA. Virology 52:456–467, 1973.

Gruber H, Finlay K, Hershberg R, Katzman S, Laikind P, Seegmiller J, Friedmann T, Yee J, Jolly D: Retroviral vector mediated gene transfer into human hematopoietic progenitor cells. Science 230:1057–1061, 1985.

Hammer RG, Palmiter RD, Brinster RL: Partial correction of murine hereditary growth disorder by germ-line incorporation of a new gene. Nature 311:65–67, 1984.

Hock R, Miller A: Retrovirus-mediated transfer and expression of drug resistance genes in human hematopoietic progenitor cells. Nature 320:275–277, 1986.

Horwich AL, Fenton W, Williams K, Kalousek F, Kraus J, Doolittle RF, Konisberg W, Rosenberg LE: Structure and expression of a complementary DNA for the nuclear coded precursor of human mitochondrial ornithine transcarbamylase. Science 224:1068–1074, 1984.

Jaenisch R, Harbers K, Schnieke A, Löhler J, Chumakov I, Jahner D, Grotkopp D, Hoffman E: Germline integration of Moloney murine leukemia virus at Mov13 locus leads to recessive lethal mutation and early embryonic death. Cell 32:209–216, 1983.

Johnson FL, Look AT, Gockerman J, Ruggeriero MR, Dalla-Pozza L, Billings FT: Bone marrow transplantation in a patient with sickle cell anemia. N Engl J Med 311:780–783, 1984.

Jolly D, Okayama H, Berg P, Esty A, Filpula D, Bohlen P, Johnson G, Shively J, Hunkapillar T, Friedmann T: Isolation and characterization of a full-length expressible cDNA for human hypoxanthine phosphoribosyltransferase. Proc Natl Acad Sci USA 80:477–481, 1983.

Jolly DJ, Willis RC, Friedmann T: Variable stability of a selected provirus after retroviral vector gene transfer into human cells. Mol Cell Biol 6:1141–1147, 1986.

Joyner A, Bernstein A: Retrovirus transduction: Generation of infectious retrovirus expressing dominant and selectable genes is associated with in vivo recombination and deletion events. Mol Cell Biol 3:2180–2190, 1983.

Joyner A, Keller G, Phillips RA, Bernstein A: Retrovirus transfer of a bacterial gene into mouse haematopoietic progenitor cells. Nature 305:556–558, 1983.

Kan YW: Thalassemias. In Stanbury JB, Wyngaarden JB, Fredrickson DS, Goldstein JL, Brown M (eds): "Metabolic Basis of Inherited Disease," 5th Ed. New York: McGraw-Hill, 1983, pp 1145–1467.

Keller G, Paige C, Gilboa E, Wagner E: Expression of a foreign gene in myeloid and lymphoid cells derived from multipotent hematopoietic precursors. Nature 318:149–154, 1985.

Kelley W, Wyngaarden J: Clinical syndromes associated with hypoxanthine-guanine phosphoribosyltransferase deficiency. In Stanbury JB, Wyngaarden JB, Fredrickson DS, Goldstein JL, Brown M (eds): "Metabolic Basis of Inherited Disease," 5th Ed. New York: McGraw-Hill, 1983, pp 1115–1143.

Krivit W, Pierpoint ME, Ayaz K et al: Bone marrow transplantation in Maroteaux-Lamy syndrome (mucopolysaccharidosis type VI): Biochemical and clinical status 24 months post-transplantation. N Engl J Med 311:1606–1611, 1984.

Kwok W, Schuening F, Stead R, Miller A: Retroviral transfer of genes into canine hemopoietic progenitor cells in culture: A model for human gene therapy. Proc Natl Acad Sci USA 83:4552–4555, 1986.

Ledley F, Grenett H, McGinnis-Shelnutt M, Woo S: Retroviral-mediated gene transfer of human phenylalanine hydroxylase into NIH 3T3 and hepatoma cells. Proc Natl Acad Sci USA 83:409–413, 1986.

Lemischka I, Raulet D, Mulligan R: Developmental potential and dynamic behavior of hematopoietic stem cells. Cell 45:917–927, 1986.

Littlefield JW: Selection of hybrids from matings of fibroblast in vitro and their presumed recombinants. Science 145:709–710, 1964.

Long G, Chandra T, Woo S, Davies E, Kurachi K: Complete sequence of the cDNA for human α-l-antitrypsin and the gene for the S variant. Biochemistry 23:4828–4837, 1984.

Mann R, Mulligan RC, Baltimore DB: Construction of a retrovirus packaging mutant and its use to produce helper-free defective retrovirus. Cell 33:153–159, 1983.

Melton D, Konecki D, Brennand J, Caskey CT: Structure, expression and mutation of the hypoxanthine phosphoribosyltransferase gene. Proc Natl Acad Sci USA 81:2147–2151, 1984.

Miller AD, Buttimore C: Redesign of a retrovirus packaging cell line to avoid recombination leading to helper virus production. Mol Cell Biol 6:2895–2902, 1986.

Miller AD, Eckner RJ, Jolly DJ, Friedmann T, Verma IM: Expression of a retrovirus encoding HPRT in mice. Science 225:630–632, 1984a.

Miller AD, Jolly DJ, Friedmann T, Verma IM: A transmissible retrovirus expressing human hypoxanthine phosphoribosyltransferase (HPRT): Gene transfer into cells obtained from humans deficient in HPRT. Proc Natl Acad Sci USA 80:4709–4713, 1983.

Miller AD, Law MF, Verma IM: Generation of helper-free amphotropic retroviruses that transduce a dominant-acting, methotrexate dihydrofolate reductase gene. Mol Cell Biol 5:431–437, 1985.

Miller AD, Ong ES, Rosenfeld MG, Verma IM, Evans RM: Infectious and selectable retrovirus containing an inducible rat growth hormone minigene. Science 225:993–998, 1984b.

Mulligan RC: Construction of highly transmissible mammalian cloning vehicles derived from murine retroviruses. In Inouye M (ed): "Experimental Manipulation of Gene Expression." New York: Academic Press, 1984.

Mulligan RC, Berg P: Expression of a bacterial gene in mammalian cells. Science 209:1422–1427, 1980.

Mulligan RC, Howard B, Berg P: Synthesis of rabbit β-globin in cultured monkey kidney cells following infection with a SV40-β-globin recombinant genome. Nature 277:108–114, 1979.

O'Reilly RJ, Brochstein J, Dinsmore R, Kirkpatrick D: Marrow transplantation for congenital disorders. Semin Hematol 21:188–216, 1984.

O'Reilly RJ, Dupont B, Pahwa S et al: Reconstitution in severe combined immunodeficiency by transplantation of marrow from an unrelated donor. N Engl J Med 297:1311–1318, 1979.

Orkin S, Daddona P, Shewach D, Markham A, Bruns G, Goff S, Kelley W: Molecular cloning of human adenosine deaminase gene sequences. J Biol Chem 258:12753–12756, 1983.

Patel PI, Framson PE, Caskey CT, Chinault AC: Fine structure of the human hypoxanthine phosphoribosyltransferase gene. Mol Cell Biol 6:393–403, 1986.

Putnam CW, Porter KA, Peters RL, Aschcavi M, Redeker AG, Starzl T: Liver replacement for α-l-antitrypsin deficiency. Surgery 81:258–261, 1977.

Rappeport JM, Ginns EI: Bone marrow transplantation in severe Gaucher's disease. N Engl J Med 311:84–88, 1984.

Santerre RF, Allen NE, Hobbs JN, Rao RN, Schmidt RJ: Expression of prokaryotic genes for hygromycin B and G418 resistance as dominant selection markers in mouse L cells. Gene 30:147–156, 1984.

Schmidtke J, Cooper D: A list of cloned human DNA sequences. Hum Genet 65:19–26, 1983 (updated Sept. 13, 1984).

Simonsen CC, Levinson AD: Isolation and expression of an altered mouse dihydrofolate reductase cDNA. Proc Natl Acad Sci USA 80:2495–2499, 1983.

Sorge J, Wright D, Erdman V, Cutting A: Amphotropic retrovirus vector system for human cell gene transfer. Mol Cell Biol 4:1730–1737, 1984.

Southern PJ, Berg P: Transformation of mammalian cells to antibiotic resistance with a bacterial gene under control of the SV40 early region promoter. J Mol Appl Genet 1:327–341, 1982.

Stanbury JB, Wyngaarden JB, Fredrickson DS, Goldstein JL Brown M (eds): "Metabolic Basis of Inherited Disease," 5th Ed. New York: McGraw-Hill, 1983.

Su TS, Bock H, O'Brien W, Beaudet A: A cloning of cDNA for argininosuccinate synthetase mRNA and study of enzyme overproduction in a human cell line. J Biol Chem 256:11826–11831, 1981.

Sures I, Goeddel D, Gray A, Ullrich A: Nucleotide sequence of human preproinsulin complementary DNA. Science 208:57–59, 1980.

Tegtmeyer P: Simian virus 40 deoxyribonucleic acid synthesis: The viral replicon. J Virol 10:591–598, 1972.

Tourian A, Sidbury J: Phenylketonuria and hyperphenylalaninemia. In Stanbury JB, Wyngaarden JB, Fredrickson DS, Goldstein JL, Brown M (eds): "Metabolic Basis of Inherited Disease," 5th Ed. New York: McGraw-Hill, 1983, pp 270–286.

Valerio D, Duyvesteyn M, van der Eb A: Introduction of sequences encoding functional human adenosine deaminase into mouse cells using a retroviral shuttle system. Gene 34:163–168, 1985.

Veres G, Craigen WJ, Caskey CT: The 5′ flanking region of the ornithine transcarbamylase gene contains DNA sequences regulating tissue-specific expression. JBC 261:7588–7591, 1986.

Walser M: Urea cycle disorders and other hereditary hyperammonemic syndromes. In Stanbury JB, Wyngaarden JB, Fredrickson DS, Goldstein JL, Brown M (eds): "Metabolic Basis of Inherited Disease," 5th Ed. New York: McGraw-Hill, 1983, pp 402–438.

Weiss R, Teich N, Varmus H, Coffin J (eds): "RNA Tumor Virus." Cold Spring Harbor, NY: Cold Spring Harbor Laboratory, 1984.

Wigler M, Silverstein S, Lee LS, Pellicer A, Cheng YC, Axel R: Transfer of purified herpes virus thymidine kinase gene to cultured mouse cells. Cell 11:223–232, 1977.

Wigler M, Sweet R, Sim GK, Wold B, Pellicer A, Lucy E, Maniatis T, Silverstein S, Axel R: Transformation of mammalian cells with genes from procaryotes and eucaryotes. Cell 16:777–785, 1979.

Williams DA, Lemischka IR, Nathan DG, Mulligan RC: Introduction of new genetic material into pluripotent haematopoietic stem cells of the mouse. Nature 310:476–480, 1984.

Williams DA, Orkin SH, Mulligan RC: Retrovirus-mediated transfer of human adenosine deaminase gene sequences into cells in culture and into murine hematopoietic cells in vitro. Proc Natl Acad Sci USA 83:2566–2570, 1986.

Williams JC, Murray AK: Enzyme replacement in Pompe disease with an α-glucosidase low density lipoprotein complex. In Desnick RJ (ed): "Enzyme Therapy in Genetic Diseases: 2." New York: Alan R. Liss, Inc., for The National Foundatiion—March of Dimes. BD:OAS XVI(1):415–423, 1980.

Willis R, Jolly D, Miller AD, Plent M, Esty A, Anderson P, Chang H-C, Jones O, Seegmiller J, Friedmann T: Partial phenotypic correction of human Lesch-Nyhan (hypoxanthine-guanine phosphoribosyltransferase-deficient) lymphoblasts with a transmissible retroviral vector. J Biol Chem 259:7842–7849, 1984.

Winslow R, Anderson WF: Hemoglobinopathies. In Stanbury JB, Wyngaarden JB, Fredrickson DS, Goldstein JL, Brown M (eds): "Metabolic Basis of Inherited Disease," 5th Ed. New York: McGraw-Hill, 1983, pp 1666–1710.

Wood WI, Capon DJ, Simonsen CC et al: Expression of active human factor VIII from recombinant DNA clones. Nature 312:330–337, 1984.

Wright S, Rosenthal A, Flavell R, Grosveld F: DNA sequences required for regulated expression of β-globin genes in murine erythroleukemia cells. Cell 38:265–273, 1984.

Yeung C, Frayne E, A-Ubaid M, Hook A, Ingolia D, Wright D, Kellems R: Amplification and molecular cloning of murine adenosine deaminase gene sequences. J Biol Chem 258:15179–15185, 1983.

Yu SF, von Rïden T, Kantoff PW, Garber C, Seiberg M, Rüther U, Anderson WF, Wagner ER, Gilboa E: Self-inactivating retroviral vectors designed for transfer of whole genes into mammalian cells. Proc Natl Acad Sci USA 83:3194–3198, 1986.

The Press Week: A Sidebar to the Short Course

Joe Mori

The March of Dimes Birth Defects Foundation, White Plains, NY 10605

THE COURSE

At the March of Dimes Birth Defects Foundation, the annual Short Course in Medical and Experimental Mammalian Genetics is usually referred to as the "Short Course" or, more often, as "Bar Harbor." The Short Course operates on an annual grant and is the longest ongoing grant given by the March of Dimes to any enterprise in its history. It is a tradition like the British Crown, though its function is hardly ceremonial or sacerdotal. A touch of the symbolic invests the Short Course. It expresses our homage to genetics and embryology as the core sciences of birth defects research, and that research as the principal architect of our future.

The stage of understanding and control of birth defects reached by 1984 was a future difficult to imagine in 1960, although the geneticists even then were already able to proclaim: "In the Beginning was the Word, and the Word was DNA."

The influence of the Short Course is largely centrifugal, and it would be difficult to evaluate its impact, but indirect evidence of Bar Harbor's success is considerable. By now, "graduates" of the course can be found in every state of the union and most provinces of Canada, scattered over North America from Hudson Bay to the Rio Grande and from sea to shining sea.

The composition of the faculty is more impressive every year, enlisting those who are on the cutting edge of research in molecular and developmental genetics among guest speakers, along with the reliable, and sometimes

Birth Defects: Original Article Series, Volume 23, Number 3, pages 323–332

scintillating, staff members of the two presenting institutions, Johns Hopkins University and The Jackson Laboratory. The professional interests of the "students" (as participants are called) has widened from a heavy concentration of medical school teachers to those in diverse biomedical areas, which one might take as a sign that genetics, along with birth defects, has taken root in the consciousness of a broad range of Americans, including epidemiologists and public health officials, and—dare I presume?—neonatologists.

One thing that has not changed is the enthusiasm of the students. They participate in after-lecture discussions, in coffee-break postmortems, in the after-classroom give and take among themselves. If you happen to sit next to their table at a restaurant, they may be discussing the semiautonomous operation of mitochondrial genes. A woman at the table reminds the men: "It all came from the mother, you know."

The most impressive exhibition of their enthusiasm, however, is manifested at the optional afternoon workshops. To take up this option and sit crowded indoors through the long, magical afternoons of Mount Desert Island, one must be driven by passion or stoicism, yet the workshops are heavily attended. In 1984, it was not surprising that there were two (nonconcurrent) workshops in molecular biology; the cytogenetics workshop was mobbed, and the devotion of the attendees was so unflagging and participatory that Sandy Miller had to lower the curtain, a reminder that they and she had shared long and exhausting hours.

Not too many people at the March of Dimes knew what went on at Bar Harbor, not firsthand anyway. Virginia Apgar, vice president for medical affairs until her death in 1974, was an exception—as she was to most rules governing ordinary humans. An ardent follower of the Short Course, she made frequent pilgrimages to Bar Harbor and was eventually awarded a diploma by Victor McKusick, with a minimum of fuss, since neither had a flair for ceremony and protocol.

THE PRESS WEEK

Others were members of the March of Dimes science information department. Very early in the history of the Short Course, the March of Dimes gained the privilege of inviting a small group of experienced science writers to Bar Harbor every other year to attend the second week of the 2-week course, to listen, query, ponder, and write about whatever their instinct and experience told them would strike a spark of interest in their public—and, of course, their editors. Essentially, they have free rein. Relations with the writers is pure Spencerian laissez-faire, though perhaps not completely laissez-passer.

The practice seems to work. The biennial press attendance at the Short Course has become an accepted part of the Bar Harbor landscape. Press week has generated its own traditions and myths. One of the traditions is that most writers stay in the same motel along with March of Dimes personnel. There is a press room with adequate facilities and another for informal gatherings. Nothing is elaborate. This en famille atmosphere is a good way of getting to know each other, but individualists can, and do, go their own ways. The press room at The Jackson Laboratory is large and very close to the auditorium where the lectures go on. The space between has a dozen or so chairs where one can see the lecturer and the blackboard on closed-circuit television.

Success or failure of press week depends on the quality of the news conferences held twice each day: at the coffee break (roughly 10:30) and at the end of the morning session, supposedly at 12:30 but usually a little later. There are about 13 to 18 writers and one, two, or three scientists at each news session. The interchange between them is one of the best—as attested to by the journalists—in this established ritual of American journalism. The bulk of reportage bearing a Bar Harbor dateline comes out of what is said in these sessions, although some originates in private interviews between individual writers and the scientists they corral on subjects of mutual interest.

The interchange is a success, whether or not it gives rise to a printed story, if it generates plenty of discussion and both sides learn something. Both sides *do* learn. Questions take the researcher deeper and deeper into his story, and suddenly another question deflects him from his course to an unexpected path that may lead to an interesting and pertinent sidelight.

Of course, he faces a highly selected—self-selected—group of writers, including some Bar Harbor veterans who usually open the questioning and carry the ball. But the young reporters are right on their heels and probably absorb more than those habituated to the place. Watching from the sidelines one can see that the scientist often enjoys the occasion as much as the reporters. Sessions may be prolonged in the heat of dialogue, the midmorning press conference almost overlapping the one at noon. We seldom get back to home base until 2:30 or 3:00 PM, and then some of the writers have to file the story of the day. The Short Course is no vacation for the reporters.

Most writers who come to the Short Course know by experience or hearsay that it is unlikely that they are going to run into a story on a cure for cancer or prevention of birth defects, the traditional "breakthroughs" editors love. What they get, along with something to write about, is a liberal education in genetics and related biologic sciences. Because these areas are moving so fast, they run into something new every year.

Gene replacement therapy is still in diapers but may soon be toilet trained and have a place in the company of other clinical therapies. With the upsurge in cancer genetics, we were inundated in 1983 and 1984 with new information on oncogenes, growth factors, receptors for these, retroviruses, long terminal repeats, enhancers, fragile sites, chromosome translocations associated with transformed cells, and those loci in the human genome at which the loss of both alleles leads to congenital tumors—what might be called antioncogenes—which brought back into fashion Knudson's two-hit theory of retinoblastoma, a topic we heard much about in the 1976 Short Course from Knudson himself.

Perhaps most spectacular has been the metamorphosis of gene mapping, a continuous companion of the Short Course for more than 10 years, into DNA mapping. We have learned to recognize, if only vaguely, what the acronym RFLP (restriction fragment length polymorphism) stands for, although at the 1984 session Allan Wilson was persuasive in his suggestion that the L should be dropped.

In any case, we know that RFP was the clue that enabled the mappers to place the Huntington disease gene on chromosome 4. Apparent omnipresence of RFPs and their mendelian inheritance make them one of the most treasured gifts of molecular biology to genetics. The writers, no less than the participants and the faculty who came to the 1984 session, have an idea of what genes are different from the people who were present at the 1976 session.

Every year, the Short Course has brought some new excitement to the habitués of the course, quite in line with explosive developments in genetics and its subdisciplines in recent times. McKusick and Roderick have been effective impresarios in inducing many stars and potential stars among investigators to perform at Bar Harbor. Dorothy Davis and I, who have tended the Short Course turf for the March of Dimes since 1968, try to make the best use of the talents brought there. The format of press week is centered around the ritual of news conferences partly because it seems an economical way of getting at what a scientist is up to, not only what he has found and how he did it but what hypothesis has emerged from it to be used for the next step toward his long-term objective and what tactics will be used to test its validity. In short, the creative side of investigation and display of a disciplined imagination.

François Jacob's brilliant Danz lectures, "The Possible and The Actual," open with a quote from Jean Rostand: "Theories pass. The frog remains." True. But the frog is important to us as an instrument to forge theories. Hypothesis is the ladder that connects the possible to the actual.

The organizers of the Short Course, besides teaching young investigators, have educated a generation of scientists and science reporters in mutual give and take. They perhaps sense this by observing the identity of interest between investigators and journalists at news conferences during press week.

MEMORIES

The old-timers of press week at the Short Course doubtless have many interesting recollections, each with a different set. I have some vivid memories, though that doesn't mean they are accurate since I believe vividness of most recollection is due to posthoc embellishments in the mind.

For example, in 1972, Jim Renwick came over from England with the story of the year, to wit, the major cause of neurotube defects is the eating of potatoes, or rather of certain fungi that grow on them. The story had wide coverage by the press, as one might expect: What villain in birth defects is more pervasive, more constantly consumed, more instantly recognizable than the potato? Why are anencephaly and spina bifida so common in the Irish but for the fact that they have had to live on potatoes since Cromwell?

Jim was a nightly visitor to the motel around happy hour, and we would have felt uneasy had an evening passed without seeing his somewhat spectral presence coming quietly through the door. Looking back, I must say his thesis based on some teratologic and a mass of epidemiologic data looks as imposing as (or more imposing than) the current idea of folic acid deficiency before the supplementation trials in high-risk mothers. The moral of this may be that, just as the butler is never guilty in a detective story, the culprit in some birth defects may be one totally unsuspected ("Murder of Roger Ackroyd") or a collusion of several individual entities ("Murder in the Calais Coach").

The 1974 press week is unique in my memory and probably also in that of others. Early in the week, March of Dimes headquarters called to say Virginia Apgar had died. Dorothy and I, over the years so close to her personally and professionally, felt the loss keenly, but we didn't have a chance to cultivate our grief in private. The final loss of American innocence was being completed that week as the last act of "Nixon Agonistes" reached its climactic crescendo and people were glued to TV. Not many Bar Harbor stories found their way into the newspapers that week.

The last example I might mention occurred in 1984. About 2 weeks before going to Bar Harbor and some 10 days before the opening of the Short Course, I called Frank Ruddle at New Haven to sound out whether he would be willing to talk to the press about the finding of possible vertebrate equivalents of certain DNA sequences (180 basepair homeobox) of *Drosophila*, which may be the key to how development begins by setting the topologic patterns of body parts.

Frank was willing. "Oh, sure," he said without hesitation. The topic was not an obvious part of his Short Course lecture on somatic cell genetics, but he sounded elated at the chance to talk about it. No wonder, since he had spent his sabbatical at the Biozentum in Basel, one of the two most active laboratories in the field, the reports from which are flooding into the journals like the first waves of a tsunami, promising more to come. Obviously, his laboratory was gearing up to comb mammalian genomes for homeotic genes. Preliminary evidence apparently shows the sequence homology to be impressive in mammalian genomes and the genealogy awesomely ancient, back to earthworms.

By the time I left the phone, I was elated, too. I concocted what seemed to me a sexy title for a news conference: "*Fushi Tarazu* and the Birth Defects Genes—From Fly to Mice and Man," ignoring that it's the *mutations* in these genes that cause birth defects, and *fushi tarazu* for purists is not quite a homeotic gene.

At our first day at the Short Course, July 23, I was wandering around the lobby and auditorium at The Jackson Laboratory trying to find several speakers we wanted for press conferences. Droves of people were coming in for the 8:30 lecture. I wasn't worried about missing Frank, since he towers over a mob like Everest in the Catskills. Then I heard that he had not been there the first week when his lecture was supposed to have been given. It didn't seem likely that he would skip Bar Harbor even if the homeotic genes cooking in his laboratory reached the boiling point. I got hold of Tom Roderick and he said, "Frank isn't coming. I talked to him last night. He is undergoing surgery right now."

For a few minutes, I looked around vacantly for others and went into the press room and told the story. After the initial flurry of shocked questioning, there was a moment of silence, not because we wouldn't meet *fushi tarazu* but because we were all wondering when we would meet Frank Ruddle again.

WHY BAR HARBOR?

Some might ask why Bar Harbor was picked for the Short Course in the first place. I've never asked; Adam must have accepted Eden, the snake and all. If the initiator of the course had been anyone but Victor McKusick, however, the chances are that he or she might have chosen somewhere else, say Woods Hole or one of the Gordon Conference spas in New Hampshire. The ultimate responsibility for the choice must be shared by the founder of The Jackson Laboratory.

In 1929, in his hegira from academia, C.C. Little went about as far east as one can in the continental U.S. to Mount Desert Island, which he already knew and loved, and where some of his rich patrons had summer palaces. There, with modest financial aid given in memory of Roscoe B. Jackson, he established a laboratory for mouse biology. He had carried stocks of mice all over the country since his Harvard days, where he acquired his taste for the genus *Mus* from his mentors W.E. Castle and E.D. Tyzzer.

Whatever his limitations, Little's instinct was right: We can learn a lot from mice and even more from inbred mice. This was at a time when many investigators believed that inbreeding could not be carried on for many generations because mice become decadent, an easy prey to microbes. Today, we accept inbreeding as the most powerful means so far known of uncovering recessive genes, and the rich array of mutant mice at The Jackson Laboratory is almost as impressive as those in *Drosophila*.

To say The Jackson Laboratory started modestly is an understatement by today's standards, but it lasted and grew, and this in spite of a fate that imposed a rebirth on the laboratory. Late in October 1947, an uncontainable fire swept the island (from west to east right to the sea). In its path was The Jackson Laboratory with all its treasures: 60,000 mice were incinerated in seconds.

The impact of the tragedy was deeply felt in the world of mammalian genetics. Many experiments had to stop in many laboratories. March of Dimes' archives have a 1947 grant application from Albert Sabin, who wanted to do comparative neuropathologic studies of inbred mouse strains infected with polio virus. The Bar Harbor fire stopped the project before it could get started. Many laboratories responded to The Jackson Laboratory's plight with generous gifts of mouse stocks they had on hand, including Jax mice. There was other assistance also, but still the rebirth after the holocaust is a heroic page in American research history.

The rise of the laboratory after the fire seemed to follow an exponential course. It steadily got bigger and more varied; growth and differentiation went together. By the time of the Short Course, Earl Green had taken the helm from Little, and everything suggests that Earl was the right man in the right place at the right time. The reputation of the Laboratory was worldwide, and so were the reputations of some of its scientists, a few of whom, such as Tibby Russell and George Snell, had come to Bar Harbor in the 1930s.

Snell, of course, had a subject that has yielded one of biology's most revolutionary revelations, mouse major histocompatibility complex (MHC). The initial work was done by Peter Gorer of Guy's Hospital, who came to The Jackson Laboratory in the 1930s to work with its inbred strains. In the historic address to the Prague Institute of Experimental Biology in 1967, Snell characterized the MHC and its enigmatic neighbor T locus as supergenes. And indeed it has remained the supernova of vertebrate gene families, like the Tyger burning bright in the forest of night. Of innumerable intellectual gifts of mouse to man, nothing has been as valuable as H-2, of which human HLA is a gross plagiarism.

So the major factor in the choice of Bar Harbor for the Short Course was not the scenery or remoteness, or lobsters and blueberries, but The Jackson Laboratory itself. The other major factor was, of course, Victor McKusick.

He was born in Parkman, Maine. I savored the juxtaposition of the two names: McKusick and Parkman. What a felicitous coincidence, I thought, that one of the most historically minded scientists I know was born in a place bearing the name of perhaps the greatest American historian, Francis Parkman. Is not "Mendelian Inheritance in Man" the "France and England in North America" of American genetic literature?

Victor spent most of his adult life outside Maine, but he is still a quintessential Downeaster. But his interests were global rather than regional; Bar Harbor probably was chosen because he was thinking of the wide world of genetics and Earl Green was sympathetic to his aims.

Sometime in the late 1940s, Victor shifted from cardiology to genetics. What prompted the move I don't know, but probably several things contributed: he had worked on Peutz-Jeghers syndrome, he may well have come upon several patients with Marfan syndrome or suspected of having it, who would naturally be of interest to a cardiovascular man, and he was sensitive to the implications of genetic advances during the decade: Beadle and Tatum's one gene–one enzyme idea was proclaimed in the early 1940s, DNA as the genetic material was reported in 1944, Neel worked out the genetics of β-thalassemia a little later, Pauling et al's "Sickle Cell Anemia, a Molecular Disease" was published in 1949, and the ever-increasing inroads of biochemistry in human genetics were yearly becoming evident.

Whatever the reason for the change in career path, it was important in the shaping of the March of Dimes' birth defects program. By 1958, Victor McKusick was one of the best known medical geneticists in the country with a base of operation in the Moore Clinic that had undergone as radical a change as the March of Dimes: from syphilis to genetics. The second edition of his "Heritable Disorders of Connective Tissue" came out about that time and was required reading for us during the March of Dimes' brief involvement with arthritis. Around the backbone of connective tissue, Victor had gathered together a half-dozen interesting and generally obscure genetic diseases, all with systemic manifestations, some (eg Marfan) with delayed onset, each introduced with a historical note. The text was interesting, but just as interesting was the fact that the author surrounded these diseases with a horizon that promised more.

The major theme of the book follows Harvey's well-known dictum that rare diseases are the windows through which we can discern the normal workings of the body. Here, the interactions of various components of connective tissue and their relation to other organ systems were discussed, and, in line with this, at the back of the book the author discussed a host of others as possible targets for future annexations to the imperium of connective tissue.

Human genetics then and now is an expanding subject, and Victor's capacity for anticipation has kept up with it. This is perhaps seen in the programming of the Short Course. The rest of us follow, panting.

THE ISLAND

The locale itself is worthy of a few words. We summer visitors are told that for many months the island is enveloped with the damp cold of the North Atlantic littoral. We suspect summer is a brief ecstacy between winter torpors. But in that very brevity perhaps lies the magic of a Bar Harbor summer.

Mount Desert (pronounced *"dessert"*) Island is a relatively large island of that Northern Aegean, the Maine Coast, so familiar to us from the work of Winslow Homer, John Marin, and many other painters—their sea, land, harbors, and, above all, boats, are what our nostalgia is made of. It's all here.

Most of the central part of the island is Acadia National Park, a rampart against rampant commercialism. Nowhere on the East Coast does the mountain come so close to the sea. Under the July–August sun, the island sparkles and the sea glitters with reflections. Save for the houses, the place has a Mediterranean air, recalling the fabled isles we dream about.

I was surprised to learn that Marguerite Yourcenar, that *rara avis*—the first and only woman elected to the Académie Francaise—had made her home on Mount Desert Island for 40 years. It was there she wrote "Memoirs of Hadrian," which is in a sense an elegy to the golden twilight of the classic Mediterranean civilization. On second thought, it is most appropriate; it was Champlain, a hero in Parkman's saga, who named the place l'Ile des Monts Deserts.

But the most beautiful scene is when the sun is behind low clouds, with mist and fog roaming the landscape, hiding, revealing, and hiding again the islands, promontories, houses, and slow-moving boats on the glistening sea in a stately striptease. No wonder Tonegawa fell in love at first sight. So would Hiroshige and the Sung Masters.

The sky, too, plays a role: gyrating gulls etched in amber clouds, the sunrise viewed from Cadillac Mountain. Very often we can spot the aurora borealis. Once a group of writers spent most of the night on a beach, watching shooting stars.

The people at the Short Course have another experience known to tourists. Theirs is the mixture of intellectual excitement and sensory delight, a potent potion. This is probably not unlike what the developmental biologists enjoy in their biennial pilgrimage to Kolymbari in Crete. The grandeur of the island scenery overlaid with the technologic myth of Daedalus must be impressive and enhances the intoxicating stimulus of hearing new embryologic legerdemain revealed for the first time.

I have watched young scientists talking excitedly on the sunny lawn of The Jackson Laboratory during Short Course coffee breaks. Their voices are happy even when they complain about something. They are happy because they are excited—about what they have just heard. This leads me to muse about what will happen to them 10 years hence. Will some of them be gripped by the Weltschmerz of a midacademic career?

Like many people, I've heard apocryphal horror stories: a minimum of 12 hours a day at lab, clinic, teaching, committees, report writing; journal scanning at odd moments—minutes here, seconds there—over lunch or while waiting for a traffic light to change. Thinking? That can be done only during sleep. Then there are the usual headaches and heartaches of institutional ladder-climbing; a methodologic block holding up your work, to be neatly solved by your competitors; the annual ordeal of composing grant requests; the flickering temptation of private practice, $200,000 a year, no?, without estranging the family.

I could see him (her, too) sneaking back to a cubbyhole of an office, sitting in a swivel chair, looking through a window at the grimy wall of the radiation department 20 feet away, and thinking, oh, God, is it worth it? I have an anodyne. If you shut your eyes, you can fish out of the well of memory the remembrance of the fortnight you spent on this island and can say, with a little liberty, I too once lived in Arcadia.

Index